MTP International Review of Science

Biochemistry
Series One

Consultant Editors

H. L. Kornberg, F.R.S. and
D. C. Phillips, F.R.S.

Publisher's Note

The MTP International Review of Science is an important new venture in scientific publishing, which is presented by Butterworths in association with MTP Medical and Technical Publishing Co. Ltd. and University Park Press, Baltimore. The basic concept of the Review is to provide regular authoritative reviews of entire disciplines. Chemistry was taken first as the problems of literature survey are probably more acute in this subject than in any other. Physiology and Biochemistry followed naturally. As a matter of policy, the authorship of the MTP Review of Science is international and distinguished, the subject coverage is extensive, systematic and critical, and most important of all, it is intended that new issues of the Review will be published at regular intervals.

In the MTP Review of Chemistry (Series One), Inorganic, Physical and Organic Chemistry are comprehensively reviewed in 33 text volumes and 3 index volumes. Physiology (Series One) consists of 8 volumes and Biochemistry (Series One) 12 volumes, each volume individually indexed. Details follow. In general, the Chemistry (Series One) reviews cover the period 1967 to 1971, and Physiology and Biochemistry (Series One) reviews up to 1972. It is planned to start in 1974 the MTP International Review of Science (Series Two), consisting of a similar set of volumes covering developments in a two year period.

The MTP International Review of Science has been conceived within a carefully organised editorial framework. The overall plan was drawn up, and the volume editors appointed by seven consultant editors. In turn, each volume editor planned the coverage of his field and appointed authors to write on subjects which were within the area of their own research experience. No geographical restriction was imposed. Hence the 500 or so contributions to the MTP Review of Science come from many countries of the world and provide an authoritative account of progress.

Butterworth & Co. (Publishers) Ltd.

BIOCHEMISTRY SERIES ONE

Consultant Editors
H. L. Kornb rg, F.R.S.
Department of Biochemistry University of Leicester and
D. C. Phillips, F.R.S., *Department of Zoology, University of Oxford*

Volume titles and Editors

1 **CHEMISTRY OF MACRO-MOLECULES**
Professor H. Gutfreund, *University of Bristol*

2 **BIOCHEMISTRY OF CELL WALLS AND MEMBRANES**
Dr. C. F. Fox, *University of California, Los Angeles*

3 **ENERGY TRANSDUCING MECHANISMS**
Professor E. Racker, *Cornell University, New York*

4 **BIOCHEMISTRY OF LIPIDS**
Professor T. W. Goodwin, F.R.S., *University of Liverpool*

5 **BIOCHEMISTRY OF CARBO-HYDRATES**
Professor W. J. Whelan, *University of Miami*

6 **BIOCHEMISTRY OF NUCLEIC ACIDS**
Professor K. Burton, F.R.S., *University of Newcastle upon Tyne*

7 **SYNTHESIS OF AMINO ACIDS AND PROTEINS**
Professor H. R. V. Arnstein, *King's College, University of London*

8 **BIOCHEMISTRY OF HORMONES**
Professor H. V. Rickenberg, *National Jewish Hospital & Research Center, Colorado*

9 **BIOCHEMISTRY OF CELL DIFFER-ENTIATION**
Professor J. Paul, *The Beatson Institute for Cancer Research, Glasgow*

10 **DEFENCE AND RECOGNITION**
Professor R. R. Porter, F.R.S., *University of Oxford*

11 **PLANT BIOCHEMISTRY**
Professor D. H. Northcote, F.R.S., *University of Cambridge*

12 **PHYSIOLOGICAL AND PHARMACO-LOGICAL BIOCHEMISTRY**
Dr. H. K. F. Blaschko, F.R.S., *University of Oxford*

PHYSIOLOGY SERIES ONE

Consultant Editors
A. C. Guyton,
Department of Physiology and Biophysics, University of Mississippi Medical Center and
D. F. Horrobin,
Department of Physiology, University of Newcastle upon Tyne

Volume titles and Editors

1 **CARDIOVASCULAR PHYSIOLOGY**
Professor A. C. Guyton and Dr. C. E. Jones, *University of Mississippi Medical Center*

2 **RESPIRATORY PHYSIOLOGY**
Professor J. G. Widdicombe, *St. George's Hospital, London*

3 **NEUROPHYSIOLOGY**
Professor C. C. Hunt, *Washington University School of Medicine, St. Louis*

4 **GASTROINTESTINAL PHYSIOLOGY**
Professor E. D. Jacobson and Dr. L. L. Shanbour, *University of Texas Medical School*

5 **ENDOCRINE PHYSIOLOGY**
Professor S. M. McCann, *University of Texas*

6 **KIDNEY AND URINARY TRACT PHYSIOLOGY**
Professor K. Thurau, *University of Munich*

7 **ENVIRONMENTAL PHYSIOLOGY**
Professor D. Robertshaw, *University of Nairobi*

8 **REPRODUCTIVE PHYSIOLOGY**
Professor R. O. Greep, *Harvard Medical School*

INORGANIC CHEMISTRY SERIES ONE

Consultant Editor
H. J. Emeléus, F.R.S.
Department of Chemistry
University of Cambridge

Volume titles and Editors

1 **MAIN GROUP ELEMENTS—HYDROGEN AND GROUPS I-IV**
Professor M. F. Lappert, *University of Sussex*

2 **MAIN GROUP ELEMENTS—GROUPS V AND VI**
Professor C. C. Addison, F.R.S. and Dr. D. B. Sowerby, *University of Nottingham*

3 **MAIN GROUP ELEMENTS—GROUP VII AND NOBLE GASES**
Professor Viktor Gutmann, *Technical University of Vienna*

4 **ORGANOMETALLIC DERIVATIVES OF THE MAIN GROUP ELEMENTS**
Dr. B. J. Aylett, *Westfield College, University of London*

5 **TRANSITION METALS—PART 1**
Professor D. W. A. Sharp, *University of Glasgow*

6 **TRANSITION METALS—PART 2**
Dr. M. J. Mays, *University of Cambridge*

7 **LANTHANIDES AND ACTINIDES**
Professor K. W. Bagnall, *University of Manchester*

8 **RADIOCHEMISTRY**
Dr. A. G. Maddock, *University of Cambridge*

9 **REACTION MECHANISMS IN INORGANIC CHEMISTRY**
Professor M. L. Tobe, *University College, University of London*

10 **SOLID STATE CHEMISTRY**
Dr. L. E. J. Roberts, *Atomic Energy Research Establishment, Harwell*

INDEX VOLUME

PHYSICAL CHEMISTRY SERIES ONE

Consultant Editor
A. D. Buckingham
Department of Chemistry
University of Cambridge

Volume titles and Editors

1 **THEORETICAL CHEMISTRY**
Professor W. Byers Brown, *University of Manchester*

2 **MOLECULAR STRUCTURE AND PROPERTIES**
Professor G. Allen, *University of Manchester*

3 **SPECTROSCOPY**
Dr. D. A. Ramsay, F.R.S.C., *National Research Council of Canada*

4 **MAGNETIC RESONANCE**
Professor C. A. McDowell, F.R.S.C., *University of British Columbia*

5 **MASS SPECTROMETRY**
Professor A. Maccoll, *University College, University of London*

6 **ELECTROCHEMISTRY**
Professor J. O'M Bockris, *University of Pennsylvania*

7 **SURFACE CHEMISTRY AND COLLOIDS**
Professor M. Kerker, *Clarkson College of Technology, New York*

8 **MACROMOLECULAR SCIENCE**
Professor C. E. H. Bawn, F.R.S., *University of Liverpool*

9 **CHEMICAL KINETICS**
Professor J. C. Polanyi, F.R.S., *University of Toronto*

10 **THERMOCHEMISTRY AND THERMO-DYNAMICS**
Dr. H. A. Skinner, *University of Manchester*

11 **CHEMICAL CRYSTALLOGRAPHY**
Professor J. Monteath Robertson, F.R.S., *University of Glasgow*

12 **ANALYTICAL CHEMISTRY —PART 1**
Professor T. S. West, *Imperial College, University of London*

13 **ANALYTICAL CHEMISTRY —PART 2**
Professor T. S. West, *Imperial College, University of London*

INDEX VOLUME

ORGANIC CHEMISTRY SERIES ONE

Consultant Editor
D. H. Hey, F.R.S.,
Department of Chemistry
King's College, University of London

Volume titles and Editors

1 **STRUCTURE DETERMINATION IN ORGANIC CHEMISTRY**
Professor W. D. Ollis, F.R.S., *University of Sheffield*

2 **ALIPHATIC COMPOUNDS**
Professor N. B. Chapman, *Hull University*

3 **AROMATIC COMPOUNDS**
Professor H. Zollinger, *Swiss Federal Institute of Technology*

4 **HETEROCYCLIC COMPOUNDS**
Dr. K. Schofield, *University of Exeter*

5 **ALICYCLIC COMPOUNDS**
Professor W. Parker, *University of Stirling*

6 **AMINO ACIDS, PEPTIDES AND RELATED COMPOUNDS**
Professor D. H. Hey, F.R.S., and Dr. D. I. John, *King's College, University of London*

7 **CARBOHYDRATES**
Professor G. O. Aspinall, *Trent University, Ontario*

8 **STEROIDS**
Dr. W. F. Johns, *G. D. Searle & Co., Chicago*

9 **ALKALOIDS**
Professor K. Wiesner, F.R.S., *University of New Brunswick*

10 **FREE RADICAL REACTIONS**
Professor W. A. Waters, F.R.S., *University of Oxford*

INDEX VOLUME

MTP International Review of Science

Biochemistry
Series One

Volume 2
Biochemistry of Cell Walls and Membranes

Edited by **C. F. Fox**
University of California, Los Angeles

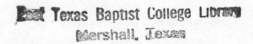

Butterworths · London
University Park Press · Baltimore

THE BUTTERWORTH GROUP

ENGLAND
Butterworth & Co (Publishers) Ltd
London: 88 Kingsway, WC2B 6AB

AUSTRALIA
Butterworths Pty Ltd
Sydney: 586 Pacific Highway 2067
Melbourne: 343 Little Collins Street, 3000
Brisbane: 240 Queen Street, 4000

NEW ZEALAND
Butterworths of New Zealand Ltd
Wellington: 26–28 Waring Taylor Street, 1

SOUTH AFRICA
Butterworth & Co (South Africa) (Pty) Ltd
Durban: 152–154 Gale Street

ISBN 0 408 70496 9

UNIVERSITY PARK PRESS

U.S.A. and CANADA
University Park Press
Chamber of Commerce Building
Baltimore, Maryland, 21202

Library of Congress Cataloging in Publication Data
Fox, C. Fred
Biochemistry of cell walls and membranes.
(Biochemistry, series one, v. 2) (MTP international review of science)
1. Cell membranes. I. Title. II. Series.
III. Series: MTP international review of science.
[DNLM: 1. Cell membrane. 2. Cell wall. W1 BI633 v.2/QH601 B613]
QP501.B527 vol. 2 [QH601] 574.1'92'08s [574.8'75]
ISBN 0-8391-1041-3 74-16253

First Published 1975 and © 1975

Typeset, printed and bound in Great Britain by
REDWOOD BURN LIMITED
Trowbridge & Esher

Consultant Editors' Note

The MTP International Review of Science is designed to provide a comprehensive, critical and continuing survey of progress in research. Nowhere is such a survey needed as urgently as in those areas of knowledge that deal with the molecular aspects of biology. Both the volume of new information, and the pace at which it accrues, threaten to overwhelm the reader: it is becoming increasingly difficult for a practitioner of one branch of biochemistry to understand even the language used by specialists in another.

The present series of 12 volumes is intended to counteract this situation. It has been the aim of each Editor and the contributors to each volume not only to provide authoritative and up-to-date reviews but carefully to place these reviews into the context of existing knowledge, so that their significance to the overall advances in biochemical understanding can be understood also by advanced students and by non-specialist biochemists. It is particularly hoped that this series will benefit those colleagues to whom the whole range of scientific journals is not readily available. Inevitably, some of the information in these articles will already be out of date by the time these volumes appear: it is for that reason that further or revised volumes will be published as and when this is felt to be appropriate.

In order to give some kind of coherence to this series, we have viewed the description of biological processes in molecular terms as a progression from the properties of macromolecular cell components, through the functional interrelations of those components, to the manner in which cells, tissues and organisms respond biochemically to external changes. Although it is clear that many important topics have been ignored in a collection of articles chosen in this manner, we hope that the authority and distinction of the contributions will compensate for our shortcomings of thematic selection. We certainly welcome criticisms, and solicit suggestions for future reviews, from interested readers.

It is our pleasure to thank all who have collaborated to make this venture possible—the volume editors, the chapter authors, and the publishers.

Leicester — H. L. Kornberg

Oxford — D. C. Phillips

Preface

The assembly of this volume was undertaken with two goals in mind. First, it was hoped that the contents would be comprehensive and give a picture of the membrane field as a whole. Second, since there are already a number of recently published review volumes on membranes, the authors were selected so that emphasis would be placed on materials not yet reviewed in detail, or on topics recently reviewed but where another viewpoint might be useful.

Though I feel that the first goal has been achieved reasonably well, the text has not been assembled in the chapter order most suitable for students or those who have not yet read extensively in the membrane area. These individuals might consider reading the chapters in the following order. Four chapters deal with the general organisation and dynamics of membranes: Chapter 5, which concentrates on the disposition of proteins in membranes; Chapter 11, which treats lipids and their physiological effects; Chapter 4, which presents the physical methodology for studying membrane dynamics; and Chapter 9, which examines models for membrane assembly and turnover. The reader might then turn to Chapters 2 and 10, which treat transport in model systems and membranes, respectively. Chapter 8 also treats a transport related topic, the involvement of the bacterial membrane in cell wall biosynthesis. Since events involving membrane fusion are apparently important in nerve impulse conduction, it might be useful to read Chapter 1 before Chapter 7. The two remaining chapters, 3 and 6, both deal with the role of the cell membrane in regulation of cell proliferation, and in the editor's opinion require more background information than the others.

With respect to the second objective, it is perhaps worthwhile to underscore the considerations that entered into the choice of some topics presented here. Though the literature is practically awash with treatments on the membranes of tumour cells, the presentations here should nevertheless offer fresh viewpoints. Pardee and Rozengurt approach the subject from the context of an understanding of regulation of events in cell division in procaryotes, and Brady and Fishman approach it from the viewpoint of the glycolipid biochemist. Transport is another topic that has of late been given a saturation treatment. The recent burst of activity in the possible role of proton gradients in energising transport, however, more than justifies inclusion of the chapter by Epstein. Marchesi has drawn heavily upon his own elegant studies to expound on the special problems faced by the protein biochemist working in the membrane field, and his treatment may offer special insight into the details that must be considered in determining the precise disposition of proteins in membranes. These impressions are representative of what I think have been the attempts made by each author to present something more than a dry rehash of what has already appeared in print elsewhere.

In closing, I thank the authors for taking time to share their ideas with the readers and me. The authors' contributions are more than adequate reward for my efforts in assembling them.

Los Angeles C. F. Fox

Contents

Biochemistry of endocytosis 1
E. D. Korn, *National Heart and Lung Institute, Bethesda, Maryland*

The ion selectivity of carrier molecules, membranes and enzymes 27
G. Eisenman and S. J. Krasne, *U.C.L.A. School of Medicine, Los Angeles*

Membranes of transformed mammalian cells 61
R. O. Brady and P. H. Fishman, *National Institute of Neurological Diseases and Stroke, Bethesda, Maryland*

Mobility of membrane components 97
F. Podo and J. K. Blasie, *University of Pennsylvania*

Molecular orientation of proteins in membranes 123
V. T. Marchesi, *Yale University*

Role of the surface in production of new cells 155
A. B. Pardee and E. Rozengurt, *Princeton University*

Membranes in nerve impulse conduction 187
C. W. Cotman and W. B. Levy, *University of California, Irvine*

The actions of penicillin and other antibiotics on bacterial cell wall synthesis 207
J. L. Strominger, *Harvard University*

Turnover of membrane proteins in animal tissues 229
R. T. Schimke, *Stanford University*

Membrane transport 249
W. Epstein, *University of Chicago*

Phase transitions in model systems and membranes 279
C. F. Fox, *University of California, Los Angeles*

Index 307

1
Biochemistry of Endocytosis

E. D. KORN
National Heart and Lung Institute, Bethesda, Maryland

1.1 INTRODUCTION

1.1.1 Definition of terms

Endocytosis is the process by which material from the medium is taken into the cell within a vesicle formed from the cell surface. When the material is particulate, the process may be called *phagocytosis;* when solutes alone are ingested, the process is *pinocytosis.* At the biochemical level, differences between pinocytosis and phagocytosis have not been established. The intracellular vesicle derived from the cell surface is an *endosome* (*phagosome* or *pinosome* in special instances). Within the cell, the endosome will fuse with other intracellular vesicles which may be (i) other endosomes, (ii) *primary lysosomes*, vesicles derived from the endoplasmic reticulum–golgi complex that contain hydrolytic enzymes, or (iii) *digestive vacuoles* (*secondary lysosomes*) that are essentially the result of previous fusions of endosomes and secondary lysosomes. In some cases, the vesicle and its contents may never leave the cell becoming an inert *residual body*, while in other instances the endosome membrane may join again with the cell surface excreting the vesicular contents. This latter process is frequently called *exocytosis*, a term also used for the separate *secretory process* by which material synthesised within the cell and packaged within vesicles may be exported. There are insufficient data to determine if the secretory process and the fusion of endosome with the cell surface are biochemically and morphologically identical. The term *intracytosis*, if it is to be used at all, might be reserved for the related phenomenon in which intracellular membranes surround cytoplasmic constituents to form a vesicle (*autophagosome*) that might then fuse with primary or secondary lysosomes.

1.1.2 Diversity of endocytosis

Endocytosis has been studied in several different cell types in each of which it may serve a different function that may be reflected in significant differences in the endocytic process. In polymorphonuclear leukocytes, macrophages and other reticuloendothelial cells, phagocytosis predominates as a mechanism for removing damaged (aged) and foreign cells and particulate matter from the circulation. Pinocytosis in such cells probably also serves largely to sequester damaged or foreign molecules. Endocytosis in these cells does not serve a nutritional function. It is likely that the phagosomes accumulate as residual bodies resulting in death or functional loss of the phagocyte. Pinocytosis also occurs in other cells of higher animals where its magnitude and function are not well studied.

Large amoebae, *Chaos chaos* and *Amoeba proteus*, are favoured experimental subjects. In such cells, phagocytosis of other living organisms is the sole known means of nutrition. Hence, unlike mammalian phagocytes, it is necessary for the products of endocytic digestion to reach the cytoplasm, and for the undigested contents of the phagosome to be excreted so that the process can continue. The large amoebae have not been successfully cultured on soluble media, probably because they are incapable of normal pinocytosis at a rate sufficient to support cell growth. Pinocytosis can be stimulated in these amoebae, usually by positively charged solute molecules, but this 'induced' pinocytosis is intermittent and morphologically distinct from 'normal' pinocytosis. Induced pinocytosis is characterised by adsorption of inducer to the cell surface, accumulation of coated membrane at the uroid ('tail') end of the amoeba and the formation of long invaginated channels from which endosomes large enough to be seen in the light microscope vesiculate internally. Subjectively, the process might seem one of removal of damaged surface membrane rather than a specific mechanism for transport of solute, and it bears at least a superficial resemblance to the 'capping' phenomenon in mammalian fibroblasts and thymocytes (Section 1.3.3.1).

Finally, small amoebae, in particular *Acanthamoeba castellanii*, have been extensively studied. In contrast to the large amoebae, *Acanthamoebae* can be grown with ease on soluble media and, therefore, pinocytosis can be studied as a natural phenomenon, the animal's only means of nutrition under these conditions since neither active transport nor facilitated diffusion of solute molecules occur. In these amoebae, the endosomes form everywhere on the cell surface as very small vesicles visible only by electron microscopy. All solute molecules are pinocytosed; inducers are not needed nor are any known. Pinocytosis is a continuing process in growing cultures of *Acanthamoebae* so that surface membrane must be continually replaced. The small amoebae are also capable of phagocytosis (in the wild, phagocytosis is probably of more importance than pinocytosis), and the process has been studied in these, and other, cells with the use of latex particles. Such experiments must be interpreted with care, however, because the latex may be treated abnormally after it is phagocytosed.

Thus, three physiologically distinct experimental systems can be distinguished: (i) specialised cells (macrophages and polymorphonuclear leukocytes) of higher animals in which endocytosis serves to protect the organism but is of no particular specific advantage to the endocytic cell; (ii) large amoebae (*Chaos chaos* and *Amoeba proteus*) for which phagocytosis of food organisms is the only known means of nutrition but pinocytosis, as usually studied, is an experimental situation induced by a restricted set of non-physiological solute molecules; and (iii) small amoebae (*Acanthamoeba castellanii*) which can live 'naturally' by either phagocytosis of food organisms or pinocytosis of solute molecules with little, if any, specificity in the uptake process.

1.1.3 General statement of the problem

The following phenomena are known or suspected to be involved in endocytosis. The endocytic cell and substrate particle or solute molecule first meet at the cell surface where, in some instances, specific interactions may occur. The nature of these interactions depends on the morphology and chemistry of the cell surface and the substrate. Either as a consequence of these interactions or as a continuing physiological process, endocytosis ensues. What initiates the process? Morphologically, endocytosis can begin as local deformations of the cell surface, as long channels that later vesiculate, or as advancing pseudopods that surround the particle. Is energy required for these movements and are structural elements such as microfilaments and microtubules involved? Eventually the endosome will close on itself, separating from the cell surface. What physical, chemical or enzymatic processes are involved in thus disrupting the plasma membrane and re-annealing the cell surface and the endosome membrane?

Within a cell, a myriad of morphological changes occur. Endosomes fuse with each other, with primary lysosomes and with digestive vacuoles to form larger vesicles from which small vesicles will bud off. What controls these events? What forces are involved in the membrane fusions? Why are the fusions restricted to a particular set of membranes not including the endoplasmic reticulum or mitochondria, for example? During this period, enzymatic reactions occur within the digestive vacuole and, at least in cells nutritionally dependent on endocytosis, the products of digestion must pass through the vacuole membrane and enter the cytoplasm. Are changes in membrane permeability necessary? Finally, how and why does the vesicle find its way back to the cell surface where its membrane can fuse with the plasma membrane?

1.1.4 Scope of this article

The previous sub-section is a better summary of what is not known about endocytosis than what is known. The remainder of this article is an assessment of the present status supported by selected data with emphasis on the biochemical aspects of the problem. For more complete discussions of all but the most recent work and for other points of view, several excellent reviews[1-6] are available. It may also be interesting to read the original descriptions of phagocytosis[7] and pinocytosis[8] if only to learn how little progress has been made since.

1.2 RECOGNITION AND DISCRIMINATION

1.2.1 The cell surface

The composition and organisation of the plasma membrane is discussed briefly in Section 1.3.2, but for the initial events of endocytosis it is probably only the external surface of the cell that is important.

1.2.1.1 Mammalian phagocytes

Mammalian cell surfaces are largely covered by carbohydrate[9,10], partially in the form of glycolipids but mainly as glycoproteins, the protein portion of which may be anchored within the lipid portion of the plasma membrane with the polysaccharide chain extending some distance from the cell surface. There are essentially no data which are specific to phagocytic cells, but in general it is the carbohydrate side-chains that give the cell surface its negative charge (sialic acid moieties) and which carry the antigenic sites, the sites for cell–cell recognition, the sites with which plant agglutins interact and probably receptors for antibodies and antigens alike. One would suspect, therefore, that interactions between the endocytic cell and its substrate would be dominated by these surface glycoproteins.

1.2.1.2 The large amoebae

The plasma membrane of the large amoebae is covered by an amorphous layer some 100–200 Å deep with hair-like projections that may extend for several thousand angstroms[3,11,12]. This material is frequently referred to as acid mucopolysaccharide, but no chemical data apparently exists, the only evidence being metachromatic staining with appropriate dyes[13]. Nevertheless, this surface coat dominates morphologically the cell surface and, with its negative charge, would be expected to dominate electrostatically as well as physically the interaction of cell and substrate.

1.2.1.3 Acanthamoeba castellannii

The cell surface of *Acanthamoeba castellanii* is being studied in detail[14-17]. Approximately one-third of the mass of isolated plasma membranes consists of a phosphonoglycan composed of three or four neutral sugars (including glucose and mannose), glucosamine, galactosamine, a large amount of two aminophosphonic acids (2-aminoethylphosphonic acid and 1-hydroxy-2-aminoethylphosphonic acid) and long chain fatty acids and 2-hydroxy fatty acids. As a consequence to the aminophosphonic acids, the lipophosphonoglycan has a very low isoelectric point. In contrast to the large amoebae, light and electron microscopy[18] of *Acanthamoebae* reveals no amorphous coat external to the plasma membrane. But since concanavalin A, a plant protein that reacts specifically with terminal glucosyl and mannosyl residues, binds specifically to the cell surface[19] (and, remarkably, to the inner surface of isolated plasma membranes as well), it is likely that the lipophosphonoglycan is anchored in the plasma membrane through hydrophobic lipid groups with its carbohydrate and aminophosphonic acid components determining the character of the cell surface.

1.2.2 Chemistry of the substrate and its interaction with the cell

There are conflicting claims of specificity and non-specificity for the substrates of endocytosis. Some of these difficulties may be referable to the methods of assay. Frequently, endocytosis has been measured not as it should be (by the amount of material taken into the cell) but by the microscopic observation of the formation of endocytic vesicles or channels. This restricts the assay to those endosomes visible in the light microscope and is not an assay amenable to quantitative kinetic analysis. Where the uptake of radioactively-labelled substrates has been measured, only infrequently have the experiments been done under conditions allowing proper quantification. If two substrates are to be compared, it is necessary that the initial rates of uptake or binding be measured as a function of the substrate concentration.

1.2.2.1 Mammalian cells

Mammalian phagocytes ingest *in vivo* and *in vitro* a range of particles[5,6] including at least erythrocytes and erythrocyte ghosts, melanin granules, bacteria, latex, quartz, india ink, chromium phosphate, colloidal iron and colloidal gold. The need for a particular substrate composition is not obvious but, although it is not the case for latex, particles are frequently more rapidly ingested by mammalian cells after their exposure to serum proteins[20]. It is generally assumed that this 'opsonisation' requires immunoglobulin, alone or with complement, for which the phagocytes (macrophages and leukocytes) have specific receptor sites[21]. On the other hand, the phagocytosis of aldehyde-treated red cells by macrophages and fibroblasts is apparently inhibited by immunoglobulin as well as by unrelated proteins[92]. This entire area needs more definitive research.

The data on substrate specificity for pinocytosis by mammalian cells are also in conflict. One group[23] reports that polyanions such as acid mucopolysaccharides, dextran sulphate and nucleic acids are excellent inducers of pinocytosis in mouse macrophages; that polyglutamic acid is active and polylysine inactive; that acidic proteins, albumin and fetuin, are active and basic proteins are inactive; that, more specifically, removal of the bound fatty acids from albumin or of the terminal sialic acid residue from the fetuin glycoprotein renders then non-inducers. (The apparent *requirement* for terminal sialic acid contrasts dramatically with recent experiments[24] in which removal of the terminal sialic acid residue has been shown to reduce drastically the circulating half-life of serum glycoproteins by inducing their pinocytosis by liver parenchymal cells.) In direct contrast to the above observations, others[25] have concluded that, at least for cells of an established tumour line, only basic macromolecules stimulate pinocytosis.

At this time it is not possible to generalise on the nature of the interaction between substrate and the mammalian endocytic cell. It may vary with cell type and better experimental methods are needed. The subject is obviously intimately related to the entire range of important cell-surface phenomena involving recognition sites. Mammalian cells do bind many plant lectins[26] which have varied specificities for sugar moieties, and these may subsequently be pinocytosed. Similarly, lymphocytes have specific receptors for antigens[27], the binding of which may also be followed by pinocytosis. Some aspects of these specific processes are discussed in Section 1.3.3.1.

1.2.2.2 The large amoebae

Early work[1-3] led to the definition of a number of inducers of pinocytosis, whose presence in the medium caused the formation of long invaginated channels from which large endosomes would vesiculate and which stimulated the uptake of non-inducers. Inducers were inorganic cations at neutral pH values, amino acids at alkaline pH values, proteins below their isoelectric points and a number of basic dyes whose effect (and charge) was independent of pH. The phenomenon was initially viewed as the specific interaction of inducer with membrane stimulating the endocytic response. More recent data[3] suggest, however, that such induction requires adsorption of inducer to almost the entire cell surface which is then removed endocytically with consequent internalisation of adsorbed molecules and, to a lesser extent, solute molecules. Cells cannot survive for long in inducing solutions and, although the process may be experimentally useful, 'induced' pinocytosis may be a mechanism for the removal of damaged cell surface rather than a stimulation of a normal transport mechanism.

In contrast, 'permanent' endocytosis in large amoebae is a more continuous process lasting for 4–5 h rather than the 30 min required for induced pinocytosis. Permanent endocytosis has no known specificity although in nature it is probably largely restricted to food organisms. The mucoid layer seems to play a major functional role in phagocytosis by interacting with the prey and preventing its movement away from the amoeba[28], but this probably does not involve specific chemical interactions.

1.2.2.3 Acanthamoeba castellanii

Pinocytosis is a continuing process in *Acanthamoeba castellanii* and cultures have been maintained for about 20 years on soluble media. No specificity for solute molecules is known, the rates of uptake of large and small molecules, neutral and charged (albumin, inulin, glucose, leucine) being indistinguishable[29]. Phagocytosis of bacteria[30] and latex particles[31, 32] is rapid. No specificity of interaction has been noted, and the amoebae, whose surface charge is negative, seem indifferent to the charge on the latex particle. In shaking cultures, the mechanical energy of the collision is probably sufficient to overcome any charge repulsion. Electron-microscopic evidence suggests that, despite the absence of a mucoid layer, *Acanthamoebae* are able to bind bacteria[30] and latex particles[32] to their surface prior to ingestion but the chemical basis of this remains unexplored.

1.3 FORMATION OF ENDOSOMES

1.3.1 Kinetics

Careful kinetic studies are of inestimable value in understanding the mechanism of endocytosis. By analogy with enzyme kinetics, one should usually determine the initial rate of uptake. Where that rate is independent of substrate concentration, specific interaction of substrate and cell is unlikely, for such a process should involve a limited number of saturable sites on the cell surface. Without pursuing too far the analogy to enzyme–substrate interactions, it is obvious that where, for whatever reason, specificity exists, the maximum amount of information will be obtained from the initial rate of endocytosis as a function of substrate concentration.

1.3.1.1 Mammalian phagocytes

The few available studies suggest that, for polymorphonuclear leukocytes, the rate of phagocytosis remains constant for not more than 10 or 20 min depending on the substrate (latex[20] and *E. coli*[33], respectively), but almost every study of phagocytosis has used longer times so that the measurement has been of cell capacity rather than of any parameter of the uptake process. Cessation of uptake is probably a consequence of membrane being removed from the cell surface more rapidly than it can be replaced.

The initial rate of pinocytosis by mammalian cells seems to be independent of the substrate concentration[34] which provides the best evidence that specific, saturable sites on the cell surface are not necessarily involved in the process. The rate of pinocytosis has in some instances been shown to be constant for as long as 10 h[35] suggesting, in contrast to phagocytosis, that internalised membrane is rapidly replaced at the cell surface. The observed rates of pinocytosis as measured by different macromolecules (albumin, inulin, horseradish peroxidase, chondroitin sulphate) by different mammalian cells (leukocytes, macrophage, Ehrlich ascites tumour, sarcoma, hamster cell line) are remarkably constant at about $0.05\,\mu l/10^6$cells h^{-1} [34-38]. The uniformity of these data, which were obtained under several different experimental conditions, strongly suggests that, at least for these macromolecules, pinocytosis is independent of substrate concentration and specificity is not a factor.

Endocytosis by mammalian cells is inhibited at low temperature and requires energy. The effect of different inhibitors depends on the metabolic source of cellular energy, whether oxidative or glycolytic. Thus, endocytosis by guinea-pig polymorphonuclear leukocytes is inhibited by iodoacetate and fluoride but not by cyanide or dinitrophenol[39] aleveolar macrophages are sensitive to any one of the four inhibitors[40, 41] and inhibition of endocytosis in Ehrlich ascites tumour cells requires the simultaneous presence of inhibitors of aerobic and anaerobic energy metabolism[33].

1.3.1.2 *Large amoebae*

Quantitative kinetic studies of phagocytosis by large amoebae have not been made. At saturating levels of protein, induced pinocytosis of ribonuclease and cytochrome c by *Chaos chaos* occurs at a rate of 3×10^{17} molecules/10^6 cells h^{-1} (*ca.* 10^6 times greater than for mammalian cells), but since endocytosis ceases after only 10 min, uptake is limited to 5×10^{16} molecules/10^6 cells[42], i.e. to an amount of protein that had been present in 50 times the volume of the cells. The fact that the process is saturated with excess substrate implies binding to the cell surface and this can be seen in the electron microscope when thorotrast[43] or ferritin[44] is the substrate. Basic proteins bind to the mucoid layer. A better measure of the volume rate of pinocytosis, therefore, its obtained from the rate of uptake on non-inducer molecules that do not adsorb to the cell surface. In the presence of inducing protein, glucose is pinocytosed at a rate equivalent to 300 μl/10^6 cells h^{-1}, but again pinocytosis stops after only 10 min when uptake of 50 μl/10^6 cells has occurred[45]. These values are not as high as might appear in comparison to mammalian cells since the volume of *Chaos chaos* is approximately 1000 times greater than the volume of the mammalian cells. The volume of medium pinocytosed is equivalent to about 2% of the volume of the amoeba. Unfortunately, there are no data on the rate of pinocytosis by large amoebae in the absence of an 'inducer' substance.

1.3.1.3 Acanthamoeba castellanii

(a) *Phagocytosis* — Phagocytosis of latex particles by *Acanthamoeba castellanii* is a highly specific process which from electron microscopic studies was shown to involve the selective uptake of the latex with the cell surface closely apposed[32]. to the particle so that very little medium is included in the endosome. For six different sizes of latex particles ranging in diameter from 0.126–2.68 μm (the diameter of the amoeba is *ca.* 20 μm), the initial rate of uptake is a function of the substrate concentration and when expressed as the mass (or volume) of latex the calculated values of K_m and V_{max} were the same for the different particle sizes[31]. These data suggested that the volume of the phagosome was approximately the same irrespective of the size of the latex particle. The conclusion drawn from the kinetic data was confirmed by electron microscopy which showed that whereas large particles tended to be ingested singly, small latex particles were accumulated at the cell surface and ingested in large numbers within one phagosome whose volume was approximately that of the larger latex particles.

The rate of phagocytosis is such that the amoebae will ingest all the particles contained in about 100 times their own volume in *ca.* 30 min. This rate compares favourably with the rate of induced pinocytosis in *Chaos chaos*. The two processes are also superficially similar in that both cease after *ca.* 15 min at saturating levels of substrate, possibly because there is insufficient surface membrane left to allow endocytosis to continue.

Quantitative kinetic studies[31] show that phagocytosis is not affected by the charge on the particle or by the pH of the medium (over a wide range), is inhibited by dinitrophenol and cyanide but not by fluoride and iodoacetate, and is also inhibited by elevated ionic strength or by inappropriate osmolality (too high or too low relative to the osmolality of the growth medium).

(b) *Pinocytosis* — Pinocytosis in *Acanthamoeba castellanii*[29] has all of the features of pure pinocytosis as originally defined, *i.e.* the continuous uptake of fluid with the consequent 'passive' ingestion of solute molecules. The rate of uptake is linear at all times. The rate of pinocytosis is independent of the substrate concentration, saturation being unobtainable. Inducers are not known; albumin, inulin, glucose and leucine are all taken in at the same rate. A radioactive molecule, then, simply serves as a tracer for the calculation of the rate of fluid uptake (as was also calculated for mammalian cells (Section 1.3.1.1)). Calculated rates are about 2.1×10^6 μl/10^6 cells h^{-1} or *ca.* 50 times greater than for mammalian cells. The rate of pinocytosis is *ca.* 22% of the volume of the cell per hour. This is about twice the maximum rate of induced pinocytosis in *Chaos chaos* which, however, can be maintained for only *ca.* 10 min while pinocytosis in *Acanthamoeba* is continuous. As with phagocytosis, pinocytosis in *Acanthamoeba* is inhibited by dinitrophenol and cyanide.

(c) *Temperature effects* — Endoctyosis in all cells is very slow or absent at 0 °C. (For some substances and some cells, binding may occur to the cell surface in the cold, but not vesiculation.) With *Acanthamoebae*, neither pinocytosis nor phagocytosis occurs[46] until the temperature reaches 16–17 °C, where there is a sharp inflection and a linear increase in the rate of endocytosis until lethal temperatures are reached. It seems unlikely that the critical temperature is related to a phase transition of membrane lipids since lipids isolated from the amoeba plasma membranes do not show a transition at that temperature[46]. It is possible that the transition temperature is related to events of vesicle formation or movement discussed elsewhere in Sections 1.3 and 1.4.

1.3.2 The plasma membrane

1.3.2.1 General considerations

Before discussing further the formation of the endosome, it is necessary to describe briefly the composition and structure of the plasma membrane. In Section 1.2.1, we dealt only with the outer surface of the cell which in some cases is superficial to, and in other cases an integral part of, the plasma membrane. All cells have in common a permeability barrier, the plasma membrane. that is *ca.* 100 Å in width consisting of protein and lipid in the approximate ratio of 1.5 : 1[47, 48] and a significant proportion of carbohydrate[9, 10]. As is true for all cell membranes, the lipid of the plasma membrane contains little glyceride[47-49] and quite a bit of phospholipid, but the plasma membrane is unusual among cell membranes in its high content of glycolipid and sterol.

It is now generally accepted[50] that most of the phospholipids and sterols of plasma membranes are in the form of a molecular bilayer orientated with the polar head groups at the inner and outer surfaces and the fatty acyl chains forming a hydrophobic interior. There is some suggestion of specific arrangements of particular phospholipids[51] and sterol molecules within the bilayer. Evidence largely derived from model systems of artificial phospholipid bilayers suggest that lipid molecules are free to move rapidly within the plane of the bilayer[52], but that they are unable to flip from one side of the bilayer to the other[53]. Theories of endocytosis generally assume (Section 1.3.4) a major role for the phospholipids in the fusion processes by which vesicles are formed initially and later unite with and split from lysosomes. Membrane sterols are thought to decrease the permeability of the plasma membrane[54], but no role has been suggested for sterols in the endocytic process.

Proteins are certainly present at the outer and inner surfaces of the plasma membrane, in particular the glycoproteins and other specific receptor sites. There is now universal agreement that proteins also lie within the hydrophobic interior of the lipid bilayer[55] and, in one instance, (the erythrocyte glycoprotein[56]), there is evidence that a single protein spans the membrane. The topographic distribution of membrane proteins is clearly a subject of great interest. A current useful working model suggests that many of the membrane proteins exist as mobile islands within the hydrocarbon sea. These are the proteins revealed as particles lying within the membrane by freeze-cleavage electron microscopy[55].

1.3.2.2 Plasma membrane of mammalian phagocytes

Plasma membranes have not been isolated from those mammalian cells, leukocytes and macrophages, most used for studies of endocytosis. In general, plasma membranes of mammalian cells contain 20–50 proteins separable by sodium dodecyl sulphate polyacrylamide gel electrophoresis whose molecular weights range between 20 000 and 250 000[57-59]. Some of these polypeptides have been speculatively related to actin and myosin, but only in one case[59] is there any direct evidence for the possible presence of proteins related to actin and myosin in association with isolated plasma membranes (see Section 1.4.1). Many enzymes are known to be present in mammalian cell plasma membranes[47-49] but none of these is presently thought to be connected to endocytosis although some of them are involved in the active transport of ions and other solute molecules. Enzymes found only in the plasma membrane may serve as useful markers of that membrane following endocytosis.

Although all of the phospholipid species are represented, the plasma membrane of mammalian cells is particularly enriched in sphingomyelin.

The only sterol is cholesterol. These two molecules may also serve as useful markers for membranes within the cell that may be related to the plasma membrane.

1.3.2.3 The amoeba plasma membrane

It is unfortunate that there have been no chemical analyses of isolated plasma membranes from *Chaos chaos* and only preliminary observations on one preparation of plasma membrane from *Amoeba proteus*[60] since so many studies of endocytosis have utilised these organisms. Much more is known about the plasma membrane of the small amoeba *Acanthamoeba castellanii*.[14-17] The isolated plasma membrane consists of about one-third protein, one-third lipid and one-third lipophosphonoglycan (Section 1.2.1). About 55% of the lipid is phospholipid, together with an equimolar amount of sterol consisting of ergosterol and the related dehydroporiferasterol. Of particular interest is the fact that 60% of the total protein is accounted for by a single polypeptide of molecular weight *ca.* 14 000. Very few enzymes are known to be present—some of those that are present will be discussed later (Section 1.3.4.1). This very simple protein composition relative to mammalian cell plasma membranes may be the consequence of a highly specialised structure for endocytosis.

1.3.3 Endosome membrane

As a by-product of the studies on the phagocytosis of latex by *Acanthamoeba castellanii*, a method has been developed for the rapid isolation of the phagosome by centrifugation of cell homogenates in an appropriate sucrose gradient[61]. The membrane of phagosomes formed during the first 15–30 min of phagocytosis was found to have the same sterol: phospholipid ratio and phospholipid composition as plasma membranes from the same cells[14, 61]. It has not been determined whether the phagosome membrane also contains the protein and lipophosphonoglycan that are characteristic of the *Acanthamoeba* plasma membrane (Section 1.3.2.3).

The same technique has recently been applied to mammalian cells[62] where it was found that the cholesterol: phospholipid ratio and the specific activity of 5′-nucleotidase (a marker enzyme for mammalian cell plasma membranes) were the same in the phagosome as in the macrophage plasma membrane. The general similarity of the endosome and plasma membrane is not incompatible with the report[63, 65] that the membrane of secondary lysosomes isolated from liver after Triton injection differ significantly in composition from the plasma membrane. It must be remembered that these secondary lysosomes are far removed in time from the initial endocytic events and are the products of multiple fusions of the endosomes with many intracellular vesicles. The experiments on the isolation of latex phagosomes were carefully designed to obtain phagosomes at the earliest practical time following their formation so as to minimise secondary fusions.

Parenthetically, it should be mentioned that there is some danger in using the latex or related techniques as a method for the isolation of 'plasma membrane'[65,119]. Even at the earliest times after internalisation, the endosome membrane is likely to have been altered. It does appear reasonable to assume, however, from these data, and from the electron-microscopic observations, that the endosome membrane is derived from the plasma membrane.

This is not to say, however, that the endosome and plasma membranes are necessarily identical before intracellular fusions have occurred. Evidence exists[66] that in both macrophages and polymorphonuclear leukocytes the rate of active transport of leucine and adenosine is undiminished following internalisation of an appreciable portion of the plasma membrane by phagocytosis of latex particles. This suggests that, in contrast to the plasma membrane 5′-nucleotidase and concanavalin A receptors[130], the transport carrier proteins of the plasma membranes, are not internalised. Such results, if confirmed, could have at least two alternative interpretations: (i) specific areas of the plasma membrane are involved in phagocytosis and these do not contain the leucine and adenosine transport proteins, whereas 5′-nucleotidase and concanavalin A receptors are distributed randomly throughout the membrane; (ii) specific endocytic areas do not exist, but during the formation of the endosome some of the membrane proteins are left behind at the cell surface.

1.3.3.1 Mobility of membrane components

There is definitive evidence that under a number of conditions some membrane proteins are mobile within the membrane[55]. Membrane lipids are generally thought of as being free to move within the plane of the membrane. Such mobility might occur during endocytosis, perhaps as a necessary concomitant of the process. It has long been known that when pinocytosis is induced in the large amoebae by Alcian Blue, the filaments of the surface coat are clumped and the clumps collect at the uroid region. It is possible that this represents movement of the external mucoid layer only, but circumstantial evidence suggests that some elements of the underlying plasma membrane may also be involved[12].

The 'capping' phenomenon of lymphocytes may be related. When B-lymphocytes are exposed to bivalent anti-immunoglobulin (or concanavalin A), the antibodies react with the diffusely-distributed immunoglobulins on the cell surface, cross-linking them and causing them to migrate to, and agglutinate at, one pole of the cell[67,68]. This capping phenomenon is frequently, but not necessarily, followed by endocytosis of the cap.

1.3.4 Mechanism of membrane fusion

1.3.4.1 Theoretical considerations

A critical stage in the endoctytic process is the moment when the vesiculating plasma membrane fuses with itself to form the endosome while retaining continuity of the cell surface. Studies at the ultrastructrual level have not yet been informative, but recent studies of a specialised example of exocytosis[69] and of pulmonary endothelial cells[119] suggest that the freeze–fracture, freeze–etch technique may be very useful.

The molecular basis of this phenomenon is obscure. The most reasonable working hypothesis is that the fusion process might involve a controlled disruption and re-formation of the phospholipid bilayer. Specifically, it has been proposed[70] that the local action of a phospholipase A2 might convert phospholipids to lysophospholipids. Lysophospholipids are known to make phospholipid bilayers unstable, which might lead to the formation of a micellar region consisting of molecules from the two fusing membranes. Controlled enzymatic re-acylation of the lysophospholipids might then allow the bilayer to form again in such a way that the endosome separates from the restored cell surface.

Direct experimental support[71] for one of the several conjectures in the above hypothesis is the observation that the fusion of avian erythrocytes can be induced by added external lysolecithin, albeit at a rather high concentration. Moreover, fusion of chicken erythrocytes can also be stimulated by 40 mmol l^{-1} Ca^{2+} ion at pH 10.5 and, under these conditions, is associated with the conversion of about 10% of the phosphatidyl choline and phosphatidyl ethanolamine to the lysophospholipids[72]. However, if the proposed mechanism is correct, the plasma membrane of endocytic cells should contain at least two enzymes: a phospholipase A2 which catalyses the hydrolytic removal of the fatty acid from the 2 position of phospholipids and an acyl CoA-lysophospholipid transferase which would catalyse the re-acylation of the lysophospholipid. Neither phospholipase A2 nor acyl CoA-lysophospholipid transferase has yet been detected in the erythrocyte membrane (but the erythrocyte is not normally regarded as an endocytic cell).

Highly purified plasma membranes isolated from *Acanthamoeba castellanii* do contain, however, both a very active phospholipase A2 and an acyl CoA-lysophospholipase transferase, in addition to a lysophospholipase[73]. The enzymes are generally not found together in plasma membranes. For example, plasma membranes of liver parenchymal cells contain a phospholipase A2 but not acyl CoA-lysophospholipid transferase[74]. The possible presence of both of these enzymes in the plasma membranes of endocytic cells other than the *Acanthamoeba* is, as yet, untested.

1.3.4.2 Radioactive tracer studies

Radioactive lysophosphatidyl choline and lysophosphatidyl ethanolamine are more rapidly acylated to phosphatidyl choline and phosphatidyl ethanolamine, respectively, in phagocytising than in resting leukocytes[75]. The source of the fatty acids appears not to be acyl CoA, however, but perhaps cellular triglyceride[76] conceivably by triglyceride–lysophospholipid transacylation. Such data provide some support for a phospholipid–lysophospholipid cycle, but since the analyses were of total cell lipids it is not certain that the reactions occurred at the plasma membrane. Similar experiments in *Acanthamoeba castellanii* have been consistently negative[77,78]. The rates of incorporation of ^{32}P-phosphate, ^{3}H-oleic acid and radioactive lysophosphatidyl choline into the phospholipids of the amoeba plasma membrane are less than their rates of incorporation into the phospholipids of the whole cell and not greater than their rates of incorporation into mitochondrial phospholipids under conditions of maximum pinocytosis (Section 1.3.5). When pinocytosis was stopped by cooling to 12 °C the inhibition of turnover of plasma membrane phospholipids was not greater than the inhibition of turnover of other cellular phospholipids.

1.3.5 Magnitude of membrane involvement

What percentage of the cell surface is involved in the endocytic process? One must know two things to answer this question: the surface area of the cell and the total surface area of the endosomes formed within the period in question.

1.3.5.1 Mammalian cells

Pertinent published data are hard to find. One very interesting paper reports the percentage of 5′-nucleotidase internalised during phagocytosis of latex particles by macrophages[62]. On the assumption that the 5′-nucleotidase is representative of the entire plasma membrane, it can be calculated that 35–50% of the surface membrane can be used to form phagosomes in 1 h.

1.3.5.2 Large amoebae

The above estimate for mammalian macrophages is not too different from the approximation[79] that about 50–100% of the surface area of *Chaos chaos* and *Amoeba proteus* can be utilised during one cycle of induced pinocytosis. These latter estimates are based on visual estimates of the relative membrane areas and cannot make adequate allowance for the numerous infoldings of the cell surface. Permanent, normal pinocytosis in a related large amoeba has been quantitatively estimated by morphometric measurement[80] to involve about 12% of the cell surface per hour. This rate can be maintained for 7–8 h.

1.3.5.3 Acanthamoeba castellanii

(a) *Phagocytosis* — The surface area of the cell has been calculated to be about 2.2×10^{-9} m^2 by microscopic measurements of amoebae rounded by exposure to dinitrophenol and confirmed by volume measurements[31,81]. Since latex particles of diameter greater than 1 μm are ingested singly within a closely apposed endosome membrane[32] and uptake of latex may be readily quantified by spectroscopic analysis[31], it can be calculated that the rate of utilisation of surface membrane is about $4\text{–}6 \times 10^{-11}$ m^2/cell min^{-1}. Thus, the entire amoeba surface would be utilised in *ca.* 40–60 min which is approximately the maximum period that phagocytosis of latex can be maintained.

These numbers do not differ very much from the previous calculations for macrophages and larger amoebae.

(b) *Pinocytosis* — *Acanthamoeba* continually pinocytose[29] at a rate equal to *ca.* 2.2×10^{-15} m^3/cell h^{-1} (Section 1.3.1.3(b)). To convert this volume of total fluid uptake to surface area of pinosome membrane, one must know the mean pinosome volume. This was done[29] by determining the size distribution of vesicles containing horse radish peroxidase within 20 s afters it addition to the medium. The results showed a sharp distribution around a mean diameter of 1200 Å, with a few larger vesicles of diameter as great as 1 μm. The mean pinosome volume would then be 9×10^{-22} m^3 which would necessitate the formation of 2.8×10^6 pinosomes/cell h^{-1} to maintain the observed rate of pinocytosis. The mean pinosome surface area would be 4.5×10^{-4} m^2 so that the total surface area of pinosomes would be 12×10^{-8} m^2/cell h^{-1}. Since this is 55 times the surface area of the amoeba, the calculated turnover rate is about 55 times per hour or once every 1.1 min. Even if one were to recalculate on the assumption that the largest observed vesicles were the initial pinosomes, the calculated turnover rate would be 5.5 times per hour or once every 11 min.

Despite its magnitude, the calculated turnover rate of cell surface as a consequence of endoctyosis in *Acanthamoeba* is not unreasonable. The cells from which these data were obtained had a mass of about 570 μg/10^6 cells. Growing in a medium containing 1.5% preteose peptone and 1.5% glucose, they would pinocytose about 66 μg/10^6 cells h^{-1} or the equivalent of their own mass in 9–10 h. This is quite compatible with their doubling time of *ca.* 20 h. More recently, culture conditions have been changed so that the doubling time of the cells has been reduced to *ca.* 12 h and the rate of pinocytosis has increased by about 50%.

1.3.6 Replacement of the surface membrane

1.3.6.1 Mammalian cells

As already mentioned, phagocytosis in mammalian cells is a cyclical phenomenon. One possible explanation is that the cell may be unable to synthesise plasma membrane at a rate equal to the rate of utilisation, so that eventually phagocytosis must stop until the surface membrane is replenished. It has been found[62] that after a lag period of 6 h following intense phagocytosis of latex beads by macrophages, synthesis of cholesterol and phospholipid is stimulated in proportion to the amount of membrane internalised. After *ca.* 10 h, the content of cholesterol and phospholipid in the cell can be doubled and only then will the cells regain the ability to phagocytose. This synthesis is in specific response to the endocytosis since no comparable increase in lipids occurs in control cells.

Other changes in lipid metabolism occur concomitantly with endocytosis, but these do not seem to reflect membrane synthesis. In polymorphonuclear leukocytes, for example (but not in alveolar macrophages), phagocytosis is associated with an increased rate of incorporation of ^{32}P-phosphate and ^{14}C-acetate into certain phospholipids, particularly phosphatidyl inositol[82-85].

The major membrane phospholipids, phosphatidyl choline, phosphatidyl serine and phosphatidyl ethanolamine, are not affected however, and it is not known whether the plasma membrane is involved in this process.

1.3.6.2 Amoebae

Induced pinocytosis in *Chaos chaos* is also discontinuous. One electron-microscopic study[12] suggests that it takes *ca.* 30 min to replace the surface membrane after maximum induction of pinocytosis by Alcian Blue and that it is probably newly synthesised membrane that appears at the cell surface, not recycled endosome membrane. Similarly, in *Acanthamoeba castellannii*, phagocytosis of latex beads continues only for a limited period presumably because the plasma membrane is not replaced at a rate equal to the rate of its utilisation[31]. Attempts using radioactive precursors to show an effect of phagocytosis on the rates of turnover of membrane lipids and proteins have all been negative[14, 26]. These results are difficult to interpret, however, because the rate of normal pinocytosis in the 'control' cells would have resulted in at least as much turnover of surface membrane as might have occurred in the experimental cells exposed to latex particles. In more carefully designed experiments, the rate of incorporation of radioactive precursors into plasma membranes was compared to the rate of incorporation into other cell membranes[78], but again with negative results (Section 1.3.4.2).

One can calculate, in fact, from the magnitude of membrane involvement in pinocytosis in *Acanthamoeba* (Section 1.3.5.2(b)) that replacement by synthesis is impossible. The plasma membrane accounts for *ca.* 3–6% of the cell mass or *ca.* 20–40 μg/10^6 cells. Since the surface membrane turns over *ca.* 55 times per hour while the amoebae ingest *ca.* 60 μg of nutrient per 10^6 cells, it is obvious that even if all of the material that was pinocytosed were used to replace the surface membrane only *ca.* 2.5–5% of the surface membrane utilised for endocytosis could be re-synthesised. Renewal of the cell surface must be a consequence of exocytosis.

1.3.7 Other metabolic effects of endocytosis

In mammalian phagocytes, endocytosis has pronounced effects on the general metabolism in addition to the specific effects on phospholipids already discussed. The rate of oxygen uptake is doubled with a concomitant shift from glycolysis to oxidation of glucose via the hexose monophosphate shunt in phagocytosing leukocytes[35, 40]. Perhaps the earliest detectable change is an increase in the activity of NADPH oxidase[87], which may be the cause of the increase in glucose oxidation by making more NADP available. These metabolic changes may not be related to the ingestion process itself, however, but to events occurring within the endosome. Leukocytes from patients with chronic granulomatous disease, for example, phagocytose bacteria normally but are unable to kill the bacteria; such cells are deficient in NADPH oxidase[88].

In the large amoebae, phagocytosis and induced pinocytosis inhibit cell locomotion[3]. It is not known whether this is due to specific interactions at the cell surface, competition for cell surface, or competition for force-generating mechanisms that may be common to both processes.

1.4 FATE OF ENDOSOMES

1.4.1 Translation of vesicles

1.4.1.1 Cell motility systems

Animal cells possess at least two distinct molecular mechanisms for the transduction of chemical energy into motion, and each of these systems has a major ultrastructural element. The better understood system is based on actin and myosin. In addition to muscle, amoebae and many mammalian cells are known to contain actin and myosin. The actin is organised in the cytoplasm as microfilaments (analogous to the thin filaments of muscle), but there is no evidence of the organisation of myosin in these cells in structures analogous to the thick myosin filaments of muscle. In all cases, the cytoplasmic myosin is an ATPase with many of the properties of muscle myosin, in particular a Mg^{2+} ATPase activity that is activated by actin. This system is probably involved in cell motility and in cytoplasmic streaming, and is speculatively associated with the endocytic process—perhaps in the formation of the endosome, perhaps in translational movements within the cell.

The second motility system is based on microtubules which are composed of the protein tubulin and which, at least in some cases, have been shown to be associated with another ATPase, dynein. Microtubules are frequently associated with the movement of vesicles within the cytoplasm in processes related to endocytosis.

(a) *Are actin or myosin membrane components?* — Actin is a protein of molecular weight 46 000, and polypeptides in this range are frequently seen in polyacrylamide gel electrophoretic patterns of membrane proteins. The presence of actin in a cell-surface membrane, however, has not been established. Some preparations of a plasma membrane from *Acanthamoeba castellanii* do contain actin[16], but this is due to associated microfilaments that may attach to the cytoplasmic surface of the plasma membrane[89]. Similar proximity of actin filaments to plasma membranes has been observed with platelets[90]. It should be emphasised that even in these two examples there is no irrefutable evidence for the specific association of the filaments with the membrane although it is certainly true that attachment of filaments to the membrane could provide the ultrastructural basis for cell motility in addition to serving a possible role in endocytosis.

There is at least one claim for the presence of myosin (actomyosin in fact) as a component of the erythrocyte membrane[91]. It is true that two molecules of appropriate molecular weight (*ca.* 220 000) are the major polypeptides in erythrocyte membranes[58]. These polypeptides seem to be sub-units of a protein called Tektin A[92] which, although an ATPase, has neither the enzymatic, or physical properties of myosin. Although it is claimed[59] that brain synaptic vesicles (vesicles which contain neurotransmitters to be exocytosed) contain myosin (which may interact with actin asssociated with the plasma membrane of the brain cell), until more rigorous enzymatic, physical and chemical criteria are employed the identification of these proteins remains in doubt. Furthermore, whether a putative actin or myosin is integrated into, associated with, or contaminants of, a plasma membrane preparation needs to be established by careful electron microscopy of whole cells and isolated membranes.

1.4.1.2 Are actin and myosin involved in endocytosis?

A circumstantial case could be argued from the following: in many cells, microfilaments lie beneath the plasma membrane and may attach to it; microfilaments are associated with the endoplasmic channel in large amoebae[3]; induced pinocytosis in large amoebae is competitive with locomotion and cytoplasmic streaming[53] which almost certainly are based on actomyosin; phagocytosis frequently resembles cell locomotion morphologically; the presumptive movement of surface components observed during induced pinocytosis of Alcian Blue in large amoebae, and the capping phenomenon in lymphocytes, resemble the movement of cell surface markers seen in locomotion of amoebae and fibroblasts. The experimental support for a role for the actomyosin system, however, depends solely on the effects of cytochalasin B.

Cytochalasin B is believed by many to interact specifically with microfilaments[94]. It has been proposed, therefore, that any process inhibited by cytochalasin B must involve microfilaments, and it has been reported that cytochalasin B inhibits phagocytosis of bacteria by leukocytes[95,96] and by macrophages[97]. On the other hand, pinocytosis of ferritin and colloidal gold is not inhibited by cytochalasin B[126]. In *Acanthamoeba castellanii* neither phagocytosis nor pinocytosis is inhibited by cytochalasin B[98], however, and it seems to inhibit only partially[89], or not at all[68], capping and related pinocytosis in lymphocytes.

Negative experiments may, of course, be explained by the possible impermeability of the cell to the reagent. On the other hand, cytochalasin B is probably not as specific in its action as was once thought, and it is possible that any effects on microfilaments are, in fact, secondary to the interaction of cytochalasin B with the plasma membrane. Many active transport mechanisms that certainly do not involve microfilaments are inhibited by concentrations of cytochalasin B too low to affect microfilaments[100-103]. At this time, the question of the possible involvement of actin and myosin in endocytosis cannot be answered.

1.4.1.3 Are microtubules involved in endocytosis?

As with the microfilaments, the evidence for the possible involvement of microtubules in endocytosis is partially circumstantial and partially based on the effects of a specific inhibitor. Microtubules frequently seem to focus at the golgi region of the cell, an area towards which endosomes often move. Also, vesicles not of endocytic origin do seem to move along the microtubules of nerve cells, and association between these vesicles and the microtubules has been observed electron microscopically.

The inhibitor studies are based on the effects of colchicine and related drugs which seem to be quite specific in their interaction with tubulin, the protein of which microtubules consist. But, as with cytochalasin B, the available data are inconsistent. Colchicine is reported to block the endocytic uptake of thyroglobulin from the follicular lumen of the thyroid gland[104], but colchicine is said not to inhibit phagocytosis of bacteria by macrophages although it interferes with the movement of the phagosomes towards the golgi region[97]. And colchicine and vinblastine seem not to inhibit phagocytosis of latex particles by leukocytes[66, 105], although they do have an interesting effect on the surface membrane. In the presence of colchicine, the transport-specific proteins of the plasma membrane that normally are *not* internalised during endocytosis (Sections 1.3.3) are lost from the cell surface in direct proportion to the extent of endocytosis. These data suggest either that microtubules are involved in the maintenance of the specific organisation of the cell surface so that the phagocytic sites and the active transport sites are kept separate, or that during the endocytic process there is a microtubule-dependent movement of the transport proteins such that they are not internalised with the endosome membrane. The possible effects of colchicine on movements of elements of the plasma membrane are not clarified by the apparently contradictory reports that colchicine blocks the capping phenomenon in cultured fibroblasts[106] but does not inhibit capping in lymphocytes[67].

From the above brief summary, it is clear that there are movements of proteins within or on the cell surface and that these are probably related to endocytosis as well as to cell motility. The dependence of these movements on either microfilaments (actin–myosin system) or on microtubules, or both, has yet to be worked out. The subsequent formation and cytoplasmic movement of endosomes may also be controlled by one or the other or both of these systems, but again the experimental facts are inadequate to allow conclusions.

1.4.2 Fusion of endosomes with lysosomes

Very soon after their formation, endosomes fuse with each other, with primary and with secondary lysosomes, and also undergo fission reactions thus forming large and small vesicles which contain a number of hydrolytic enzymes[29, 61, 108, 109]. The origin of these enzymes is presumably the endoplasmic reticulum via the golgi region of the cell, but the myriad of morphological transitions occurring in the cell have made it impossible to determine in any particular instance the relative contribution of primary and secondary lysosomes to the final mix within the digestive vacuole. The general aspects of these phenomena have been exhaustively and competently reviewed[4, 109, 110] but the molecular basis of the internal fusions and the selectivity with which they occur remain obscure. It is presumably the lipid or protein composition of the membranes that allows the fusions to occur and that discriminates between those membranes that can fuse with one another and those which cannot. Until much more is known about the composition and organisation of the individual membranes, there is little point in speculation.

In most cases, the evidence for the presence of hydrolytic enzymes in the digestive vacuoles comes from cytochemical techniques that allow localisation of specific enzymes (usually acid phosphatase) by electron microscopy. In some cases, the secondary lysosomes have been isolated from homogenates taking advantage of specific-density characteristics conferred by their content of Triton, of latex particles, or of paraffin emulsions that the cells had endocytosed. This latter technique allows quantitative, kinetic studies.

Thus, it was found[61] that within 30 min after *Acanthamoebae* were exposed to latex particles the isolated phagosomes contained about 25% of the total cellular acid phosphatase and β-glucosidase. When polymorphonuclear leukocytes were exposed to paraffin emulsions[111], the isolated phagosomes contained hydrolytic enzymes within 15 min (the earliest time studied), but whereas the acid and alkaline phosphatase levels continued to increase for at least 1 h, the levels of NADH oxidase, β-glucuronidase and peroxidase did not. The differential rate of appearance of the two groups of enzymes within the phagosomes confirms other evidence that they are present in different vesicles within the cell.

1.4.3 Utilisation of endosome contents

Very little information has been obtained on the digestion and utilisation of the material within the endosome. The process seems to be very slow in mammalian macrophages where the digestion of albumin (probably to amino acids) has a half-time of *ca.* 5 h[35] and the digestion, of haemoglobin and horse radish peroxidase a half-time of *ca.* 25 h[34, 112]. Even the inactivation of the enzymic activity of horse radish peroxidase was very slow with a half-life of 7–9 h[34]. The rates of these processes are in agreement with the hypothesis that endoctyosis in these cells serves a protective, not a nutritional, function; once the foreign material has been removed from the circulation, the organism is protected. This conclusion is further supported by the observation that bacteria are 'killed' by polymorphonuclear leukocytes within 30 min, by which time only a small fraction of the bacterial phospholipids has been degraded[113].

Available data suggest that digestion in amoeba, which are nutritionally dependent on endocytosis, may be much more rapid. About 50% of the radioactivity of ingested [^{131}I]albumin is lost from *Acanthamoebae* within 15 min after endocytosis[29], in contrast to the hours required for mammalian macrophages. Despite the importance of the process, little more is known about the utilisation of the endocytic contents. How do the hydrolytic products reach the cytoplasm of the cell? Does the permeability of the endosome membrane differ from the permeability of the plasma membrane?

1.4.4 Re-utilisation of endosome membrane

Exocytosis is a well-documented phenomenon in specific secretory processes such as synthesis of plasma proteins in the liver, pancreatic enzymes, numerous hormones and neurotransmitters, plant cell walls, etc. But it is not known to what extent, and how, the endosome membrane is returned to the cell surface. It has already been suggested here, and in other reviews, that in some circumstances the endocytic membrane is not re-utilised (phagocytosis of latex particles by amoebae, or endocytosis in many mammalian phagocytes in which secondary lysosomes accumulate). This may even be exaggerated in certain disease states where hydrolytic enzymes are deficient and the contents of the endosomes are in consequence indigestible.

In contrast, it was calculated in Section 1.3.6.2 that the rate of utilisation of the *Acanthamoeba* cell surface by endocytosis was so great that replacement by re-synthesis was impossible and that the internalised surface membrane must ultimately find its way back to the cell surface. Little direct evidence for this process exists, however. The electron-microscopic observations on *Chaos chaos* and the radioactive tracer experiments with *Acanthamoeba castallanii*, both of which have already been discussed (Sections 1.3.4.2 and 1.3.6.2), would seem to be more closely connected to the growth-related synthesis of plasma membrane than to the replacement of membrane ingested by endocytosis. Also, there is strong electron-microscopic, chemical and enzymatic evidence that as a consequence of digestive and fusion processes the membranes of secondary lysosomes differ significantly from plasma membranes. Thus, if the endosome membrane is ultimately to rejoin the plasma membrane it must be reconverted to its former state.

1.5 CONCLUDING REMARKS

The importance of endocytosis to protozoa has long been apparent; its importance to man is being increasingly recognised. In addition to the more obvious phagocytic defence mechanisms, endoctyosis now seems to be a necessary event in eliciting the immune response and in processing macromolecules such as thyroglobulin. Also a growing list of hereditary diseases appear to be the consequence of degenerate endocytic processes. Although membrane lesions may also be involved, these pathological states are frequently polysaccharide and lipid–storage diseases characterised by a deficiency of one or more lysosomal enzymes. Such diseases might be alleviated by inducing pinocytosis of exogenously supplied enzymes. Precedent for this lies in experiments[114] in which the toxic effects of sucrose accumulated by pinocytosis in the lysosomes of macrophages was reversed by the subsequent pinocytosis of invertase, an enzyme lacking in macrophages. Similarly, functional pinocytosis of uricase has been obtained with alveolar macrophages[115]. In fact, clinical attempts of treating glycogen storage disease by α-glycosidase, the deficient enzyme, preceded these experimental models[116].

An interesting ramification of this theme is the recent observation[117] that fibroblasts cultured from patients with I (inclusion)-cell disease, in which at least seven lysosomal enzymes are deficient, secrete these enzymes into the medium. The enzymes appear to be 'faulty' in that the I-cells cannot pinocytose them as well as they pinocytose their normal counterparts. The inference is drawn that, contrary to the accepted theory that lysosomes are formed with their enzymatic content in the golgi region, fibroblast lysosomes obtain their enzymatic content by pinocytosis of enzymes secreted by other cells. Whatever the cause of the disease, pinocytosis of normal enzymes remains possible therapy. In fact, pinocytosis may provide a mechanism for entry of functional macromolecules other than proteins into the cell should it prove possible for them to get through the endosome membrane and into the cytoplasm.

Although much of this review has been concerned with pinocytic processes that seem to lack substrate and cell specificity, there is increasing evidence, in addition to that discussed in the preceding paragraph, for the existence of specific pinocytic mechanisms in mammalian systems. Thus, B-lymphocytes have specific receptors on the cell surface for antigenic molecules[27], and liver parenchymal cells apparently have specific receptor sites for a terminal β-galactosyl moiety on a glycoprotein[24]. Electron microscopy has revealed morphologically-distinct regions on the surface of many cells, referred to as coated pits, which may be involved specifically in pinocytosis of proteins[121-125]. The possibility should be considered that coated pits are the ultrastructural counterpart of chemically-specific receptor sites on the cell surface.

It will be obvious from this review how little is really known of the biochemical mechanisms of endocytosis. The molecular basis of substrate recognition, of endosome formation, of membrane fusion, of translational activities and of digestive and transport functions within the lysosome are all relatively obscure. What is clear is that the biochemistry of endocytosis is intimately involved with the yet more general problems of the biochemistry of cell membranes and the biochemistry of cell motility.

Speculatively, a picture is beginning to develop in which membrane fusion may be a consequence of specific interactions of particulate proteins within the membrane (based on very fragmentary electron-microscopic evidence[69, 119]) and a micellisation and re-formation of the phospholipid bilayer as a consequence of enzymatic deacylation and re-acylation. Might enzymes be associated with the intramembrane particles? The movement of membrane proteins that may accompany endocytosis and/or the vesiculation and subsequent translation of the endocytic vesicles may involve microfilaments and/or microtubules. It has taken a number of years to reach this point, but now that the problems are better defined and many experimental systems are available to study endocytosis and membrane fusion (for example, viral-induced cell fusions[72, 127], membrane fusions *in vitro*[128] and phospholipid liposomes in addition to those discussed in this review), perhaps we can anticipate more rapid progress.

References

1. Holter, H. (1959). *International Review of Cytology*, Vol. 8, 481 (G. H. Bourne and J. F. Danielli, editors) (New York: Academic Press)
2. Holter, H. (1965). *Symposia of the Society for General Microbiology*, 15, 89
3. Stockem, W. and Wohlfarth-Bottermann (1969). *Handbook of Molecular Cytology* (A. Lima de-Faria, editor) (New York: Wiley-Interscience)
4. De Duve, C. and Wattiaux, R. (1966). *Ann. Rev. Physiol.*, **28,** 435
5. Gropp, A. (1963). *Cinematography in Cell Biology*, 279 (G. G. Rose, editor) (New York: Academic Press)

6. Jacques, P. J. (1969). *Lysosomes in Biology and Pathology*, Vol. 2 (H. Fell and J. Dingle, editors) (Amsterdam: North Holland)
7. Metchnikoff, E. (1883). *Biol. Zbl.*, **3,** 560
8. Lewis, W. H. (1931). *Johns Hopkins Hospital Bulletin*, **49,** 17
9. Cook, G. M. W. (1968). *Biol. Revs.*, **43,** 363
10. Hughes, R. C. (1973). *Progress in Biophysics and Molecular Biology*, Vol. 26, 189 (J. A. V. Butler and D. Noble, editors) (Oxford: Pergamon)
11. Pappas, G. D. (1959). *Ann. N.Y. Acad. Sci.*, **78,** 448
12. Nachmias, V. T. (1966). *Exp. Cell. Res.*, **43,** 583
13. Rustad, R. C. and Rustad, L. C. (1961). *Biol. Bull.*, **121,** 377
14. Ulsamer, A. G., Wright, P. L., Wetzel, M. G. and Korn, E. D. (1971). *J. Cell. Biol.*, **51,** 193
15. Korn, E. D. and Olivecrona, T. (1971). *Biochem. Biophys. Res. Commun.*, **45,** 90
16. Korn, E. D. and Wright, P. L. (1973). *J. Biol. Chem.*, **248,** 439
17. Korn, E. D., Dearborn, D. G., Fales, H. M. and Sokoloski, E. A. (1973). *J. Biol. Chem.*, **248,** in press
18. Bowers, B. and Korn, E. D. (1968). *J. Cell. Biol.*, **39,** 95
19. Bowers, B. and Korn, E. D. (1973). Submitted for publication
20. Michell, R. H., Pancake, S. J., Noseworthy, J. and Karnovsky, M. L. (1969). *J. Cell Biol.*, **40,** 216
21. Rabinovitch, M. (1969). *Exp. Cell Res.*, **54,** 210
22. Rabinovitch, M. (1970). *Exp. Cell Res.*, **59,** 272
23. Cohn, Z. A. and Parks, E. (1967). *J. Exp. Med.*, **125,** 213
24. Ashwell, G. and Morell, A. G. (1971). *Glycoproteins of Blood Cells and Plasma*, 173 (G. A. Jamieson and T. J. Greenwalt, editors) (Philadelphia: J. B. Lippincott)
25. Ryser, H. J. P. (1968). *Science*, **159,** 390
26. Sharon, N. and Lis, H. (1972). *Science*, **177,** 949
27. Pierce, C. W., Asofsky, R. and Solliday, S. M. (1973). *Fed. Proc.*, **32,** 41
28. Christiansen, R. G. and Marshall, J. M. (1965). *J. Cell Biol.*, **25,** 443
29. Bowers, B. and Olszewski, T. E. (1972). *J. Cell Biol.*, **53,** 681
30. Ray, D. L. (1959). *J. Exp. Biol.*, **118,** 443
31. Weisman, R. A. and Korn, E. D. (1967). *Biochemistry*, **6,** 485
32. Korn, E. D. and Weisman, R. A. (1967). *J. Cell. Biol.*, **34,** 219
33. Roberts, J. and Quastel, J. H. (1963). *Biochem. J.*, **89,** 150
34. Steinman, R. M. and Cohn, Z. A. (1972). *J. Cell. Biol.*, **55,** 186
35. Ehrenreich, B. A. and Cohn, Z. A. (1967). *J. Exp. Med.*, **126,** 941
36. Ryser, H. J.-P. (1963). *Lab. Invest.*, **12,** 1009
37. Berger, R. R. and Karnovsky, M. L. (1966). *Fed. Proc.*, **25,** 840
38. Saito, H. and Uzman, B. G. (1971). *Exp. Cell. Res.*, **66,** 90
39. Sbarra, A. J. and Karnovsky, M. L. (1959). *J. Biol. Chem.*, **234,** 1355
40. Oren, R., Farnham, A. E., Saito, S., Milofsky, E. and Karnovsky, M. L. (1963). *J. Cell Biol.*, **17,** 487
41. Cohn, Z. A. (1966). *J. Exp. Med.*, **124,** 557
42. Schumaker, V. N. (1958). *Exp. Cell. Res.*, **15,** 314
43. Hausmann, E. and Stockem, W. (1972). *Cytobiologie*, **5,** 281
44. Nachmias, V. T. and Marshall, J. M. (1961). *Biological Structure and Function*, Vol. 2, 605 (T. W. Goodwin, and O. Lindberg, editors) (New York: Academic Press)
45. Chapman-Andresen, C. and Holter, H. (1964). *Compt. Rend. Trav. Lab. Carlsberg*, **34,** 211
46. Bowers, B. and Korn, E. D., unpublished observations
47. Korn, E. D. (1969). *Ann. Rev. Biochem.*, **38,** 263
48. Korn, E. D. (1969). *Theoretical and Experimental Biophysics*, Vol. 2, 1 (A. Cole, editor) (New York: Marcel Dekker)
49. Korn, E. D. (in the press). *The Inflammatory Process*, Vol. 1 (B. W. Zweifach, L. Grant and R. T. McClusky, editors) (New York: Academic Press)
50. Hendler, R. W. (1971). *Physiol. Rev.*, **51,** 66
51. Bretscher, M. (1972). *J. Mol. Biol.*, **71,** 523
52. Kornberg, R. D. and McConnell, H. M. (1971). *Proc. Nat. Acad. Sci. USA*, **68,** 2564
53. Kornberg, R. D. and McConnell, H. M. (1971). *Biochemistry*, **10,** 1111

54. Demel, R. A., Bruckdorfer, K. R. and van Deenen, L. L. M. (1972). *Biochim. Biophys. Acta*, **255,** 321
55. Singer, S. J. and Nicholson, G. L. (1972). *Science*, **175,** 720
56. Bretscher, M. (1971). *J. Mol. Biol.*, **59,** 351
57. Neville, D. M. and Glossmann, H. (1971). *J. Biol. Chem.*, **246,** 6335
58. Fairbanks, G., Steck, T. L. and Wallach, D. F. H. (1971). *Biochemistry*, **10,** 2606
59. Berl, S., Puszkin, S. and Nicklas, W. J. (1973). *Science*, **179,** 441
60. O'Neill, C. H. (1964). *Exp. Cell Res.*, **35,** 477
61. Wetzel, M. G. and Korn, E. D. (1969). *J. Cell. Biol.*, **43,** 90
62. Werb, Z. and Cohn, Z. A. (1972). *J. Biol. Chem.*, **247,** 2439
63. Henning, R., Kaulen, H. D. and Stoffel, W. (1970). *Hoppe-Seyler's Z. Physiol. Chem.*, **351,** 1191
64. Kaulen, H. D., Henning, R. and Stoffel, W. (1970). *Hoppe-Seyler's Z. Physiol. Chem.*, **351,** 1555
65. Heine, J. W. and Schnaitman, C. A. (1971). *J. Cell Biol.*, **48,** 703
66. Tsan, M. and Berlin, R. D. (1971). *J. Exp. Med.*, **134,** 1016
67. Raff, M. C. and de Petris, S. (1973). *Fed. Proc.*, **32,** 48
68. Karnovsky, M. J. and Unanue, E. R. (1973). *Fed. Proc.*, **32,** 55
69. Satir, B., Schooley, C. and Satir, P. (1973). *J. Cell Biol.*, **56,** 153
70. Lucy, J. A. (1970). *Nature (London)*, **227,** 815
71. Poole, A. R., Kowell, J. I. and Lucy, J. A. (1970). *Nature (London)*, **227,** 810
72. Toister, Z. and Loyter, A. (1973). *J. Biol. Chem.*, **248,** 422
73. Victoria, E. and Korn, E. D., unpublished observations
74. Victoria, E. J., van Golde, L. M. G., Hostelter, K. Y., Scherphof, G. L. and van Deenen, L. L. M. (1971). *Biochim. Biophys. Acta*, **239,** 443
75. Elsbach, P. (1968). *J. Clin. Invest.*, **47,** 2217
76. Elsbach, P. and Farrow, S. (1969). *Biochim. Biophys. Acta*, **176,** 438
77. Ulsamer, A. U., Smith, F. R. and Korn, E. D. (1969). *J. Cell Biol.*, **43,** 105
78. Simmons, S. and Korn, unpublished observations
79. Chapman-Andresen, C. (1961). *Progress in Protozoology*
80. Hausmann, E., Stockem, W. and Wohlfarth-Bottermann, K. E. (1972). *Z. Zellforsch.*, **127,** 270
81. Bowers, B. and Korn, E. D. (1969). *J. Cell. Biol.*, **49,** 786
82. Sbarra, A. J. and Karnovsky, M. L. (1960). *J. Biol. Chem.*, **235,** 2224
83. Karnovsky, M. L. and Wallach, D. F. H. (1961). *J. Biol. Chem.*, **236,** 1895
84. Berger, R. R. and Karnovsky, M. L. (1966). *Fed. Proc.*, **25,** 840
85. Sastry, P. S. and Hokin, L. E. (1966). *J. Biol. Chem.*, **241,** 3354
86. Goodall, R. J., Lai, Y. F. and Thompson, J. E. (1972). *J. Cell. Sci.*, **11,** 569
87. Zatti, M. and Rossi, F. (1965). *Biochim. Biophys. Acta*, **99,** 557
88. Baehner, R. L. and Karnovsky, M. L. (1968). *Science*, **162,** 1277
89. Pollard, T. D. and Korn, E. D. (1973). *J. Biol. Chem.*, **248,** 4682
90. Zucker-Franklin, D. (1970). *J. Cell Biol.*, **47,** 293
91. Guidotti, G. (1972). *Ann. Rev. Biochem.*, **41,** 731
92. Clarke, M. (1971). *Biochem. Biophys. Res. Commun.*, **45,** 1063
93. Nachmias, V. T. (1968). *Exp. Cell. Res.*, **51,** 347
94. Wessels, N. K., Spooner, B. S., Ash, J. F., Bradley, M. I., Luduena, M. A., Taylor, E. L., Wrenn, J. T. and Yamada, K. M. (1971). *Science*, **171,** 135
95. Davis, A. T., Estensen, R. and Quie, P. G. (1971). *Proc. Soc. Exp. Biol., Med.*, **137,** 161
96. Malawista, S. E., Gee, J. B. L. and Bensch, K. G. (1971). *Yale J. Biol. Med.*, **44,** 286
97. Allison, A. C., Davies, P. and de Petris, S. (1971). *Nature (London)*, **232,** 153
98. Weihing, R. R. and Korn, E. D., unpublished observations
99. Taylor, R. B., Duffus, W. P., Raff, M. C. and de Petris, S. (1971). *Nature (London)*, **233,** 225
100. Sanger, J. W. and Holtzer, H. (1972). *Proc. Nat. Acad. Sci. USA*, **69,** 253
101. Kletzien, R. F. and Perdue, J. F. (1973). *J. Biol. Chem.*, **248,** 711
102. Estensen, R. D. and Plagemann, P. G. W. (1972). *Proc. Nat. Acad. Sci. USA*, **69,** 143
103. Mizel, S. B. and Wilson, L. (1972). *J. Biol. Chem.*, **247,** 4102
104. Williams, J. A. and Wolff, J. (1972). *J. Cell Biol.*, **54,** 157
105. Ukena, T. E. and Berlin, R. D. (1972). *J. Exp. Med.*, **136,** 1
106. Edidin, M. and Weiss, A. (1972). *Proc. Nat. Acad. Sci. USA*, **69,** 2456

107. Cohn, Z. A., Fedorka, M. E. and Hirsch, J. H. (1966). *J. Exp. Med.*, **123,** 757
108. Hirsch, J. G., Fedorka, M. E. and Cohn, Z. A. (1968). *J. Cell. Biol.*, **38,** 629
109. Novikoff, A. B. (1963). *Colloq. Intern. Centre Nat. Rech. Sci.* (*Paris*), **115,** 68
110. Novikoff, A. B., Essner, E. and Quintana, N. (1964). *Fed. Proc.*, **23,** 1023
111. Stossel, T. P., Pollard, T. D., Mason, R. J. and Vaughan, M. (1971). *J. Clin. Invest.*, **50,** 1745
112. Ehrenreich, B. A. and Cohn, Z. A. (1968). *J. Cell. Biol.*, **38,** 244
113. Patriarca, P., Beckerdite, S., Pettis, P. and Elsbach, P. (1972). *Biochim. Biophys. Acta*, **280,** 45
114. Cohn, Z. A. and Ehrenreich, B. A. (1969). *J. Exp. Med.*, **129,** 201
115. Theodore, J., Acevedo, J. C. and Robin, E. D. (1972). *Science*, **178,** 1302
116. Hug, G. and Schubert, W. (1967). *J. Cell. Biol.*, **35,** C1
117. Hickman, S. and Neufeld, E. F. (1972). *Biochem. Biophys. Res. Commun.*, **49,** 992
118. Ryan, J. W. and Smith, U. (1971). *Biochim. Biophys. Acta*, **249,** 177
119. Smith, U., Ryan, J. W. and Smith, D. S. (1973). *J. Cell. Biol.*, **56,** 492
120. Lutton, J. D. (1973). *J. Cell Biol.*, **56,** 611
121. Roth, T. F. and Porter, K. R. (1904). *J. Cell Biol.*, **20,** 313
122. Kanaseki, T. and Kadota, K. (1969). *J. Cell Biol.*, **42,** 202
123. Friend, D. S. and Farquhar, M. G. (1967). *J. Cell Biol.*, **35,** 352
124. Douglas, W. W., Nagasawa, J. and Schulz, R. A. (1971). *Nature* (*London*), **232,** 340
125. Nagasawa, J., Douglas, W. W. and Schulz, R. A. (1971). *Nature* (*London*), **232,** 341
126. Wills, E. J., Davies, P., Allison, A. C. and Haswell, A. D. (1972). *Nature* (*London*), **240,** 58
127. Baker, R. F. (1972). *J. Cell. Biol.*, **53,** 244
128. Kaspar, C. B. and Kubinski, H. (1971). *Nature* (*New Biol.*), **231,** 124

2
The Ion Selectivity of Carrier Molecules, Membranes and Enzymes*

G. EISENMAN and S. J. KRASNE
U.C.L.A. School of Medicine, Los Angeles

* Based in part on a lecture presented to the Symposium on Membrane Structure and Function, IVth International Biophysics Congress, Moscow, USSR, August, 1972.

2.1 INTRODUCTION

One of the salient characteristics of all biological membranes[1] and many enzymes[2] is their sensitivity to, and ability to distinguish amongst, such chemically similar cations as Li^+, Na^+, K^+, Rb^+ and Cs^+, as well as others such as NH_4^+ and Tl^+. Although this ion selectivity is dependent on ion size, even for the alkali metal cations it is not a simple monotonic function of the radius of the ions being selected, as has been discussed elsewhere[1,3-6]. For example, the permeability of the 'K^+ channel' of frog nerve[7], as well as the resting permeability of the squid axon[8], to alkali metal cations is characterised by the selectivity sequence (K > Rb > Cs > Na > Li), while the activation of crab nerve adenosine triphosphatase[9,10] can occur either in the sequence Na > Li > K > Cs > Rb when activated by magnesium alone or in the sequence K > Rb > Cs > Na > Li when activated by sodium as well as magnesium[4,9,10].

Recently, a variety of cyclic molecules has been isolated from natural sources[11-14], as well as prepared synthetically[15], which exhibit selective ion complexation[16,17] and act as ion carriers in artificial lipid bilayer membranes[18-24]. Molecular models of the naturally-occurring carriers, nonactin,

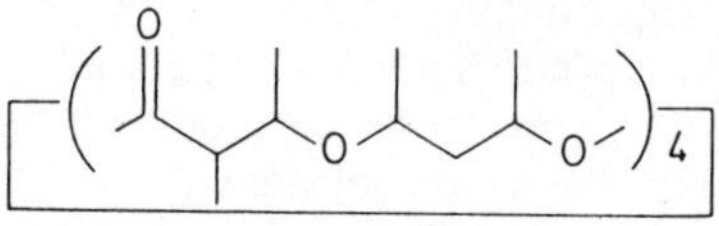

and valinomycin, cyclo-(D-Val-L-Lac-L-Val-D-HyIv)$_3$, as well as the synthetic carrier hexadecavalinomycin, cyclo-(D-Val-L-Lac-L-Val-D-HyIv)$_4$, synthesised by the Shemyakin Institute[25] are illustrated in Figure 2.1. These molecules share the common property of combining a lipophilic exterior with a polar interior consisting of oxygen ligands (e.g. ethers, ester carbonyls, amide carbonyls) energetically suitable for replacing all or part of the normal hydration shell of cations. The molecular structures of the ion-selective 'sites'

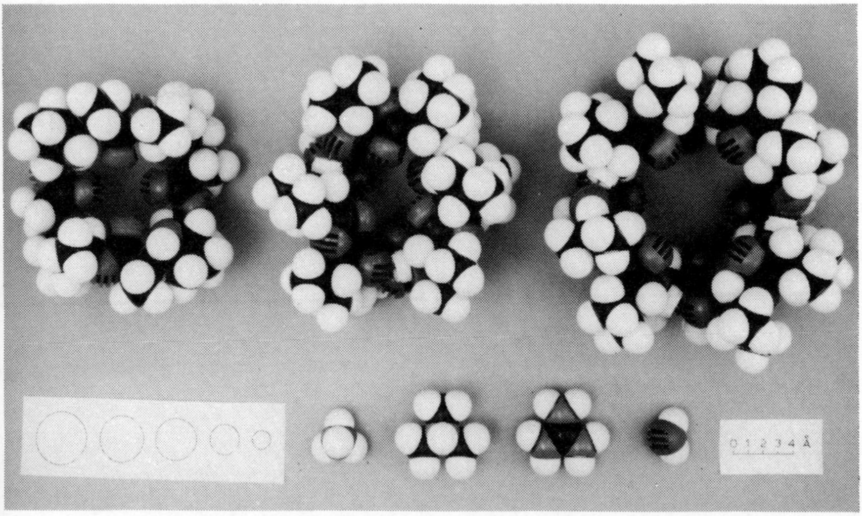

Figure 2.1 CPK models of three typical carriers and representative ions which they carry. From left to right above, the carriers are nonactin, valinomycin, and hexadecavalinomycin. All of these are in the configurations inferred for their cation complexes[17,23,25-31]. From left to right below are presented the circumferences of Cs^+, Rb^+, K^+, Na^+, and Li^+, CPK models of NH_4^+, $(CH_3)_3NH^+$ and hydrazinium$^+$, and a CPK model of the water molecule. Notice that in nonactin and valinomycin the cavity size is about appropriate to the middle-sized alkali cations K^+ or Rb^+ whilst it is considerably larger for hexadecavalinomycin. Also notice that nonactin is characterised by an approximately cubic coordination, with the four tetrahedrally-orientated carbonyl ligands appearing more prominent and the four tetrahedrally-orientated ether ligands less clearly visible in this view. The six octahedrally-arrayed ester carbonyls of valinomycin are easily seen, as are the eight ester carbonyl ligands of hexadecavalinomycin which are arrayed at the corners of a square-antiprism. Comparing the geometric array of the eight ligands in nonactin with that in hexadecavalinomycin, it can be seen that there is no tetrahedral array in the latter, in contrast to the situation in nonactin.

of membranes and enzymes are of great interest but presently unknown, whereas for these cyclic molecules not only are the primary structures completely known but so are the conformations of their ion complexes in appropriate solvents[17, 23, 25, 26] or crystals[27-31]. These molecules contain the same ligands as those presumably present in enzymes or membranes and utilise natural molecular components (e.g. amino acids); they therefore provide unique model systems in which to study at the molecular level the elementary factors giving rise to ion selectivity.

A sin the biological examples cited above, the alkali cation selectivity of these ion complexers (or carriers) is not a simple monotonic function of the

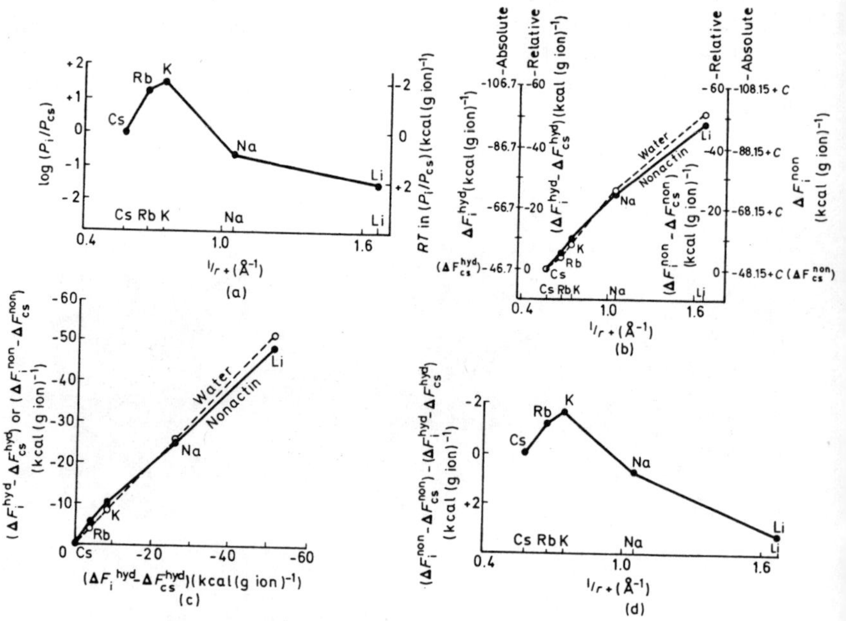

Figure 2.2 (a) Non-monotonic dependence on ion size of the selectivity of nonactin.

The logarithm of the permeability ratios relative to Cs^+ determined from zero-current potential measurements of a bilayer membrane[24] in the presence of nonactin is ploted versus the reciprocal of the cation radius[39]. The right-hand ordinate gives the equivalent in energy units.

(b) Monotonic dependence of the ion-binding energy to nonactin and of the ion-hydration energy on the reciprocal cation radius.

The left-hand ordinate pertains to the ion-hydration energy (open circles, dashed line), and the right-hand ordinate pertains to the ion-binding energy of nonactin (closed circles, solid line). Each ordinate has two scales. The external scales, labelled 'absolute', are absolute values of free energy and the internal scales, labelled 'relative, are values of free energy relative to those for Cs^+.

(c) The asymmetry between nonactin and water.

The ordinate is the relative free energy of binding to nonactin (closed circles, solid lines) or the relative free energy of hydration (open circles, dashed lines) of the alkali cations and the abscissa is the relative free energy of hydration for the alkali cations.

(d) The selectivity of nonactin.

The ordinate is the difference between the relative free energy of binding of the alkali cations to nonactin and the relative free energy of hydration of the alkali cations and the abscissa is the reciprocal ion radius.

sizes of the ions being selected[32-38]. For example, for nonactin the sequence of permeability ratios (K > Rb > Cs > Na > Li) has a maximum[24] at the intermediate-sized K^+ ion as is illustrated in Figure 2.2(a), where the logarithm of the experimentally-observed permeability ratio has been plotted as a function of the reciprocal of the naked (Pauling[39]) cation radius. Other non-monotonic sequences are characteristic of different molecules; for example, a sequence (Rb > K > Cs > Na > Li) in which the maximum occurs at the Rb^+ ion is observed for valinomycin[18, 19] and hexadecavalinomycin[40], while for cyclic polyethers a variety of further sequences is seen[41-44].

The detailed mechanisms by which these complexers act as carriers to enable ions to permeate bilayers has been discussed elsewhere. In particular, it has been demonstrated that over a wide range of conditions these non-monotonic selectivities reflect well-defined chemical equilibria[24, 35, 45, 47]. We therefore restrict our present considerations to equilibrium interactions without becoming involved in the greater complications attendant when kinetic factors become important, as they certainly do in the 'kinetic domain'[35, 45] of experimental conditions even for the present carriers[46-51] or in the kinetic aspects of membrane permeation through channels.

The lack of a monotonic dependence on ion size for complexer selectivity might seem paradoxical since the binding energy between an ion and the ligands of the binding molecule is expected to increase monotonically with decreasing ion size (and therefore decreasing distance of closest approach)*. However, the equilibrium selectivity of a complexer reflects not just the binding energy' of the complexer to each of the ions but represents the balance of the energy involved in the process of taking one ion out of water and putting it inside of the complexer versus that for the ion (this is because the aqueous solution is generally chosen as the reference state in defining selectivity)†. Thus, the selectivity reflects the *difference* between the 'binding energy' of the ions to the complexer and the 'binding energy' of the ions to water (i.e. the ion-hydration energy). We shall now show that, although each of these binding energies is a monotonic function of ion size, the difference between these energies will have a non-monotonic dependence on ion size if the two energies are 'asymmetrical' as defined below.

2.2 'ASYMMETRICAL' INTERACTIONS OF IONS WITH WATER MOLECULES AND WITH OTHER MOLECULES AS THE BASIS OF ION SELECTIVITY AND THE VARIOUS THEORIES ON THE PHYSICAL ORIGINS OF THESE ASYMMETRIES

2.2.1 Monotonic dependence on ion size of binding and hydration energies

Figure 2.2(b) illustrates that both the free energy of ion hydration and the free energy of ion binding to nonactin are functions which increase monotonically relative to reciprocal ion size. This is true whether we plot the free

* This holds not only for ion—monopole interactions but also for more complex attractions (e.g. ion—multipole, induction forces, dispersion forces, etc.).

† In this paper we will use the term 'selectivity' in this narrow sense, i.e. with reference to water.

energy for each ionic species or the differences of free energies relative to Cs^+, both of which yield the same data points when plotted against the appropriate ordinate. Thus, the left-hand ordinate pertains to ion hydration and has two scales of free energy. The scale labelled 'absolute' is the absolute free energy of ion hydration for the individual ion, ΔF_i^{hyd} (referred to vacuum[52]), for which we have plotted as open circles Salomon's[53] values calculated from thermochemical data. The scale labelled 'relative' is for the free energy of ion hydration relative to that of Cs^+, $(\Delta F_i^{hyd} - \Delta F_{Cs}^{hyd})$, which it will be useful to call 'relative hydration energies'. Notice that such a plot of the data may be superimposed with the absolute values since the two scales on the left-hand ordinate differ in value only by an additive constant, the free energy of hydration of Cs^+ ($\Delta F_{Cs}^{hyd} = 46.7$ kcal (g ion)$^{-1}$). The right-hand ordinate pertains to the ion-binding energy of nonactin, and again both 'absolute' and 'relative' scales are provided. We have reconstructed the absolute free energy of ion binding to nonactin (referred to vacuum), ΔF_i^{non}, to within the value of an additive constant, C, from the nonactin salt-extraction equilibrium constants for the alkali cations using a classical Born–Haber thermodynamic cycle (for details see Ref. 37). These values are plotted as filled circles. The scale labelled 'relative' plots the free energy of ion binding to nonactin relative to that of Cs^+ $(\Delta F_i^{non} - \Delta F_{Cs}^{non})$, which we will call the 'relative binding energy'. Notice again that data points for 'relative' and 'absolute' energies so plotted are identical since these two scales differ only by an additive constant ($48.15 + C$ kcal (g ion)$^{-1}$ [37]).

2.2.2 'Asymmetrical' and 'symmetrical' interactions

It is useful to distinguish two classes of interactions between ions and the molecules which bind (or solvate) them. The first class of interactions, which we shall call 'symmetrical', is capable of producing only monotonic selectivity sequences. In contrast, the second class of interactions, which we shall call 'asymmetrical', is capable of producing non-monotonic selectivity sequences as well.

Symmetrical and asymmetrical interactions can be defined from Figures 2.2(b) and 2.2(c). Although the free energies underlying both binding and hydration in Figure 2.2(b) are seen to be monotonic functions of decreasing ion radius, they are not identical functions. If the curve for hydration energy is expressed as a particular empirical function of ion radius,

$$\Delta F_i^{hyd} = f(r_+) \tag{2.1}$$

then the curve for ion-binding energy, ΔF_i^{bind}, which will in general be a different empirical function of ion radius, is defined as being *symmetrical* (to the hydration energy) when

$$\Delta F_{i(sym)}^{bind} = B \times f(r_+) + C \tag{2.2}$$

and as being *asymmetrical* (to the hydration energy) when

$$\Delta F_{i(asym)}^{bind} \neq B \times f(r_+) + C \tag{2.3}$$

In equations (2.2) and (2.3), B and C are constants characteristic of the particular binding molecule(s).

Whether the ion-binding energy is symmetrical or asymmetrical to the hydration energy cannot be easily seen from a plot such as Figure 2.2(b), but this distinction becomes apparent on plotting the relative binding energies of ions with nonactin ($\Delta F_i^{non} - \Delta F_{Cs}^{non}$) and with water ($\Delta F_i^{hyd} - \Delta F_{Cs}^{hyd}$) against the relative binding energy to water ($\Delta F_i^{hyd} - \Delta F_{Cs}^{hyd}$) as is done in Figure 2.2(c). For water we obtain, of course, a straight line of unit slope passing through the origin, while for nonactin we see a distinct curvature. (Note that we would reach the same conclusion had we plotted the absolute binding energies instead of the relative binding energies.) Generalising the above definitions of symmetrical and asymmetrical interactions to this form of plot leads to

$$(\Delta F_i^{bind} - \Delta F_{Cs}^{bind})_{(sym)} = B(\Delta F_i^{hyd} - \Delta F_{Cs}^{hyd}) \tag{2.4}$$

and

$$(\Delta F_i^{bind} - \Delta F_{Cs}^{bind})_{(asym)} \neq B(\Delta F_i^{hyd} - \Delta F_{Cs}^{hyd}) \tag{2.5}$$

Thus, if the interactions of ions with the binding molecule(s) are symmetrical to their interactions with water, their relative binding energies to the molecule(s) will be proportional to their relative hydration energies and will produce a straight line on a plot such as Figure 2.2(c) (cf. the dashed line for water). In contrast, if the interactions of ions with the binding molecule(s) are symmetrical to their interactions with water, their relative binding energies will not be proportional to their hydration energies and will therefore produce a curved line when plotted against the relative hydration energy as in Figure 2.2(c) (cf. the curve for nonactin). We shall therefore define any binding which produces a curvature on such a plot as being asymmetrical, whilst any binding which yields a straight line when so plotted will be defined as being symmetrical.

The curvature in Figure 2.2(c) for the plot of nonactin versus water reflects a difference in the relative ion-binding energies of nonactin and water which attains a maximum for K^+ and, of course, it is this energy difference which corresponds to the selectivity of nonactin. This energy difference is plotted in Figure 2.2(d) versus the reciprocal cation radius and can be almost identically superimposed on the curve of the nonactin selectivity deduced from bilayer permeability ratios (cf. Figure 2.2(a)). The similarity between these two curves is no surprise since both the bilayer permeability ratios and the salt-extraction ratios reflect the same underlying equilibrium selectivity of nonactin (as has been discussed elsewhere[24, 35, 45, 54]).

The pattern of selectivity of Figure 2.2(d) is one of the non-monotonic types of selectivity sequences (i.e. sequence IV of Eisenman[1, 55]) which result from asymmetrical interactions. It is worth noting that for carriers showing symmetrical interactions relative to water the selectivity will be a monotonic function of ion size (either yielding the sequences Li > Na > K > Rb > Cs, Cs > Rb > K > Na > Li, or the degenerate sequence Li = Na = K = Rb = Cs).

2.2.3 Theories on the physical origin of asymmetries between ion carriers and water

There are two principal classes of theories as to the factors which produce the asymmetries observed for carriers or ion-binding 'sites' compared to water.

One class, due to Eisenman[1,3,55] and Ling[56], emphasises the asymmetries expected in the interations between cations and the individual ligand groups of the carrier or site (cf. Refs. 32, 35 and 37). The other, 'steric', class of theories, propounded by Eigen and his colleagues[33,57] as well as by Simon and Morf[34,36], starts with the implicit assumption that the individual ligand groups are symmetrical to water molecules and then relegates any observed asymmetries of the carriers or sites to constraints on the packing of the ligands inside the carrier or site.

Although it seems likely that the observed selectivity among 'noble gas' cations of Group IA for a particular carrier molecule containing multidentate ligands should involve a contribution from both an asymmetry inherent in the individual ligands and an asymmetry arising from steric constraints in the packing of these ligands, we shall confine our considerations here to those properties of individual ligand groups which produce asymmetries. Indeed, we will now show that the solvation of Group IA cations by ligands of appropriate solvents can mimic the selectivity seen in the 'solvation' of Group IA cations by various ion carriers, which implies that the special structural features of the macrocyclic carriers are not necessarily required for their selectivity among the alkali metal cations. It will further be shown in Section 2.5.1.2 that the anomalous selectivities in carriers, membranes and enzymes for Tl^+ compared to the similar sized Rb^+ ion cannot be accounted for if one assumes that the ligands are themselves symmetrical to water, and that selectivity is being determined only on a 'best-fit' basis. Moreover, in Section 2.8, we will present evidence favouring the 'individual ligand' basis of selectivity over the steric 'fit' basis even for valinomycin from the similarity in the selectivity of valinomycin with that of hexadecavalinomycin which has a much less sterically-restricted cavity.

2.3 ENERGIES OF INTERACTIONS WITH MODEL SOLVENTS AS PROTOTYPES FOR ION BINDING TO CARRIERS

The solvation of ions by appropriate solvents can serve as a useful prototype for the 'solvation' of ions by macrocyclic carriers. In particular, solvents exist which seem likely to solvate ions through the same ligand groups as those in ion carriers (e.g. ester carbonyls, amide carbonyls and ether oxygens);

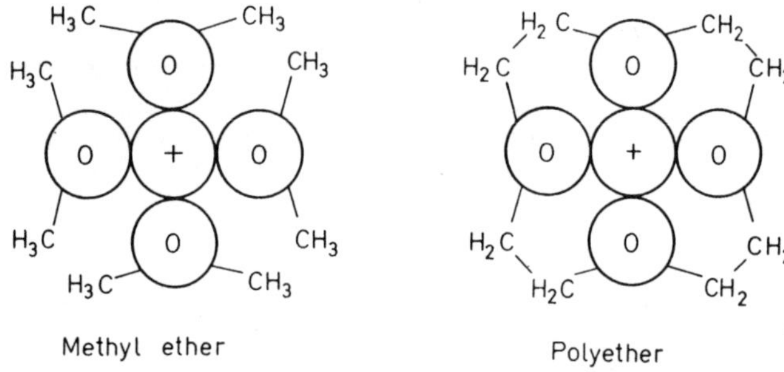

Figure 2.3 Ion solvation by methyl ether compared to 'ion solvation' by a polyether

and the intrinsic selectivities of these ligand groups can serve as excellent prototypes for the selectivity of the same groups inside of a carrier molecule. Even the constraints upon the packing of solvent molecules might provide useful prototypes for the packing of such groups within certain carriers. The analogy we are suggesting is illustrated schematically in Figure 2.3 where an ion solvated by four oxygens of a cyclic polyether is compared with an ion solvated by four methyl ether molecules. In this representation, the principle difference between the carrier and the solvent is the presence of backbone constraints on the carrier.

2.3.1 Cation selectivity of amide solvents

Somsen and Weeda's studies[58,59] on ion solvation in formamide, n-methyl formamide and dimethyl formamide provide the essential data for such a comparison*. Figure 2.4(a) plots the enthalpies of ion solvation in amide solvents as a function of the reciprocal cation radius. For comparison, the hydration enthalpies are also presented. First, it should be noted that the solvation energies of the amides and of water are seen to be comparable in magnitude (the small increases in enthalpy of ion solvation with increasing methylation of the amide solvents will be discussed later). Second, it should be noted that although the enthalpy of solvation is a monotonically increasing function of reciprocal cation radius for each of these solvents, there is an asymmetry compared to water, as revealed in Figure 2.4(b), which follows the procedure set forth in Figure 2.2(c) and plots the enthalpy of ion solvation for each solvent versus the enthalpy of hydration. The asymmetries of these amide solvents are revealed by their curvatures in this plot, signifying that these solvents are selective. Their selectivity can be compared with that for a typical carrier by plotting the data in the manner of Figure 2.4(c) and comparing this with Figure 2.2(a).

The data of Figure 2.4(c) illustrate the existence of a maximum in selectivity for K^+ for each of these amide solvents. Indeed, the selectivity sequence observed for each of these solvents ($K > Rb > Cs > Na > Li$) is the same as that illustrated for nonactin in Figures 2.2(a) and 2.2(d). Note also that the magnitudes of the selectivities in the solvent model, although not quite as large as those exhibited by nonactin, are certainly comparable. From these results, it should be apparent that the carbonyl ligands of these solvent molecules suffice to produce an ion selectivity comparable to that of a carrier system. Therefore, none of the special structural features of the macrocyclic carriers (e.g. the primary or secondary structure of the polymeric ring) are required for such ion selectivity.

2.3.2 Effects of increasing methylation

An interesting detail in the data of Figure 2.4(a) is the effect of increasing methylation on the enthalpy of solvation of the various cations. This in-

* Unfortunately, no data for ester carbonyl solvents are available as yet to use as models for carriers (e.g. nonactin, valinomycin) containing ester carbonyl ligands, and we are therefore forced to use data for amide solvents here as models for carriers containing ester ligands as well as those containing amide ligands.

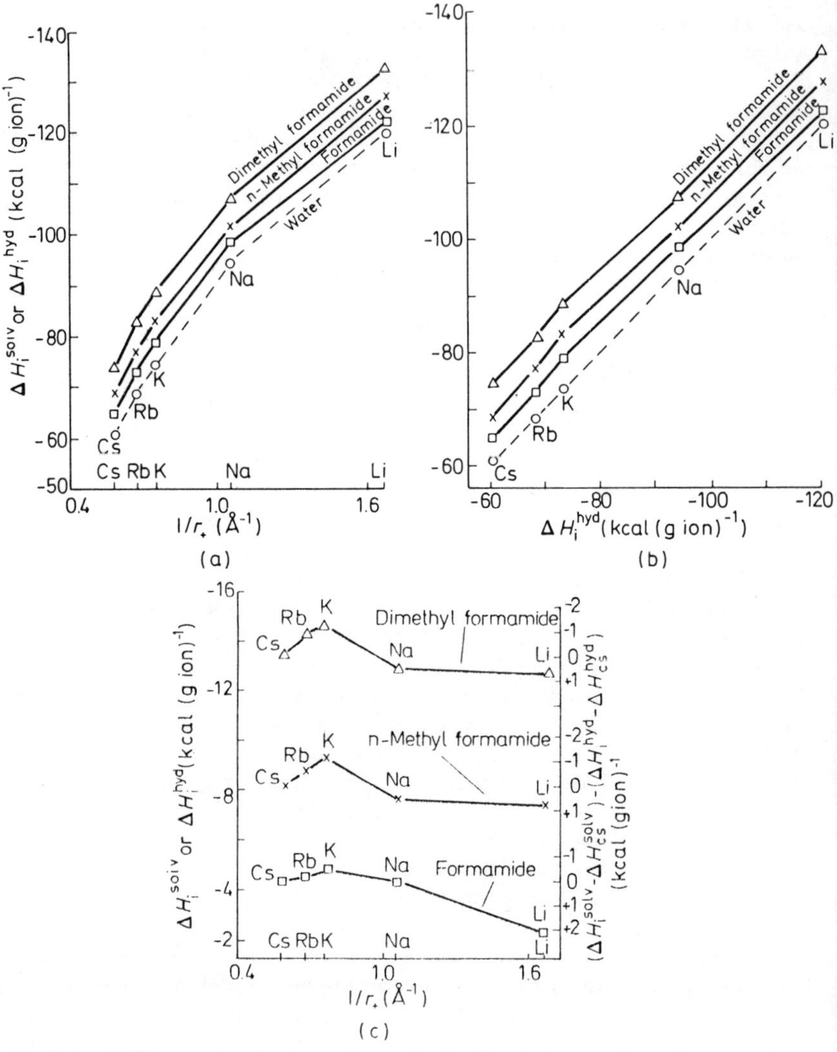

Figure 2.4 (a) The monotonic dependence of solvation enthalpy on the reciprocal ion radius.

The ordinate is the solvation enthalpy in amide solvents and in water of the alkali cations as measured by Somsen and Weeda[58,59] and the abscissa is the reciprocal ion radius.

(b) The asymmetry between amide solvents and water.

The ordinate is the absolute enthalpy of ion solvation in the amide solvents and in water and the abscissa is the absolute enthalpy of ion hydration. The curvatures of the lines for formamide, n-methyl formamide and dimethyl formamide in this plot illustrate that the interactions of these solvents with the alkali cations are asymmetrical to the interactions of water with these ions.

(c) The selectivity of amide solvents.

The left-hand ordinate is the difference between the enthalpy of solvation in each amide solvent and in water of the alkali cations and the right-hand ordinate is this difference for each cation relative to that for Cs^+. The abscissa is the reciprocal ion radius. Note that the selectivity sequence for each of the amide solvents is the same as that observed for nonactin in Figures 2.2(a) and 2.2(d) (i.e. K > Rb > Cs > Na > Li).

crease in methylation is expected to have three effects on these amide solvents: (a) a decrease in the hydrogen-bonded structure of the solvent is expected with increasing methylation[59]; (b) an increase in the bulkiness of the solvent molecules will result with increasing methylation; and (c) a change in the dipole moment of the molecules is expected from the inductive effects[60] of the added methyl groups, which for these molecules correlates with an observed[59] increase in dipole moment with increasing methylation*. Each of these effects of methylation on the amide solvents is in turn expected to influence the ion-solvation enthalpies in each of the solvents and can be a prototype for effects of varied methylations of carrier molecules.

Thus the increase in the absolute enthalpy of solvation seen in Figure 2.4(a), as one goes from water to dimethyl formamide, is expected to result from the effects of methylation on hydrogen bonding between the solvent molecules and on the dipole moment of the solvating ligand groups. The hydrogen-bonding effect occurs because fewer hydrogen bonds are broken in the more methylated solvent in the process of solvating a cation, as has been pointed out by Somsen and Weeda[59]. Thus, provided that approximately the same number of molecules are involved in solvating the ion in each of these solvents, the enthalpy on ion solvation will increase with increasing methylation. The dipole moment effect would increase ion binding due to electrostatic interactions and thereby also lead to an increase in the solvation enthalpy with methylation.

Changes in the selectivities among ions are also expected from 'field-strength' considerations[32] (as discussed below) as methylation is increased in this series of solvents and are indeed observable in the data of Figure 2.4(c). Thus, although each of these solvents exhibits the selectivity sequence K > Rb > Cs > Na > Li, increasing methylation appears to slightly decrease the selectivity for larger ions relative to K^+ (i.e. the Cs^+/K^+ ratio is smaller for dimethyl formamide than for formamide) and to slightly increase the selectivity for smaller ions relative to K^+ (i.e. the Li^+/K^+ ratio is larger for dimethyl formamide than for formamide). Such an effect could be due to the higher dipole moments of the more methylated amide carbonyls brought about by the inductive effects of the methyl substituents or, alternatively, to a change in packing about the ions with the increasing bulkiness of the solvent molecules caused by greater methylation. A change in selectivity due to increasing the dipole moment is expected from selectivity calculations for a discrete two-charge model of a dipolar ligand[32], for which increasing the 'field strength' (which corresponds to an increase in dipole moment) yields a decrease in selectivity for larger ions relative to K^+ and an increase for smaller ions relative to K^+. Therefore, the dipole effect is consistent with the observed changes in selectivity. In contrast, the restraints in packing about the ions due to the increased bulkiness of the solvent molecules would be expected to discriminate more against the smaller ions (e.g. Li^+) than against the larger ions (e.g. Cs^+) whereas, in fact, just the opposite trend in selectivity is observed, making it seem unlikely that this steric factor is responsible for the changes in selectivity observed with increasing methylation of the amide solvents.

* Formamide, n-methyl formamide and dimethyl formamide have dipole moments of 3.71, 3.84 and 3.86 debyes, respectively[59].

Interestingly, the effect of methylation on the selectivity of the amide solvents is similar to that previously observed for the ion carriers nonactin, monactin, dinactin and trinactin (each of which has one more methyl group than the preceding one) and suggested to be due to an inductive increase in the 'field strength' of the ligand oxygens[45]. Indeed, the 'pattern' of ionic selectivity derived for this series of structurally-similar molecules[35] follows the expectations of the 'field-strength' theory[3, 32]. As description of such patterns goes beyond the scope of the present paper, the reader is referred elsewhere (cf. Section V of Eisenman *et al.*[35]).

2.4 THEORETICAL CONSIDERATIONS OF THE ASYMMETRY IN ELECTROSTATIC INTERACTIONS OF IONS WITH INDIVIDUAL CARBONYL LIGANDS AND INDIVIDUAL WATER MOLECULES

We have deduced above that a sufficient condition for a carrier or site to be selective among cations is for the interaction energy between a cation and the carrier or site to have a different dependence on ion radius than for the interaction energy of the cation with water; and we have just shown that this asymmetry exists for model solvents. We now illustrate by a theoretical calculation that a single ligand with a different charge array than that characteristic of a water molecule, will have asymmetrical interactions with ions*.

2.4.1 Calculated selectivity for an individual carbonyl ligand versus an individual water molecule

As an example, we have used the discrete charge-array representations of model water and carbonyl groups pictures in Figure 2.5 to illustrate the selectivity which occurs upon replacing a single water molecule with a single ligand group having a different charge distribution from that of water†. The asymmetry implied by the differences in charge arrays of the model ligand versus the model water is illustrated in Figure 2.6 where Figure 2.6(a) (to be compared to Figures 2.2(b) and 2.4(a))‡ is a plot of the internal energy of ion

* Although differences in other variables (e.g. the dependence of the coordination number for water on ion size versus that for the solvents and carriers discussed above) will also contribute to asymmetrical interactions of ions with solvents and carriers, we shall not discuss these additional variables here. Similarly, we shall not consider additional contributions to the intermolecular forces between ions and ligands (for example, p. 25ff of Hirschefelder *et al.*[61]) beyond the simple electrostatic interactions discussed in this section and some interactions for 'noble gas' ions presented below, although asymmetry in these additional types of interactions will also contribute to the selectivities of carriers and solvents.

† Talekar[64] has based a calculation of the K^+-binding energy to valinomycin on a discrete charge-array representation by assaying the partial charges on the backbone atoms of valinomycin. This 'discrete charge-array' representation is not to be confused with 'point multipolar' representations in which the charge distribution on a molecular assemblage is represented as a sum of point multipoles. Certain significant differences between these two types of representations (and their misuse (see for example Ref. 65)) are discussed elsewhere[37].

‡ The differences in the magnitudes of the energies in Figure 2.6 as compared to Figures 2.2 and 2.4 arise to a large degree from the fact that we are considering interactions with single ligands in Figure 2.6 whereas the empirical values in Figures 2.2. and 2.4 reflect multiple coordination.

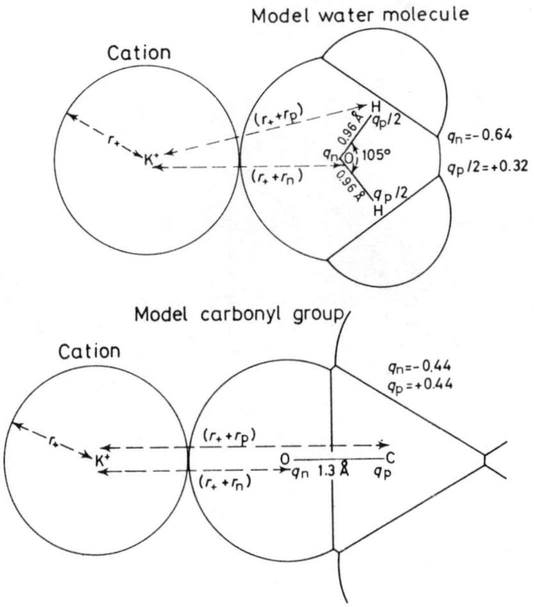

Figure 2.5 The charge distributions in a model water molecule and in a model carbonyl group.

The top figure represents a cation (in this illustration, K^+) in contact with a single water molecule and the bottom figure represents a cation in contact with a single carbonyl group. The outlines of these ligands are drawn to scale using CPK models (the van der Waal's radii therefore being determined by these models). The van der Waal's radius of the oxygen is 1.4 Å, this being the value of r_n in this figure. The values of the partial charges, the bond lengths and the bond angle of the model water molecule are those proposed by Rowlinson[62], with the value of partial charge at the oxygen atom, q_n, being -0.64 electron units and that at each of the two hydrogen atoms, $q_p/2$, being $+0.32$ electron units, with the O—H distance being 0.96 Å and with the H—O—H angle being 105 degrees. The C = O bond length of the model carbonyl group, 1.3 Å, is that proposed by Pauling[39]. The value of partial charge at the oxygen atom, q_n, is -0.44 electron units and the value of partial charge on the carbon atom, q_p, is $+0.44$ electron units. This value of partial charge was chosen so that the bond length multiplied by the partial charge would produce a dipole moment (2.75 debye) consistent with those observed[63] for typical carbonyl-containing molecules (e.g. acetaldehyde (2.72 debye) and acetone (2.89 debye). However, the precise value of partial charge was chosen to produce the particular selectivity sequence K > Rb > Cs > Na > Li so that Figure 2.6 could be appropriately compared to the analogous figures in this section. (Note that values of partial charge which are larger in magnitude will produce 'higher field strength' selectivity sequences and values lower in magnitude will produce 'lower field strength[3]' selectivity sequences. The radius of the cation is designated r_+. The distances from of charge of the cation to the centres of negative charge in the model water and carbonyl ligands are designated $(r_+ + r_n)$. The distances from the centre of charge of the cation to the centres of positive charge in the model water and carbonyl ligands are designated $(r_+ + r_p)$

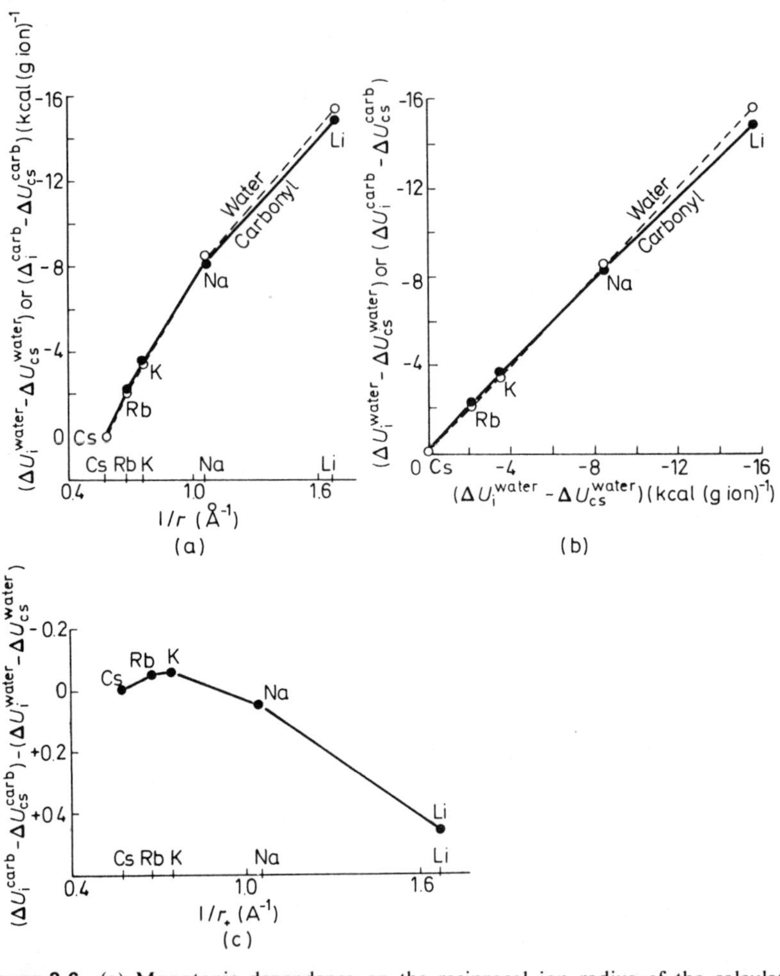

Figure 2.6 (a) Monotonic dependence on the reciprocal ion radius of the calculated internal energy of ion binding to a model water molecule and to a model carbonyl group.

The ordinate is the potential energy of interaction between the model carbonyl group of Figure 2.5 and the alkali cations (filled circles, solid line) and between the model water molecule of Figure 2.5 and the alkali cations (open circles, dashed line), both relative to that with Cs^+ and calculated using equation (2.6). The abscissa is the reciprocal ion radius.

(b) The asymmetry between the calculated internal energy of ion binding to a model carbonyl group and that to a model water molecule.

The ordinate is the calculated internal energy of binding to the alkali cations (relative to that of Cs^+) of the model carbonyl group (filled circles, solid lines) and of the model water molecule (open circles, dashed lines) and the abscissa is the calculated internal energy of binding to the alkali cations (relative to that of Cs^+) of the model water molecule. The assymmetry between the model carbonyl group and the model water molecule is reflected by the curvature of the line for the model carbonyl group in this plot.
(c) The calculated selectivity of a single model carbonyl group (referred to a single model water molecule).

The ordinate is the difference between the calculated internal energy of binding to the alkali cations (relative to that of Cs^+) of the model carbonyl group and of the model water molecule. The abscissa is the reciprocal ion radius binding (relative to Cs^+) to the model water molecule and to the model carbonyl molecule as a function of the reciprocal ion radius, calculated using Coulomb's law for the internal energy U (in kcal mol^{-1}) between an ion of charge q_+ and radius r_+ and the molecules illustrated in Figure 2.5,

$$U = 332q_+\left(\frac{q_n}{r_+ + r_n} + \frac{q_p}{r_+ + r_p}\right) \tag{2.6}$$

the parameters being defined in the figure legend.

Although Figure 2.6(a) shows that the calculated interaction energies with the ions are monotonically-increasing functions of decreasing ion size for both the ligand and water molecules, the interaction energies for the carbonyl ligand are *asymmetrical* to those with the model water molecule as can be seen from the curvature when these are plotted in Figure 2.6(b). This asymmetry gives rise to the non-monotonic selectivity sequence ($K > Rb > Cs > Na > Li$) seen in Figure 2.6(c). The reason for this asymmetry between the model water and the model ligand has been examined elsewhere[37] and shown to arise from differences in the relative value of $(r_+ + r_n)$ and $(r_+ + r_p)$, as discussed briefly in the next paragraph.

2.4.2 Physical basis of asymmetry

Since the particular functional dependence of U on cation size depends upon the relationship between r_+, r_n and r_p, a ligand with different values of r_n and r_p than water (i.e. different 'discrete-charge arrays') will have a different functional dependence of U on r_+ than is characteristic of the water molecule. The interactions of ions with the ligand and with the water molecule will therefore be asymmetrical. Comparing the distances $(r_+ + r_p)$ and $(r_+ + r_n)$ for the model water and model carbonyl in Figure 2.5, we see that the difference between $(r_+ + r_p)$ and $(r_+ + r_n)$ is much greater for the carbonyl than for water (water is a ligand with an especially small distance between its positive and negative centres of charge compared to the other ligands (e.g. carbonyls) with which we are concerned). This geometrical difference in charge distribution can account for the asymmetry between these ligands and water*.

2.5 CONSIDERATIONS FOR NON-'NOBLE GAS' MONOATOMIC AND POLYATOMIC CATIONS

Up to this point we have considered cations with 'noble gas' electronic configurations. Now we consider non-'noble gas' monoatomic ions such as Ag^+ and Cu^+, whose outer orbitals contain ten D electrons, and Tl^+ and In^+,

* It may help the reader to note that for ions for which r_+ is very small compared to r_n and r_p, the value of U in equation (2.6) will be independent of r_+, whereas for ions for which r_+ is very large compared to r_n and r_p, the value of U will be dependent upon the second power of r_+.

whose outer most orbitals contain two s electrons as well[66]. We also consider polyatomic ions (e.g. NH_4^+) which have an intrinsic 'shape', as of course do those monoatomic ions with outer d^{10} and $d^{10}s^2$ orbitals.

2.5.1 Non-'noble gas' monoatomic cations and ligand type

It will be shown below that non-'noble gas' cations have interactions with ligands which are above and beyond those of the Group IA cations. Further, we shall demonstrate that these excess interactions vary for different ions according to the particular ligand and thus can be diagnostic of the ligand type.

2.5.1.1 Non-'noble gas' monoatomic cations having excess binding energies to ligands

For non-'noble gas' cations, such as Tl^+ and Ag^+, excess ion–ligand interaction energies, as compared to the 'noble gas' cations of Group IA, can be shown empirically to exist and are illustrated in Figure 2.7. Figure 2.7(a) demonstrates that the energies of hydration for Tl^+ and Ag^+ ions are much higher than those of alkali cations of comparable crystallographic radii. Figure 2.7(b) illustrates that this is also the case with the energy for the solvation of these ions by the carbonyl ligands of the solvent propylene carbonate. Indeed, excess energies for Tl^+ and Ag^+ are the general rule, being apparent also in their interactions with carriers having variously orientated carbonyl and ether ligands (tetrahedral carbonyl plus tetrahedral ether is typified by nonactin in Figure 2.7(c), octahedral carbonyls by valinomycin in Figure 2.7(d), carbonyls in a square-antiprism[25] by hexadecavalinomycin in Figure 2.7(e) and planar sixfold coordinate ethers by the cyclic polyether dicyclohexyl-18-crown-6[42, 68] in Figure 2.7(f)*.

The explanation for these excess energies lies in the fact that the electron distributions for these ions cannot be represented as simple spheres (as, to first approximation, can those of the alkali cations); rather, these non-noble gas ions have electron distributions which can be influenced by their environment. In the presence of appropriate ligands, these ions can distort so that their electron clouds have a minimum overlap with those of the ligands[69], or they may maximise the overlap to produce partial covalency with the ligands[70]. The nature of the distortion and the excess energies produced by such distortions are expected to depend upon the types of ligands[70] and on their orientations[71], and to occur over and above the simple electrostatic interaction energies expected for spherical cations of comparable crystallographic radii.

Molecules having excess energies such as these will be shown in Section

* The ion-binding energies for nonactin, valinomycin, hexadecavalinomycin and the polyether relative to Cs^+ have been calculated in the same manner as for nonactin in Figure 2.2(b), but in Figures 2.7(c), (d) and (e) bilayer permeability ratios rather than salt extraction data were used to estimate the ratio of complexation constants in water whereas in Figure 2.7(f) the aqueous complexation constants were used.

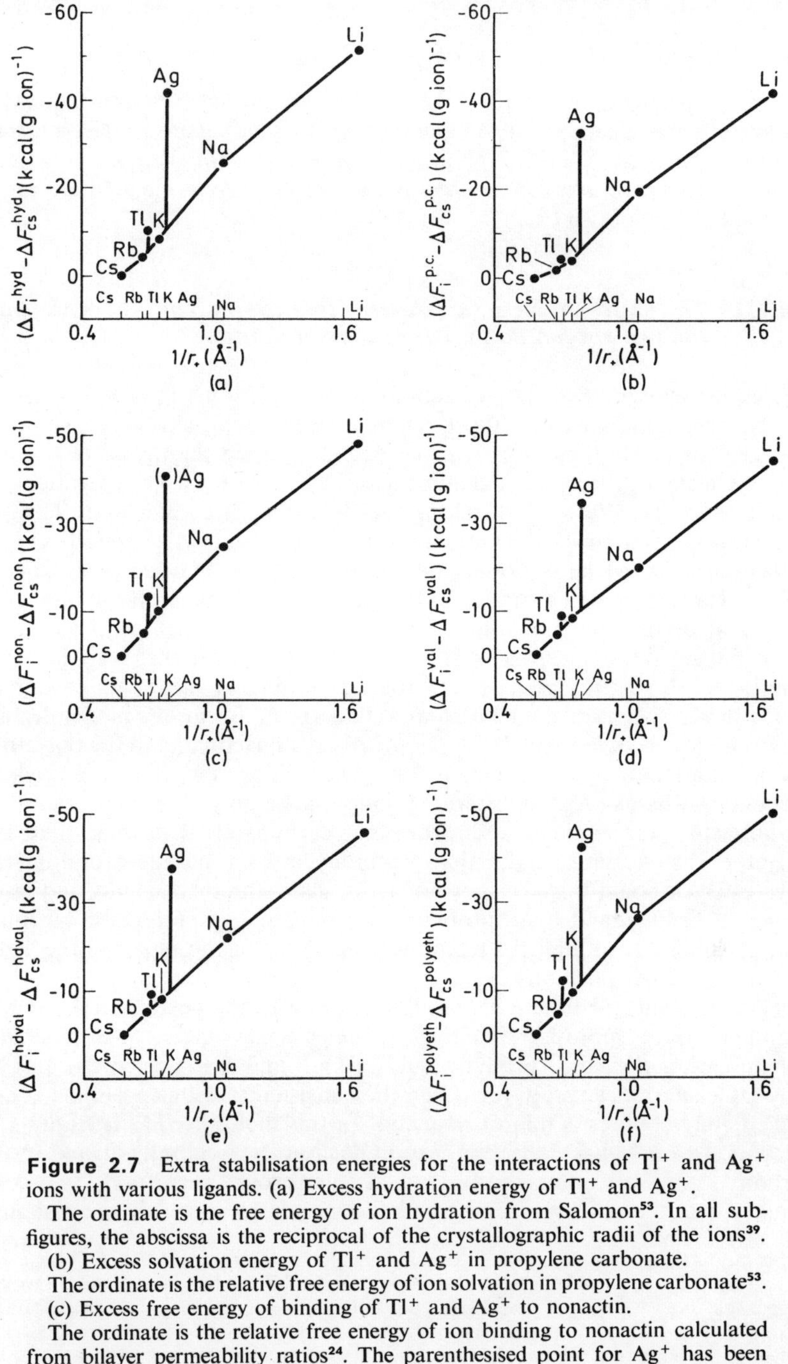

Figure 2.7 Extra stabilisation energies for the interactions of Tl^+ and Ag^+ ions with various ligands. (a) Excess hydration energy of Tl^+ and Ag^+.

The ordinate is the free energy of ion hydration from Salomon[53]. In all sub-figures, the abscissa is the reciprocal of the crystallographic radii of the ions[39].

(b) Excess solvation energy of Tl^+ and Ag^+ in propylene carbonate.

The ordinate is the relative free energy of ion solvation in propylene carbonate[53].

(c) Excess free energy of binding of Tl^+ and Ag^+ to nonactin.

The ordinate is the relative free energy of ion binding to nonactin calculated from bilayer permeability ratios[24]. The parenthesised point for Ag^+ has been extrapolated for nonactin from the value measured by Laprade[67] for the trinactin–Ag^+ complex.

(d) Excess free energy of binding of Tl^+ and Ag^+ to valinomycin.

The ordinate is the relative free energy of ion binding to valinomycin calculated from bilayer permeability ratios[35].

(e) Excess free energy of binding of $Tl^+$and Ag^+ to hexadecavalinomycin.

The ordinate is the relative free energy of ion binding to hexadecavalinomycn calculated from bilayer permeability ratios[40].

(f) Excess free energy of binding of Tl^+ and Ag^+ to the polyether dicyclohexyl-18-crown-6.

The ordinate is the relative free energy of ion binding to dicyclohexyl-18-crown-6 calculated from aqueous binding constant measurements[42,43].

2.5.1.2 to produce diagnostic selectivity features which are useful for distinguishing particular ligands (e.g. ether versus ester versus amide oxygens), and in Sections 2.6.1, 2.6.2, 2.7.1 and 2.7.2 it will be demonstrated how these same selectivity features can be used in assessing the types of ligand groups in molecules of unknown chemical structure.

2.5.1.2 *The 'sub-IA' and 'supra-IA' selectivity for Tl^+ and Ag^+ ions of molecules with known ligands*

The excess energies for the interactions of Tl^+ and Ag^+ ions with these ligands, when compared with those for their interactions with water, show up as a selectivity for these ions which, depending upon the ligand (e.g. ether versus ester), is either greater than or less than that for Group IA cations of comparable size. We will call selectivities less than or greater than those of comparably-sized Group IA cations, 'sub-IA' and 'supra-IA', repectively.

In Figure 2.8, we have plotted the selectivities for the Group IA cations as well as for Tl^+ and Ag^+ (and NH_4^+, see below) of one solvent and of a variety of carriers of known ligand types. We refer to such plots as 'selectivity fingerprints' since, as is shown here and below, they can be used empirically to identify the nature of the cation-binding site, much as a complex infrared spectrum can be used as a 'fingerprint' for an organic molecule. In this figure, it is seen that Tl^+ is 'supra-IA' in nonactin and in the polyether but is 'sub-IA' in valinomycin, in hexadecavalinomycin and in propylene carbonate, whereas Ag^+ is 'supra-IA' only in the polyether, being 'sub-IA' in nonactin, valinomycin and propylene carbonate*. Knowing that the polyether has ether ligands, that valinomycin and hexadecavalinomycin have ester carbonyl ligands, that nonactin has both ether ligands and ester carbonyl ligands and that propylene carbonate has carbonate carbonyl ligands, allows us to infer the relationship between ligand type and the 'sub-IA' or 'supra-IA' positions of Tl^+ and Ag^+. In particular, all of the 'selectivity fingerprints' in Figure 2.8 are consistent with the postulates that ether ligands produce 'supra-IA' Tl^+ and Ag^+ selectivities whereas ester and carbonate carbonyl ligands produce 'sub-IA' Tl^+ and Ag^+ selectivities†. The only molecular 'fingerprint' for which the consistency of this postulate is not immediately obvious is that of nonactin. In this molecule, Tl^+ is 'supra-IA' whereas Ag^+ is 'sub-IA'. Since nonactin has four tetrahedrally-arrayed ester carbonyls and four tetrahedrally-arrayed ether groups, we can see that consistency with the postulate demands that Ag^+ can contact only the carbonyl

* This means that, compared to the interaction energies with water, Tl^+ and Ag^+ are gaining relatively more (supra-IA) or less (sub-IA) energy with these ligand groups than would comparable sized 'noble gas' ions. To make this completely explicit, the curve for the selectivity of a water-like ligand (for which water is of course the appropriate model) is plotted at the bottom of Figure 2.8, where all of the points for these ions are shown to lie exactly on the horizontal line labelled 'water', corresponding to zero selectivity.

† Notice that we have only been able to examine a limited spectrum of carbonyl ligands so far. Thus, for example, lack of equilibrium data on model compounds has prevented us from characterising Tl^+ for amide carbonyl and for charged ligands such as carboxylate groups, although the enzyme activation effects to be discussed in Section 2.6.1 may be taken as involving these ligands (see also the footnote to Section 2.7.1).

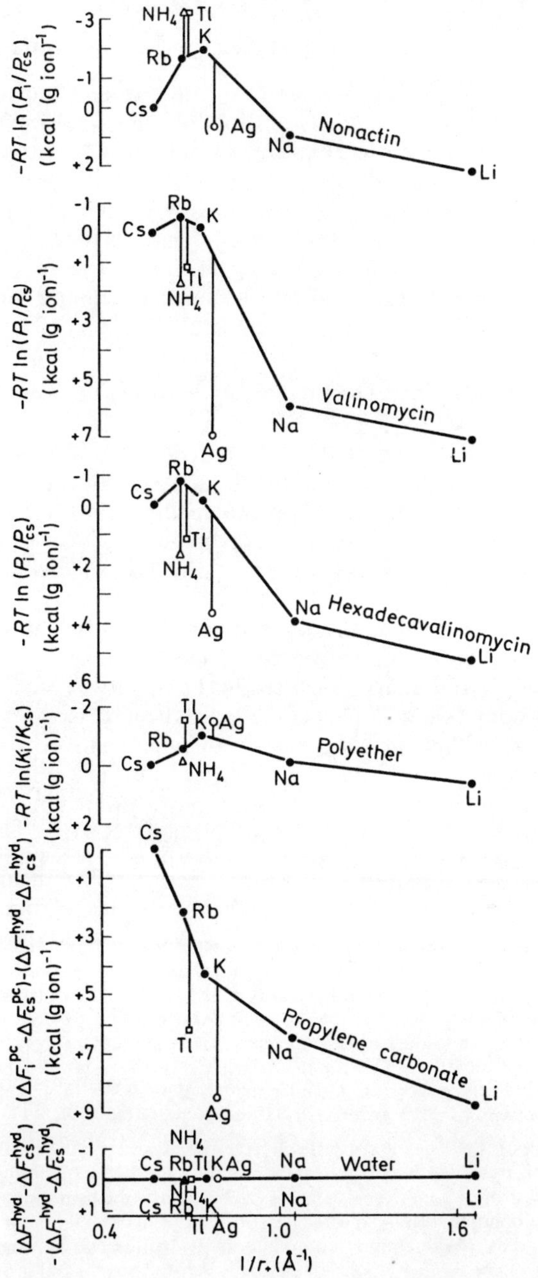

Figure 2.8 Selectivity fingerprints of model ion-binding systems of known chemical composition.

For each sub-figure, the selectivities are plotted in the same manner as in Figures 2.2.(d), 2.4(c) and 2.6(c) but also including data for Tl^+, Ag^+ and NH^+. The ordinate is the energy equivalent of the selectivity for each ion compared to Cs^+. The abscissa is the reciprocal of the crystallographic ion radius[39]. The nonactin[24], valinomycin[35] and hexadecavalinomycin[40] data are calculated from bilayer permeability ratios, whereas the polyether data is calculated from aqueous binding constant measurements[42,43]; the propylene carbonate data is taken directly from Salomon[53].

ligands whereas Tl^+ is able to interact with the ether groups as well. In fact, this is precisely the ion–ligand coordination one would expect both from inspection of CPK models (cf. Figure 2.1) and by analogy to the x-ray crystal data on the Na^+–nonactin[29] and K^+–nonactin[27, 28] complexes*.

In Sections 2.6.1, 2.6.2. 2.7.1 and 2.7.2, we shall demonstrate how the establishment of a relationship between ligand type and the 'sub-IA' or 'supra-IA' selectivities for non-'noble gas' cations can be used to infer the types of ligands in systems of unknown molecular compositions.

2.5.2 Polyatomic ion 'shape' and the spatial array of ligands

We have previously suggested[35, 54, 72] that polyatomic ions with particular 'shapes', such as NH_4^+, may be valuable in assessing the coordination geometry of the ligands in sites of unknown structure. Thus, the tetrahedral NH_4^+ ion might be expected to have a more favourable interaction energy with four tetrahedrally-arrayed oxygens than with six octahedrally-arrayed oxygens, even if the electric potential produced by the ligands at the centre of the NH_4^+ ion is the same for both distributions of ligands. This is because NH_4^+, rather than having its net positive charge located strictly at the centre of the ion, has some positive charge localised in the vicinity of each of the hydrogen atoms[39]†. Although we shall only cite NH_4^+ in the next sections to demonstrate how the shape of an ion can be used to infer ligand coordination, other monoatomic ions and molecular ions may be just as useful. One example is Hille's[7, 73] ingenious use of NH_4^+ derivatives with different sizes, shapes and charge distributions (e.g. $CH_3NH_3^+$, $HONH_3^+$, $H_2NNH_3^+$, $(NH_2)_2C = NH_2^+$) to infer the spatial array and type of ligands in the Na^+ channel of frog nerve‡.

* The crystallographic radius of Ag^+ lies between that of Na^+ and that of K^+. In the K^+–nonactin complex[27, 28], the carbonyl oxygens and the ether oxygens are almost equidistant from the K^+ ion, whereas in the smaller cavity of the Na^+–nonactin complex[29], the carbonyl oxygens are significantly closer to the Na^+ than are the ether oxygens (cf. Figure 2.1), so contact between Ag^+ and the carbonyl oxygens but not the ether oxygens is to be expected. In contrast to this situation, Tl^+, having a larger crystallographic radius than K^+, could easily contact all eight oxygens (as is approximately the case for K^+), and thus the expected 'supra-IA' selectivity of ether groups for Tl^+ could manifest itself.

† Pauling[39] deduces that each hydrogen will have approximately 0.18 partial positive charges and that the nitrogen will have about 0.28 partial positive charges. Additionally, when the possibility of hydrogen bonding exists between the ion and the ligands, an extra stabilisation from hydrogen bonding is also expected. A corollary of hydrogen bonding is a reduction of the 'effective radius' of the ammonium ion compared to the non-hydrogen-bonded ammonium ion.

‡ We have recently found that such large polyatomic ions are also well carried by hexadecavalinomycin[40] (see Table 2.1 and p. 55). This extends considerably the possibilities for the method of 'fingerprinting' biological channels and sites described below. It should also be noted that certain monoatomic ions with non-'noble gas' electron configurations, such as the transition metals, should be usable in probing the coordination geometry of the ligands since different ligand geometries split the energy levels of the electron orbitals differently thereby giving rise to additional stabilisation energies (e.g. as treated by ligand-field theory) (cf. p. 660ff of Cotton and Wilkinson[66]). This problem is analogous to the situation in crystals, which has been masterfully stated by Dunitz and Orgel[71].

2.5.2.1 *The ammonium ion as a probe for tetrahedrally-arrayed ligands*

We shall now consider the postulate that NH_4^+, through its tetrahedral shape, gives us some clue as to the orientation of the ligands. Included in the 'selectivity fingerprints' of Figure 2.8 are the selectivities of the ion carriers for NH_4^+. Whereas nonactin has a 'supra-IA' selectivity for NH_4^+, valinomycin, hexadecavalinomycin and the polyether all have 'sub-IA' selectivities for NH_4^+. Nonactin has four tetrahedrally-directed carbonyl ligands and four tetrahedrally-directed ether ligands, whereas neither valinomycin, hexadecavalinomycin, nor the polyether has tetrahedrally-directed ligands, valinomycin having six octahedrally-directed carbonyls, hexadecavalinomycin having eight carbonyls orientated in a square-antisprim[25] (see the molecular models of Figure 2.1) and the polyether having six planar ether ligands[42, 68, 75].

Since in Figure 2.8 nonactin is the only carrier with a tetrahedral array of ligands and also is the only carrier with a 'supra-IA' selectivity for NH_4^+, we feel safe in postulating that a 'supra-IA' selectivity for NH_4^+ implies a tetrahedral orientation of the ligands. We shall use this postulate in Sections 2.6.1 and 2.6.2 to infer the array of ligands in ion-selective biological systems of unknown geometry.

2.6 ION SELECTIVITY IN PRESUMABLY EQUILIBRIUM BIOLOGICAL PHENOMENA—ION BINDING AND ENZYME ACTIVATION

The ion selectivities for the model systems we have discussed thus far reflect well-defined equilibrium properties of these systems. Certain of the ion selectivities encountered in biological systems are also likely to reflect the equilibrium selectivities of the ion-binding sites. Two such cases are presented below, that for enzyme activation (Section 2.6.1) and that for ion binding to ion-exchange sites in *Chlorella* (Section 2.6.2).

2.6.1 Enzyme activation

Subsequent to the discovery by Boyer, Lardy and Phillips in 1942 that K^+ activated the reaction catalysed by pyruvate kinase[76], monovalent cations have been identified as obligatory co-factors in activating a number of important enzymes[2, 6, 77]. Their selectivity in such enzyme activation can be interpreted as reflecting their equilibrium affinity for enzyme sites and is amenable to being 'fingerprinted' in a way entirely analogous to that just presented for solvents and ion carriers.

Figure 2.9 illustrates such 'fingerprints' for the effects of monovalent cations in activating those enzymes for which we could find sufficiently complete data in the literature. The data are plotted in the same way as in Figure 2.8, except we now use Michaelis–Menten 'affinity' constants (K_m) as the measure of selectivity instead of permeability ratios.

The uppermost enzyme is diol dehydratase, for which the Michaelis–Menten constants of Manners, Moralee and Williams[78] have been plotted (these data are in close agreement with those of Toraya *et al.*[79]). The selectivity 'fingerprint' of this enzyme can be seen to be quite similar to that manifested by nonactin in Figure 2.8, not only as regards the 'supra-IA' selectivity for both

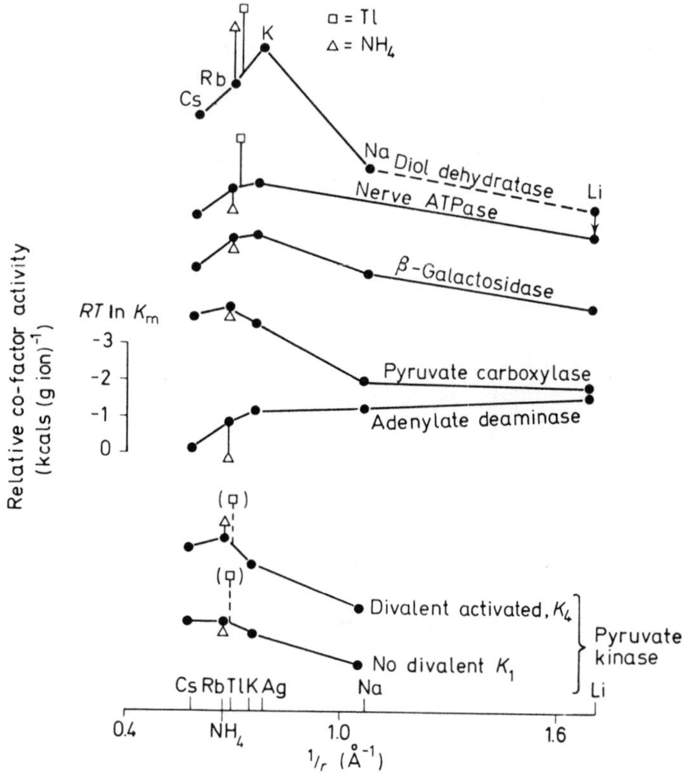

Figure 2.9 Selectivity 'fingerprints' for the monovalent cation activation of various enzymes.

These data are plotted in the same manner as for Figure 2.8 except in this figure the ordinate is the selectivity calculated from the Michaelis–Menten 'affinity' values K_m. From top to bottom, the 'fingerprints' are for diol dehydratase[78,79], nerve ATPase[10,80,81], β-galactosidase[82], pyruvate carboxylase[83], adenylate deaminase[84], pyruvate kinase activated by divalent ions[6,76,78,88] and pyruvate kinase not activated by divalent ions[6,76,78,88]. Notice that the position of Tl^+ is invariably 'supra-IA' whereas the position of NH_4^+ varies between 'sub-IA' and 'supra-IA'.

NH_4^+ and Tl^+, but also as to the sequence (K > Rb > Cs > Na > Li) and magnitude of the selectivity among the alkali metal cations.

The next 'fingerprint' is for the (Mg^{2+} + Na^+)-activated ATPase of nerve[10,80], for which data for Tl^+ have been added using Inturrisi's measurements[81]. This enzyme is seen to be 'supra-IA' for Tl^+ but 'sub-IA' for NH_4^+, resembling the polyether of Figure 2.8 in this regard.

The next three enzymes plotted are β-galactosidase[82], pyruvate carboxylase and adenylate deaminase[84], for none of which are Tl^+ data available. In the first two of these enzymes, NH_4^+ can be seen to be slightly 'sub-IA' in its effect, while in the last it is clearly 'sub-IA' (note that the data for adenylate deaminase are not K_m values but 'relative activities'[84]).

The bottom two sets of data are for pyruvate kinase, the first enzyme for which a monovalent requirement was described[76]. Group IA and NH_4^+ data from Suelter's[6] Table 4 have been plotted for two conditions: K_4, when the enzyme is activated to the maximum extent by divalent cations and K_1 in the absence of divalent activation. Data for Tl^+ of Kayne[85] and Manners, Moralee and Williams[78] have been included tentatively on both of these plots. The change of NH_4^+ effect from 'sub-IA' in the absence of divalent activation to 'supra-IA' when the enzyme is activated to its maximum extent is striking.

Scrutinising all of these data, it can be seen that Tl^+ is invariably 'supra-IA' in its action, whereas NH_4^+ can be variably 'supra-IA' or 'sub-IA'. It we apply the arguments of Section 2.5.1 and 2.5.2 in using Tl^+ to infer the ligand type and NH_4^+ to infer the ligand coordination, we can deduce from the invariably 'supra-IA' Tl^+ selectivity that for all of the enzymes the ligands are of the same 'ether-like' type, whereas from the variable NH_4^+ selectivity we deduce that the enzymes differ in the coordination geometry of their ligands—diol dehydratase and the divalent activated state of pyruvate kinase having tetrahedrally-orientated ligands as judged from the 'supra-IA' selectivity for NH_4^+, and the other enzymes lacking a tetrahedral orientation of their ligands as judged from their 'sub-IA' selectivities for NH_4^+. Interestingly, the change in NH_4^+ effectiveness relative to the Group IA cations upon activation of pyruvate kinase by divalent ions suggests a change in the conformation of the monovalent binding site as a consequence of this interaction.

The inference that the ligands of these enzymes are 'ether-like' requires clarification. Clearly these enzymes, being pure polypeptides, contain no ether groups *per se*. Rather, it seems likely to us that the ligands of the ion-binding site in these polypeptides consist chiefly (perhaps entirely) of amide carbonyls derived from the polypeptide backbone together with a single carboxylate group from acidic side-chains*. Since, as was mentioned previously, we have not yet been able to assess the equilibrium selectivity 'fingerprints' for such ligands in molecules of known structure, we cannot bring independent data to support this postulate (but see footnote to Section 2.7.2). However, Stroud, *et al*,[89,90] and Drenth *et al.*[91] have recently obtained results which at least support the existence of such a proposed structure for the site. Both in Subtilisin and Trypsin a Tl^+ binding site is observed crystallographically which has the structure postulated above, namely a number of amide carbonyl ligands from the polypeptide backbone and a single carboxylic group from a side chain. Moreover, the site in subtilisin binds ions in the probable 'supra Ia' sequence $Tl^+ > K^+ > Na^+$.

* This carboxyl group could either be in contact with the cation as an actual ligand, as seen in the ion carrier nigericin[86], or it could merely be in the vicinity of the cation, as exemplified in the carrier monensin[87]. There is evidence that carboxyl groups *per se* manifest 'supra-Ia' selectivity for Tl^+. For example, Hille[87] (p. 654) has pointed out that Tl^+ forms more stable complexes with carboxylate anions (e.g. acetates and polyacetates) than do any of the alkali cations, which of course implies that the carboxylate group is 'supra Ia' in present terminology.

Combining the structural data for subtilisin and the selectivity data for the enzymes, we therefore postulate that amide carbonyls, possibly together with a carboxylate group, are 'ether-like' in leading to 'supra-IA' Tl^+ selectivities.

2.6.2 Ion binding to Chlorella

In addition to the enzyme selectivities, there is one reasonably unambiguous example of equilibrium ion selectivity for a cell known to us, namely Cohen's careful experiments[92] on the ability of cations to displace radioactive Rb from both killed and living algal cells. These experiments demonstrate the existence of a set of ion-binding sites which survive heat- and freeze-killing and alcohol extraction, and for which he has measured equilibrium selectivity for all of the ions of present interest. The selectivity for heat-killed and freeze-killed cells from Cohen's Table I[92] has been plotted in Figure 2.10 using his concentrations for 50% displacement of radioactive Rb to calculate[4] the values of the equilibrium constants K_i for an ideal ion exchanger.

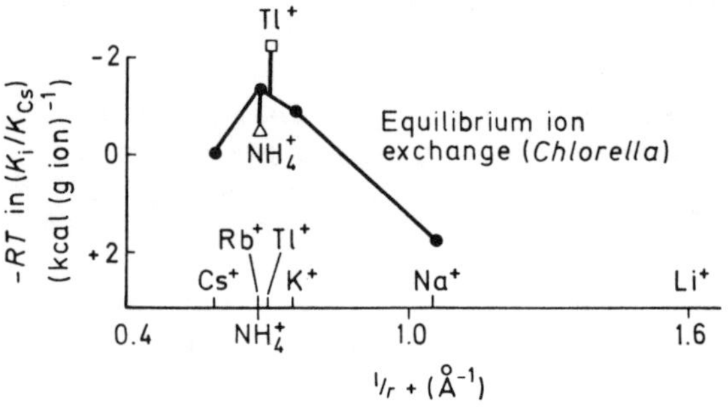

Figure 2.10 Selectivity 'fingerprint' for the cation-binding sites of heat- and freeze-killed *Chlorella* cells.

These data are plotted in the same manner as for Figure 2.8 except that in this figure instead of using permeability ratios the selectivities have been calculated from the value of equilibrium constant K_i for ion exchange for each species, which has been derived from Cohen's data[92] for the concentrations of each ion needed for 50% displacement of Rb^{86}

The ion Tl^+ is seen to be 'supra-IA' whereas NH_4^+ is 'sub-IA' in its binding to *Chlorella*, resembling nerve ATPase (and the 'K^+ channel' of nerve (cf. Figure 2.11)) in this respect. Thus, by analogy to the previous discussion, we can conclude that the ligands forming this ion-binding site are likely to be of the same 'ether-like' type as those in enzymes and are not tetrahedrally arrayed.

2.7 ION SELECTIVITY IN PRESUMABLY NON-EQUILIBRIUM BIOLOGICAL SYSTEMS—ION PERMEATION THROUGH MEMBRANE 'CHANNELS'

In contrast to the previous situation for the selectivity of enzyme activation and ion binding by *Chlorella*, which presumably reflect simple *equilibrium* interactions, the permeation of cations through membranes could involve important *non-equilibrium* factors since such permeation appears to take place through ion-selective 'channels'[7,73,93], rather than utilising simple 'carriers' like nonactin or valinomycin*. Non-equilibrium factors arise because, in general, ion permeation through channels should depend also on kinetic factors (e.g. relative ion mobilities[94]) and therefore cannot be taken as reflecting solely the equilibrium selectivity for the ion-binding sites of these channels. Despite the complications expected from kinetic effects, we shall proceed in Sections 2.7.1 and 2.7.2 to analyse the selectivity 'fingerprints' for the K^+ and Na^+ 'channels' of nerve using the same criteria which we established above for equilibrium selectivity. Before doing so, we would like to point out that although neglecting mobility effects cannot be justified except under special circumstances such as when mobility ratios depend on equilibrium energetics in a particularly simple way or differ much less than equilibrium free energies from ion to ion, there are two sets of experimental data which support the usefulness of extending this procedure to 'channels'.

First, it has been observed (as will be noted in the footnote to p. 52) that a single 'biological selectivity pattern' for alkali metal cations appears to be shown[1,4] by biological phenomena representing purely equilibrium selectivity (e.g. ion binding and enzyme activation) as well as by biological phenomena presumably complicated by kinetics (e.g. membrane permeation). Similarly, it has been observed (cf. Figures 8A and 8B of Eisenman[55]) that nearly the same quantitative selectivity *pattern* exists for selectivities derived from measurements of permeabilities in a variety of aluminosilicate glasses as for selectivities derived from measurements of equilibrium-binding affinities in solid ion exchangers composed of aluminosilicate minerals†. There is at least a formal similarity between ion permeation through a hydrated glass membrane and through a membrane 'channel', in that permeability measurements in both systems reflect contributions from ion mobilities as well as from equilibrium constants. Indeed, diffusion of ions through the interstices of a glass has been suggested[94] to be fundamentally similar to diffusion through a narrow channel‡.

* If this permeation should turn out to involve carriers, the reader should recall that the appearance of equilibrium selectivity factors directly in the selectivity of carrier-mediated permeation has been clearly demonstrated for the nonactin-type[24,35,45,46] and valinomycin-type[35,46,72] antibiotics over a considerable range of experimental conditions.

† There are small quantitative differences between the permeability and equilibrium patterns which are due to mobility effects and which have been discussed elsewhere[95,96].

‡ Even such molecular details as the expectations that in glass only one of the (approximately six) oxygen ligands surrounding each cation will bear a net charge (the others being 'bridging oxygens' of the matrix or water molecules) bears a close resemblance to the situation proposed by Hille[7] or Bezanilla and Armstrong[93] for the channels of nerve.

2.7.1 The 'K^+ channel' of nerve[7,8]

In Figure 2.11 are illustrated the selectivity 'fingerprints' observed for the 'K^+ channel' of frog nerve based on permeability ratios measured by Hille[7], and for the resting permeability of the squid axon based on permeability ratios measured by Hagiwara[8]. Both quantitatively and qualitatively, these two 'fingerprints' are almost identical, implying the same underlying molecular composition and structure for the two channels. The 'supra-IA' behaviour of Tl^+ and the 'sub-IA' behavour of NH_4^+ are consistent with the characteristics

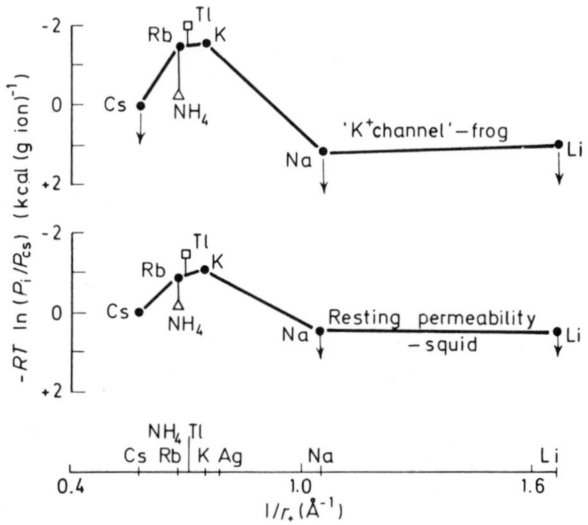

Figure 2.11 Selectivity 'fingerprints' for permeation through the 'K^+ channel' of frog node and for the resting permeation in squid axon.

The data of this figure are plotted in the manner of Figure 2.8. The data for the 'K^+ channel' in the frog node of Ranvier are from Hille[7] and those for the resting permeability of squid axon are from Hagiwara *et al.*[8]. Notice that the 'fingerprints' for both systems are almost identical.

of the ion-binding sites in enzymes, and the 'fingerprint' is virtually identical to that of nerve ATPase and *Chlorella**.

By analogy to the molecules discussed in Sections 2.5.1 and 2.5.2, the 'sub-IA' selectivity for NH_4^+ and the 'supra-IA' selectivity for Tl^+ suggest a

* The similarity of the 'fingerprint' for nerve ATPase in Figure 2.9, for *Chlorella* in Figure 2.10 and those for the 'K^+ channel' in Figure 2.11 is striking. In all systems, Tl^+ is 'supra-IA' and NH_4^+ is 'sub-IA'. Moreover, the selectivities are of comparable magnitude, the principle difference being that for the binding to *Chlorella* a sequence III[55] (Rb > K > Cs > Na) is seen instead of IV[55] (K > Rb > Cs > Na) for the permeability channel' and for activation of nerve ATPase. These data can indeed be located on a common 'biological selectivity pattern' not only for the alkali cations, as is already known[1,4], but even as regards NH_4^+[35] and Tl^+. Further discussion of the significance of such a pattern goes beyond the scope of the present chapter and will be deferred to the future.

non-tetrahedral array of the ligands* in these channels and imply that the ligands are 'ether-like' which, by analogy to the enzyme data, suggests that the channels are lined with amide carbonyls and possibly some carboxylate groups†.

2.7.2 The 'Na^+ channel' of frog nerve[73]

Figure 2.12 presents the 'fingerprint' characteristic of the 'Na^+ channel' of frog nerve and is based on the permeability ratios measured by Hille[73] in his pioneering studies of this system. Notice that the selectivity sequence among Group IA cations is different here than for the 'K^+ channel', being Na > Li > K > Rb > Cs instead of K > Rb > Cs > Na = Li (i.e. sequence X[55] instead of IV[55]). Despite this difference, Tl^+ is markedly 'supra-IA' in the 'Na^+

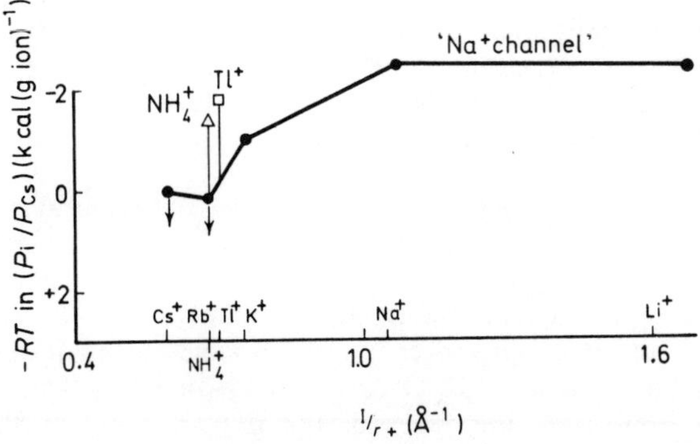

Figure 2.12 Selectivity 'fingerprints' for permeation through the 'Na^+ channel' of frog node.

The data of this figure, which are taken from permeability measurements made by Hille[73] for the action potentials recorded in the frog node of Ranvier, are plotted in the manner of Figure 2.8

* The precise non-tetrahedral structure of the ligands in the 'K^+ channel' cannot yet be specified. For example, it might contain six octahedrally-coordinated ligands, analogous to the array in valinomycin, or it could equally well contain eight ligands arrayed in a square-antiprism, analogous to the array in hexadecavalinomycin, both of the structures producing 'sub-IA' NH_4^+ selectivities. Since it might be possible to distinguish between a valinomycin-like structure and a hexadecavalinomycin-like structure by using molecular cations[74], we are presently trying to do so (cf. the data of Table 2.1).

† In addition to the evidence presented above for the involvement of amide carbonyls and carboxylate groups in the ion-binding sites of enzymes, we have recently obtained evidence[40] based on bilayer permeability ratios measured in the presence of the synthetic (low molecular weight) polyamino acid, poly(L-leucine), that the combination of a carboxylate group and amide carbonyls from a polypepeide backbone produce 'supra-IA' selectivities for Tl^+. This linear polypeptide, which appears to form ion-conducting channels in bilayer membranes, shows substantial selectivities among monovalent cations, and has the approximate permeability ratios (relative to K^+) at alkaline pH values: Tl, 2.45; K, 1.0; NH_4, 0.77; Rb, 0.5; Cs, 0.105; Na, 0.034; Li, 0.027, with Tl^+ and NH_4^+ being 'supra-IA'[49].

channel' as it was in the 'K^+ channel'. However, in the 'Na^+ channel' NH_4^+ is also 'supra-IA' in contrast to its 'sub-IA' behaviour for the 'K^+ channel'. From the same line of reasoning as before, we infer that the ligands are of the same type in the 'Na^+ channel' as in the 'K^+ channel' but that they have a component of tetrahedral array in the Na^+ channel unlike the K^+ channel.

2.7.3 Conclusion

Whether or not the particular speculations put forth above turn out to be correct, we hope we have shown how the size, shape and electronic conformation of cations can be used to infer significant details of the molecular composition and arrangement of the ligands of a variety of ion-binding 'sites' and even 'channels' not as yet accessible to direct characterisation.

2.8 EFFECTS PREDOMINANTLY DUE TO INDIVIDUAL LIGANDS—DISCUSSION OF RELATIVE CONTRIBUTIONS TO SELECTIVITY OF INDIVIDUAL LIGAND ASYMMETRIES AND STERIC FACTORS

In Section 2.2.3, two principal classes of theories for the origin of ion selectivity were mentioned, one attributing selectivity to steric factors (e.g. cavity size) the other attributing it to properties of the individual ligands (e.g. 'field strength'). One approach to evaluating the relative roles of each of these factors in ion selectivity has been made possible through the pioneering synthesis and characterisation by Ovchinnikov, Ivanov and their colleagues of a number of important analogues of the dodecadepsipeptide, valinomycin[17, 23, 25].

2.8.1 A comparison of valinomycin and hexadecavalinomycin

Table 2.1 compares the permeability ratios for valinomycin and hexadecavalinomycin[40], and these permeability ratios are plotted as selectivity 'fingerprints' in Figure 2.8. The most noteworthy feature of these 'fingerprints' is their striking similarity. Indeed the Group IA selectivity sequence (Rb > K > Cs > Na > Li) is identical for valinomycin and for hexadecavalinomycin with the only significant difference in the selectivities of these two molecules residing in a 2 kcal mol^{-1} decrease in the spread of the selectivity for hexadecavalinomycin as compared with valinomycin; in addition, the 'supra-IA' behaviour for Tl^+ and 'sub-IA' behavour of NH^+ are the same for these two molecules*.

* An interesting detail of Figure 2.8 is that NH_4^+ is 'sub-IA' in both valinomycin and hexadecavalinomycin. Although hexadecavalinomycin has eight ligands, and in this regard is similar to nonactin for which NH_4^+ was 'supra-IA', the ligands in hexadecavalinomycin are arrayed in a square-antiprism[25], a geometry which cannot provide the tetrahedral arrangement of ligands which we have suggested in nonactin produces 'supra-IA' selectivity for NH_4^+.

Table 2.1 A comparison of the permeability ratios for valinomycin and hexadecavalinomycin (glyceryl oleate-decane membranes)[40]

	Valinomycin P_i/P_k	*Hexadecavalinomycin* P_i/P_k
Li^+	0.0000042	0.00011
Na^+	0.000036	0.001
K^+	1.0	1.0
Rb^+	1.8	3.0
Cs^+	0.76	0.74
Ag^+	0.0000056	0.0016
Tl^+	0.105	0.11
NH_4^+	0.039	0.045
$CH_3NH_3^+$	0.00039	0.21
$(CH_3)_3NH^+$	$\leqslant$0.0000046	0.32
Choline$^+$	$\leqslant$0.000031	1.3
Acetylcholine$^+$	0.000016	4.5
Formamidinium$^+$	$\leqslant$0.0059	0.052
Guanidinium$^+$	$\leqslant$0.00072	0.09

If the cavity size is the determining factor in the selectivity of valinomycin among monovalent cations[33, 36, 97], it is difficult to see why hexadecavalinomycin should show an identical selectivity sequence among Group IA cations and such a similar quantitative 'fingerprint' (recall that the 'sub-IA' behaviour of Tl^+ in valinomycin has now been ascribed to steric factors). In contrast, this result is consistent with the view that the 'field strength'[3, 32, 55] of the individual ligand determines the selectivity since the ligands in these two molecules are chemically identical.

In addition, inspection of molecular models suggests that the cations may be partially hydrated in hexadecavalinomycin. This raises the question of the effect of hydration on selectivity. If the major difference between valinomycin and hexadecavalinomycin is the hydration state of the ions in the two molecules then, by analogy to the 'free-solution model' for hydrated ion exchangers (see pp. 296–301 of Eisenman[55]), we would expect that the selectivity sequence might be determined solely by the nature of the ligand whereas the degree of hydration of the ions within the complex would effect only the magnitude of selectivity.

2.8.2 Situations in which steric factors are particularly important

Whereas we have considered cases of ion carriers for which the selectivities appear to be governed more by ligand type than by cavity size, clearly it is possible to find ions which are sufficiently large that they will not fit inside the cavity of a carrier, just as it is possible to make a molecule which is sufficiently rigid that the backbone constraints override the selectivity contribution of the individual ligands. This is illustrated by the discrimination against polyatomic cations by valinomycin and the favouring of such species by hexadecavalinomycin[40, 74], seen in Table 2.1. A case in which such steric constraints in the complexer appear to be important even for monoatomic cations is that of the 1:1 complexes of the smaller cyclic polyethers[15, 42, 75, 98], where the polyethers with the smallest cavities tend to prefer the smaller cations over the larger cations and polyethers with somewhat larger cavities tend to prefer larger cations over smaller ones[75]. However, once the ring is large enough to have some flexibility, the asymmetries of the individual ligands appear to outweigh those attributable to steric factors*. For example, the polyether dicyclohexyl-24-crown-8, containing eight ether oxygens, prefers Cs^+ over the other alkali cations[98], whereas the somewhat larger polyether dicyclohexyl-30-crown-10, containing 10 ether oxygens, prefers K^+ and Rb^+ over the other alkali cations[40, 41].

It is worth noting, in addition, that even though a molecule may have a backbone which is sufficiently rigid that it maintains a relatively inflexible cavity size, the equilibrium selectivity may *still* not be determined by the cavity size. Thus the cyclic polyether, t-butyl dicyclohexyl-18-crown-6, which contains six ether oxygens, has a cavity size which is optimal for K^+ and indeed has been found in salt-extraction experiments[42, 98] and in bilayers at low concentrations[44] to show the selectivity sequence $K > Rb > Na > NH_4 > Cs > Li$ for the 1:1 ion-carrier complex. However, in bilayers at higher concentrations of this polyether, the selectivity changes to $Cs > Rb > K > NH_4 > Na > Li$, reflecting the selectivity of the energetically more favourable complex between one ion and three polyether molecules, this complex, of course, having a very different packing of the polyether ligands around the ion.

2.9 SUMMARY

The ion-selective conductances and permeabilities of bilayers in the presence of certain ion carriers can often be understood simply from the equilibrium selectivity of these carriers without having to consider their kinetics. We have shown that the selectivity of such carriers can be comprehended in terms of the equilibrium interactions between the ions and the individual ligand groups composing the ion-binding structure. We have also demonstrated, using thermochemical data, that solvents having comparable ligands (e.g. carbonyl oxygens, ether oxygens) to those of appropriate ion carriers show selectivities comparable to those seen for these ion carriers, without having to have their special structural features. Further, we have shown theoretically that the ion selectivity of individual ligands can arise simply from differences in the electrostatic interactions between ions and individual ligand groups compared to those between ions and water molecules. We have illustrated how the ionic selectivities of known molecules (e.g. carriers and appropriate solvents) for Tl^+, NH_4^+ and the Group IA cations can be used to infer the types of ligands and their coordination for such biological ion-specific systems which are not yet amenable to direct structural analysis as the 'channels' in the cell membrane and the 'sites' in enzymes. Lastly, the striking similarities in selectivity between two molecules of identical ligand type but

* Even for the small polyethers, there are manifestations of the interactions between ions and the individual ligands as is illustrated, for example, in Figure 2.8 where Tl^+ is seen to be more selected than an alkali cation of comparable radius.

very differing cavity sizes (the dodecadepsipeptide, valinomycin, and its hexadecadepsipeptide analogue) emphasise the importance of asymmetry (at the molecular level) in the interactions between ions and individual ligands versus individual water molecules as a central factor in the origin of non-monotonic selectivity sequences.

Acknowledgements

This work has been made possible by the support of the National Science Foundation (Grant GB 30835) and the National Institutes of Health (Grant NS 09931–04). We are particularly grateful to Professors Yuri Ovchinnikov and Vadim Ivanov for their gift of the priceless hexadecadepsipeptide, hexadecavalinomycin, synthesised in the Shemyakin Institute of the Soviet Academy of Sciences.

References

1. Eisenman, G. (1965). *Proc. 23rd Int. Congr. of Physiol. Sci.: Exerpta Medica*, **87,** 489
2. Suelter, C. H. (1970). *Science*, **168,** 1789
3. Eisenman, G. (1961). *Symposium on Membrane Transport and Metabolism*, 163 (A. Kleinzeller and A. Kotyk editors) (New York: Academic Press)
4. Eisenman, G. (1963). *Bol. Inst. Estud. Med. Biol.* (*Mex.*), **21,** 155
5. Diamond, J. M. and Wright, E. (1969). *Ann. Rev. Physiol.*, **31,** 581
6. Suelter, C. H. (1972). *Metal Ions in Biology*, Vol. 3. (H. Segel editor) (New York: Marcel Dekker) in press
7. Hille, B. (1973). *J. Gen. Physiol.*, **61,** 669
8. Hagiwara, S., Eaton, D. C., Stuart, A. E. and Rosenthal, N. P. (1972). *J. Memb. Biol.*, **9,** 373
9. Skou, J. C. (1957). *Biochim. Biophys. Acta*, **23,** 394
10. Skou, J. C. (1960). *Biochim. Biophys. Acta*, **42,** 6
11. Kinsky, S. C. (1961). *J. Bacteriol.*, **82,** 889
12. Lampen, J. O., Arnow, P. M. and Safferman, R. S. (1961). *J. Bacteriol.*, **80,** 200
13. Moore, C. and Pressman, B. C. (1964). *Biochem. Biophys. Res. Commun.*, **15,** 562
14. Lardy, H. A., Graven, S. N. and Estrada-O, S. (1967). *Fed. Proc.*, **26,** 1355
15. Pedersen, C. J. (1967). *J. Amer. Chem. Soc.*, **89,** 7017
16. Stefanac, Z. and Simon, W. (1967). *Microchem. J.*, **12,** 125
17. Shemyakin, M. M., Ovchinnikov, Y. A., Ivanov, V. I., Antonov, V. K., Vinogradova, E. I., Shkrob, A. M., Malenkov, G. G., Evstratov, A. V., Laine, I. A., Melnik, E. I. and Ryabova, I. D. (1969). *J. Memb. Biol.*, **1,** 402
18. Mueller, P. and Rudin, D. O. (1967). *Biochem. Biophys. Res. Commun.*, **26,** 398
19. Lev, A. A. and Buzhinsky, E. P. (1967). *Tsitologiya*, **9,** 102
20. Pressman, B. C., Harris, E. J., Jagger, W. S. and Johnson, J. H. (1967). *Proc. Nat. Acad., Sci. U.S.A.*, **58,** 1949
21. Andreoli, T. E., Tieffenberg, M. and Tosteson, D. C. (1967). *J. Gen. Physiol.*, **50,** 2527
22. Eisenman, G., Ciani, S. M. and Szabo, G. (1968). *Fed. Proc.*, **27,** 6
23. Ovchinnikov, Y. A., Ivanov, V. T., Evstratov, A. V., Bystrov, V. F., Abdullaev, N. D., Popov, E. M., Lipkind, G. M., Arkhipora, S. F., Efremov, E. S. and Shemyakin, M. M. (1969). *Biochem. Biophys. Res. Commun.*, **37,** 668
24. Szabo, G., Eisenman, G. and Ciani, S. (1969). *J. Memb. Biol.*, **3,** 346
25. Ovchinnikov, Yu. A. (1971). *23rd Int. Congr. of Pure and Appl. Chem.*, Vol. 2, 121
26. Prestegard, J. H. and Chan, S. I. (1970). *J. Amer. Chem. Soc.*, **92,** 4440
27. Kilbourn, B. T., Dunitz, J. D., Pioda, L. A. R. and Simon, W. (1967). *J. Mol. Biol.*, **30,** 559
28. Dobler, M., Dunitz, J. D. and Kilbourn, B. T. (1969). *Helv. Chim. Acta*, **52,** 2573

29. Dunitz, J. D., personal communication
30. Pinkerton, M., Steinrauf, L. K. and Dawkins, P. (1969). *Biochem. Biophys. Res. Commun.*, **35,** 512
31. Dobler, M., Dunitz, J. D., Krajewski, J. (1969). *J. Mol. Biol.*, **42,** 603
32. Eisenman, G. (1969). *Ion-Selective Electrodes*, Special Publication 314, 1–56 (R. A. Durst editor) (Washington: National Bureau of Standards)
33. Diebler, H., Eigen, M., Ilgenfritz, G., Maas, G. and Winkler, R. (1969). *Pure Appl. Chem.*, **20,** 93
34. Morf, W. E. and Simon, W. (1972). *Symposium on Molecular Mechanisms of Antibiotic Action on Protein Biosynthesis and Membranes* (E. Munoz, F. Garcia-Ferrandiz and D. Vasquez editors) (Amsterdam: Elsevier Publishing Co.), 523
35. Eisenman, G., Szabo, G., Ciani, S., McLaughlin, S. G. A. and Krasne, S. (1973). In *Progress in Surface and Membrane Science*, Vol. 6, 139 (J. F. Danielli, M. Rosenberg, D. Cadenhead editors) (New York: Academic Press)
36. Simon, W. and Morf, W. E. (1973). In *Membranes—A Series of Advances*, Chapt. 4, Vol. 2, (G. Eisenman editor) (New York: Marcel Dekker), 329
37. Krasne, S. and Eisenman, G. (1973). In *Membranes—A Series of Advances*, Section V of Chapt. 3, Vol. 2 (G. Eisenman editor) (New York: Marcel Dekker), 277
38. Eisenman, G. and Krasne, S. (1972). Lecture to Symposium on Membrane Structure and Function, *IVth Int. Biophys. Congr. Moscow, USSR*, August, 1972
39. Pauling, L. (1960). *The Nature of the Chemical Bond* (Ithaca, New York: Cornell Univ. Press)
40. Krasne, S. and Eisenman, G. (1973). Unpublished data
41. Chock, P. B. (1972). *Proc. Nat. Acad. Sci. USA*, **69,** 1939
42. Frensdorff, H. K. (1971). *J. Amer. Chem. Soc.*, **93,** 600
43. Izatt, R. M., Nelson, D. P., Rytting, J. H., Haymore, B. L. and Christensen, J. J. (1971). *J. Amer. Chem. Soc.*, **93,** 1619
44. McLaughlin, S. G. A., Szabo, G., Ciani, S. and Eisenman, G. (1972). *J. Memb. Biol.*, **9,** 3
45. Eisenman, G., Ciani, S. M. and Szabo, G. (1969). *J. Memb. Biol.*, **1,** 249
46. Szabo, G., Eisenman, G., Laprade, R., Ciani, S. and Krasne, S. (1973). In *Membranes—A Series of Advances*, Chapt. 3, Vol. 2 (G. Eisenman editor) (New York: Marcel Dekker), 179
47. Ciani, S. M., Eisenman, G., Laprade, R. and Szabo, G. (1973). In *Membranes—A Series of Advances*, Chapt. 2, Vol. 2 (G. Eisenman editor) (New York: Marcel Dekker), 61
48. Markin, V. S., Kristalik, L. I., Liberman, E. A. and Toplay, V. P. (1969). *Biofizika*, **14,** 256
49. Lauger, P. and Stark, G. (1970). *Biochim. Biophys. Acta*, **211,** 458
50. Stark, G. and Benz, R. (1971). *J. Memb. Biol.*, **5,** 133
51. Grell, E., Funck, Th. and Eggers, F. (1972). *Symposium on Molecular Mechanisms of Antibiotic Action on Protein Biosynthesis and Membranes* (E. Munoz, F. Garcia-Ferrand and D. Vasquez editors) (Amsterdam: Elsevier Publishing Company), 646
52. Bockris, J. O'M. and Reddy, A. (1970). *Modern Electrochemistry*, Vol. 1, Chapt. 2,
53. Salomon, M. (1970). *J. Phys. Chem.*, **74,** 2519
54. Eisenman, G., Szabo, G., McLaughlin, S. G. A. and Ciani, S. M. (1973). *Bioenergetics*, **4,** 93
55. Eisenman, G. (1962). *Biophys. J.*, **2,** Part, 2, 259
56. Ling, G. N. (1962). *A Physical Theory of the Living State: The Association—Induction Hypothesis* (New York: Blaisdell)
57. Eigen, M. and Winkler, R. (1971). *Neurosci. Res. Prog. Bull.*, **9,** 330
58. Somsen, G. and Weeda, L. (1971). *J. Electroanalyt. Chem.*, **29,** 375
59. Somsen, G. and Weeda, L. (1971). *Rec. Trav. Chim.*, **90,** 81
60. Gould, E. S. (1959). *Mechanism and Structure in Organic Chemistry*, (New York: Holt, Rinehart and Winston)
61. Hirschfelder, J. O., Curtiss, C. F. and Bird, R. B. (1954). *Molecular Theory of Gases and Liquids* (New York: John Wiley and Sons)
62. Rowlinson, J. S. (1951). *Trans. Faraday Soc.*, **47,** 120
63. Hodgkin, C. D., Weast, R. C. and Selby, S. M. (1960). *Handbook of Chemistry and Physics*, 2536–7 (Cleveland, Ohio: The Chemical Rubber Company)

64. Talekar, S. V. (1970). *Ph.D. Thesis*, All-India Institute of Medical Sciences
65. Rein, R., Rabinowitz, J. R. and Swissler, T. J. (1972) *J.. Theoret. Biol.*, **34,** 215
66. Cotton, F. A. and Wilkinson, G. (1967). *Advanced Inorganic Chemistry*, 2nd Ed. (New York: Interscience)
67. Laprade, R. (1972), personal communication
68. Fenton, D. E., Mercer, M. and Truter, N. R. (1972). *Biochem. Biophys. Res. Commun.*, **48,** 10
69. Deverell, C. and Richards, R. E. (1966). *Mol. Phys.*, **10,** 551
70. Bloor, E. G. and Kidd, R. G. (1968). Can. *J. Chem.*, **46,** 3425
71. Dunitz, J. D. and Orgel, L. (1960). *Adv. Inorg. Chem. Radiochem.*, **2,** 1
72. Eisenman, G., Szabo, G., McLaughlin, S. and Ciani, S. (1972). *Symposium on Molecular Mechanisms of Antibiotic Action on Protein Biosynthesis and Membranes* (E. Munoz, F. Garcia-Ferrandiz and D. Vasquez editors) (Amsterdam: Elsevier Publishing Company), 545
73. Hille, B. (1971). *J. Gen. Physiol.*, **58,** 599
74. Eisenman, G. and Krasne, S. (1973). *Biophysical Society Abstracts*, 244a
75. Christensen, J. J., Hill, J. O. and Izatt, R. M. (1971). *Science*, **174,** 459
76. Boyer, P. D., Lardy, H. A. and Phillips, P. H. (1942). *J. Biol. Chem.*, **146,** 673
77. Williams, R. J. P. (1970). *Quart. Rev. Chem. Soc.*, **24,** 331
78. Manners, J. P., Morallee, K. G. and Williams, R. J. P. (1970). *Chem. Commun.*, 965
79. Toraya, T., Sugimoto, K., Tanado, Y. and Shimizu, S. (1971). *Biochemistry*, **10,** 3475
80. Bader, H. and Sen, A. K. (1966). *Biochim. Biophys. Acta*, **118,** 116
81. Inturrisi, C. E. (1969). *Biochim. Biophys. Acta*, **173,** 567
82. Cohn, M. and Monod, J. (1951). *Biochim. Biophys. Acta*, **7,** 153
83. McClure, W. R., Lardy, H. A. and Kneifel, H. P. (1971). *J. Biol. Chem.*, **246,** 3569
84. Setlow, B. and Loewenstein, J. M. (1967). *J. Biol. Chem.*, **242,** 607
85. Kayne, F. J. (1971). *Arch. Biochem. Biophys.*, **143,** 232
86. Steinrauf, L. K. and Pinkerton, M. (1968). *Biochem. Biophys. Res. Commun.*, **33,** 29
87. Agtarap, A., Chamberlin, J. W., Pinkerton, M. and Steinrauf, L. (1967). *J. Amer. Chem. Soc.*, **89,** 5737
88. Hill, B. (1972). *J. Gen. Physiol.*, **59,** 637
89. Stroud, R. M., Kay, L. M. and Dickerson, R. E. (1971). *Cold Spring Harbor Symp. Quant. Biol.*, **36,** 125
90. Stroud, R. M., personal communication
91. Drenth, J., Hol. W. G. J., Jansonius, J. N. and Koekoek, R. (1971). *Cold Spring Harbor Symposium on Quantitative Biology*, **36,** 107
92. Cohen, D. (1962). *J. Gen. Physiol.*, **45,** 959
93. Bezanilla, F. and Armstrong, C. M. (1972). *J. Gen. Physiol.*, **60,** 588
94. Eisenman, G., Sandblom, J. P. and Walker, J. L., Jr. (1967). *Science*, **155,** 965
95. Eisenman, G. (1968). *Ann. N.Y. Acad. Sci.*, **148,** 5
96. Eisenman, G. (1967). *Glass Electrodes for Hydrogen and Other Cations: Principles and Practice*, Chapt. 5, 133 (G. Eisenman editor) (New York: Marcel Dekker)
97. Mayers, D. F. and Urry, D. W. (1972). *J. Amer. Chem. Soc.*, **94,** 77
98. Pederson, C. J. (1968). *Fed. Proc.*, **27,** 1305

3
Membranes of Transformed Mammalian Cells

R. O. BRADY and P. H. FISHMAN
National Institute of Neurological Diseases and Stroke, Bethesda, Maryland

---

3.1 INTRODUCTION

A number of the properties of a cell are changed when it is transformed by an oncogenic virus. These alterations include: (a) modification of the morphology of the cell, usually towards a smaller and more rounded shape; (b) loss of contact inhibition of growth and movement in culture; (c) increased growth rate; (d) acquisition of the ability to grow in suspension; (e) change in the chromosome number; (f) increased tumourigenicity in appropriate animal hosts; and (g) the appearance of new, specific antigens in the nucleus and on the surface of the virus-transformed cell. It seems reasonable to suspect that all of these phenomena are inter-related, but we are still too unsophisticated to assemble these various physiological and biochemical aspects into a coherent series of inter-related events. Furthermore, at the present time, our thinking seems partially blocked by the simplistic view that the relatively small amount of genetic information in the tumourigenic DNA virus genomes seem to be fairly well-accounted for, thus leaving little room for templates for the synthesis of additional molecules which are responsible for the various phenomena listed above. If this is true, then it is obvious that we must direct our thinking and imagination to mechanisms which permit the formulation of novel concepts and the design of experiments to elucidate the cause of these alterations. In order to accomplish this intellectual quantum jump, it would be extremely helpful to find a consistent set of differences in the biochemical make-up of transformed cells which are related to and a consequence of tumourigenic virus transformation. Perhaps an even more fortunate situation would be forthcoming if this change or changes resulted in an alteration of the surface properties of the transformed cell, since it seems very likely that the social behaviour of cells must be related to a large extent to events which occur on the membrane of cells exposed to other cells and other factors in the environment. Such a change has been well documented within the last few years regarding the composition and synthesis of acidic glycolipids called gangliosides which are highly concentrated in the plasma cell membrane. These discoveries provide an important basis for investigating biochemical events which follow the transformation of cells with tumourigenic viruses. Whilst we do not wish to naïvely restrict our thinking to the idea that these changes are a *sine qua non* of oncogenic virus-related neoplastic disease, the consistency with which these alterations occur, and their obligatory link to the process of tumourigenic virus transformation, make them pre-eminently worthy of exploration. It is therefore extremely important to try to understand why these manifestations occur and how they are related to the insertion of the oncogenic virus genome into the genetic apparatus of a normal cell. We will review these findings in order to acquaint the reader with these recent developments, and attempt certain extrapolations of this information which appear reasonable at this time. Finally, we shall indicate further avenues of investigation which seem to hold promise for providing an insight into viral carcinogenesis.

3.2 ALTERATION OF GLYCOLIPID COMPOSITION IN VIRUS-TRANSFORMED CELLS

3.2.1 Ganglioside patterns in normal, SV40 and polyoma virus-transformed mouse cell lines

Gangliosides are acidic glycolipids which have a hydrophobic portion of their structure in common. This moiety, which is called ceramide, consists of sphingosine $[CH_3{\cdot}(CH_2)_{12}{\cdot}CH{:}CH{\cdot}CH(OH){\cdot}CH(HN_2){\cdot}CH_2OH]$ and a long-chain fatty acid joined by an amide link to the nitrogen atom on C-2 of sphingosine. In addition to ceramide, gangliosides contain hexoses, hexosamines and sialic acid. The simplest ganglioside found in most mammalian cells contains one molecule each of ceramide, glucose galactose and *N*-acetylneuraminic acid (or *N*-glycolylneuraminic acid). This monosialoganglioside is called G_{M3}. Normal contact-inhibited mouse cell lines contain a family of gangliosides ranging from G_{M3} to the disialotetrahexosylceramide G_{D1a} (Figure 3.1). These gangliosides usually appear as doublets on analytical thin-layer chromatograms due to the slightly slower migration of the *N*-glycolylneuraminic acid derivative than the corresponding *N*-acetylneuraminic acid compounds. Some minor difference in mobility may also be due to variation in the length of the fatty-acid chain. When mouse cells are

(G_{M3}) Hematoside: ceramide–glucose–galactose–*N*-acetylneuraminic acid

(G_{M2}): ceramide–glucose–galactose–*N*-acetylgalactosamine
|
N-acetylneuraminic acid

(G_{M1}): ceramide–glucose–galactose–*N*-acetylgalactosamine–galactose
|
N-acetylneuraminic acid

(G_{D1a}): ceramide–glucose–galactose–*N*-acetylgalactosamine–galactose
| |
N-acetylneuraminic acid *N*-acetylneuraminic acid

Figure 3.1 Structures of gangliosides of normal mouse cell lines

Table 3.1 Distribution of gangliosides in AL/N and Swiss mouse cells line and in polyoma and SV40 virus-transformed derivative lines*

Cell type	*Growth in culture*	*Gangliosides* (nmol mg^{-1} protein) G_{D1a}	G_{M1}	G_{M2}	G_{M3}	*Total*
N-AL/N	Contact-inhibited	1.8	1.5	0.8	0.6	4.7
SVS-AL/N	High density	0.16	0.22	0.1	1.9	2.4
Py-AL/N	High density	0.1	0.15	0.2	1.8	2.3
Swiss 3T3	Contact-inhibited	2.4	2.6	1.8	4.0	10.8
Py11	High density	0.1	0.2	0.05	3.2	3.5
SV101	High density	0.05	0.1	0.05	3.5	3.7

* From Refs. 1 and 2

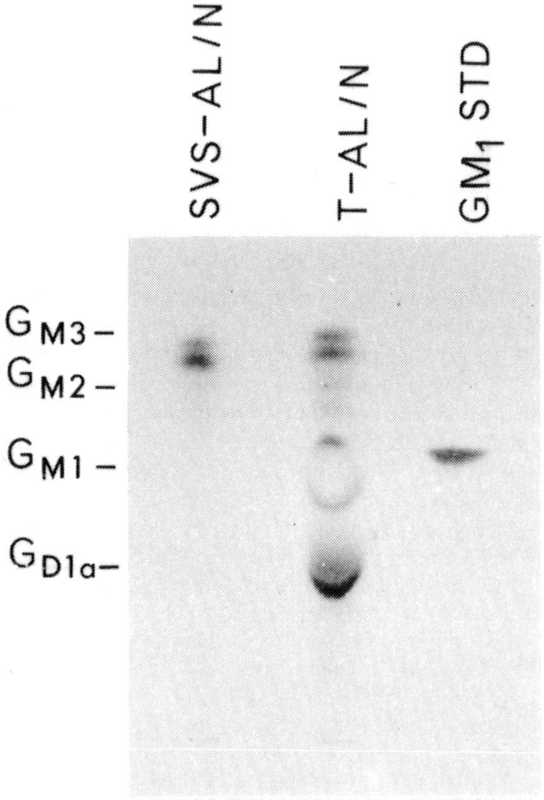

Figure 3.2 Ganglioside patterns in AL/N mouse cell lines. The family of gangliosides in the spontaneously-transformed T-AL/N cell line is essentially the same as that in contact-inhibited AL/N cells and is in sharp contrast with only G_{M3} in the Simian virus 40 (SVS-AL/N) transformed cell line. G_{M1} is a reference standard. (From Mora *et al.*[1], by courtesy of the National Academy of Sciences, U.S.A.)

transformed by tumourigenic DNA viruses such as Simian virus 40 (SV40) or polyoma virus, there is a dramatic alteration in the ganglioside composition in the transformed cell and clonal derivative lines[1,2] (Figure 3.2). The concentration of all of the ganglioside homologues larger than G_{M3} is drastically reduced (Table 3.1). This alteration in ganglioside composition after virus transformation has been verified by analyses of the gangliosides by gas–liquid chromatography[3].

3.2.2 Gangliosides in tumourigenic RNA virus-transformed mouse cell lines

Once the discovery of a shift in the ganglioside pattern in tumourigenic DNA virus-transformed cells was established, the question immediately arose regarding the universality of this phenomenon. In particular, it was extremely important to learn whether this alteration occurred in tumourigenic RNA virus-transformed cells. We have recently shown that there is a similar change in gangliosides in Swiss mouse cells which are transformed with Moloney sarcoma virus (MSV)[4].

3.2.3 Ganglioside patterns in cultured cells from other species

An early report dealt with a change in ganglioside in a polyoma virus transformant of a baby hamster kidney cell line (BHK21)[5]. The only ganglioside in BHK21 cells was found to be G_{M3} and the quantity of this glycolipid was said to be decreased in the polyoma-transformed cells. We have recently carried out an investigation of gangliosides in BHK cells and, in keeping with the observation of Hakomori and Murakama, G_{M3} appears to be the only ganglioside in these cells. However, in three polyoma-transformed BKH cell lines that we have examined, no significant decrease occurred in the quantity of this ganglioside in the transformants[6].

Another established hamster cell line, NIL, also contains only the G_{M3} ganglioside[7]. In some spontaneously- and virally-transformed sub-lines, there is a reduction in this sialoglycolipid but the changes are associated with the density of the cells in culture (which will be discussed further in Section 3.4.2).

Chick embryo secondary fibroblasts following infection with Rous sarcoma virus (RSV) have been reported to have reduced amounts of their major gangliosides G_{M3} and G_{D3} (disialylactosylceramide)[8]. The decreases follow the morphological transformation induced by this virus; no changes were observed when the cells were infected with the helper virus, Rous associated virus which is a non-transforming virus. However, no changes in ganglioside composition have been observed in chick embryo fibroblasts transformed either with RSV or temperature-sensitive RSV[9]. The latter findings are in agreement with our observations. No decrease has been observed in chick embryo secondary cells transformed by another strain of RSV or its temperature-sensitive variant[10].

Established rat kidney cells (NRK) also contain G_{M3} as their only sialoglycolipid. Cells transformed by MSV have reduced levels of this ganglioside.

In addition, NRK cells carrying temperature-sensitive MSV genome, when grown at the non-permissive temperature for transformation (39 °C) have normal amounts of G_{M3}, but when grown at the transforming temperatures (31 °C) have reduced amounts[11].

3.2.4 Neutral glycolipids

Most mammalian cells contain several neutral glycolipids in addition to the acidic gangliosides. The neutral sphinogoglycolipids range from glucosylceramide to the pentahexosylceramide called the Forssman hapten (Figure 3.3). In the report of decreased G_{M3} in polyoma-transformed BHK cells, it was stated that ceramide lactoside (Gl–2) was increased in the transformants[5]. A later report from the same laboratory indicated that the augmentation of Gl–2 was not observed in subsequent studies, but glucosylceramide (Gl–1) was present in greater amounts than that seen in the controls[12]. We have not observed consistent changes in the neutral glycolipids in our studies with DNA virus-transformed mouse cell lines[1].

A subsequent report has indicated that the quantity of ceramide trihexoside (Gl–3 in Figure 3.3) is diminished in transformed hamster cells[13]. We have not found evidence for a consistent change of Gl–3 in transformed mouse cells. However, we have recently observed that the concentration of Gl–3 is diminished in cultured rat kidney (NRK) cells after MSV virus-transformation. This alteration depends on the phenotypic expression of the transformed state which is affected by the temperature at which the temperature-sensitive transformed mutant cells available to us are grown[11]. When these cells are cultured at 31 °C, the characteristics of transformed cells outlined in Section 3.1 are apparent. If the cells are grown at a higher temperature, such as 39–40°C, they regain the morphological and other characteristics of contact-inhibited non-transformed cells. The cells grown at the higher temperature have the same amount of Gl–3 as the non-transformed parent cells, but when the cells are grown at the permissive temperature of 31 °C, the quantity of Gl–3 is decreased similarly to cells transformed by wild type MSV. Thus, the amount of this glycolipid is modulated. It must now be determined if this change is due to an alteration of catabolism or synthesis of Gl–3 in these cells under the various conditions. In addition, a change in Gl–3 level has been reported to depend on the density of cells in culture[14]. We shall discuss this aspect further in Section 3.4.2.

3.3 METABOLISM OF COMPLEX LIPIDS IN TRANSFORMED CELLS

3.3.1 Catabolic studies

It appeared important to determine the nature of the metabolic alteration responsible for the altered ganglioside pattern in virus-transformed mouse cells. One possibility was an increase in the rate of catabolism of the higher ganglioside homologues G_{M2}, G_{M1} and G_{D1a} after transformation (see

Glucosylceramide (G1–1): ceramide —— glucose

Ceramidelactoside (G1–2): ceramide —— glucose —— galactose

Ceramidetrihexoside (G1–3): ceramide —— glucose —— galactose —— galactose

Globoside (G1–4): ceramide —— glucose —— galactose —— galactose —— *N*-scetylgalactosamine

Forssman hapten (G1–5): ceramide —— glucose —— galactose —— galactose —— *N*-acetylgalactosamine —— *N*-acetylgalactosamine

Figure 3.3 Structures of neural glycolipids in clutured mammalian cells

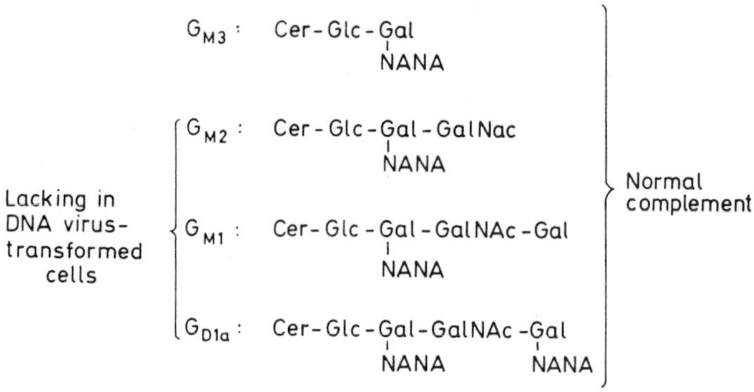

Figure 3.4 Gangliosides of normal or tumourigenic virus-transformed mouse cell lines. Abbreviations: Cer = ceramide; Glc = glucose; Gal = galactose; NANA = *N*-acetylneuraminic acid; GalNAc = *N*-acetylgalactosamine

Figure 3.4) and this concept has been examined by determining the rate of enzymatic hydrolysis of gangliosides in control and virus-transformed cells.

The principal gangliosides in the normal and transformed cells are G_{D1a} and G_{M3}, respectively. Catabolism of both of these lipids is initiated by the cleavage of the molecule of *N*-acetylneuraminic acid linked to the terminal molecule of galactose in the respective glycolipids (reactions (3.1 and 3.2)):

$$\text{Cer-Glc-Gal-(NeuNAc)-GalNAc-Gal-NeuNAc} + H_2O \xrightarrow{\text{neuraminidase}} \text{Cer-Glc-Gal-(NeuNAc)-GalNAc-Gal} + \text{NeuNAc} \quad (3.1)$$

$$\text{Cer-Glc-Gal-NeuNAc} + H_2O \xrightarrow{\text{neuraminidase}} \text{Cer-Glc-Gal} + \text{NeuNAc} \quad (3.2)$$

where Cer = ceramide, Glc = glucose, Gal = galactose, NeuNAc = *N*-acetylneuraminic acid and GalNAc = *N*-acetylgalactosamine.

In order to measure the catabolism of these gangliosides, they were exclusively labelled in the *N*-acetylneuraminic acid moieties[15]. Using these labelled substrates, it has been found that the catabolism of these compound is essentially the same in the control and in the virus-transformed cells[16]. Thus the possibility that excessively rapid catabolism of gangliosides is the cause of the alteration in ganglioside composition in the transformed cell is excluded.

3.3.2 Ganglioside synthesis in transformed cells

3.3.2.1 Ganglioside synthesis in DNA virus-transformed mouse cells

Gangliosides in mouse cells will become labelled if appropriate radioactive precursors are added to the culture medium. We have investigated ganglioside synthesis by including *N*-acetyl-[^{3}H]mannosamine in the medium because it

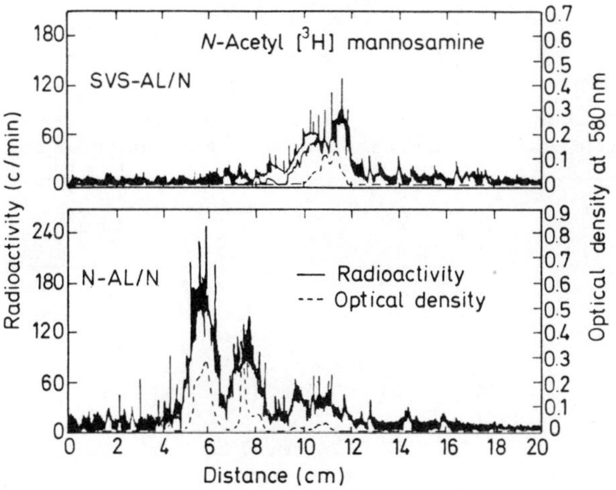

Figure 3.5 Scans of absorbance at 580 nm and radioactivity of thin-layer chromatograms of gangliosides extracted from contact-inhibited N-AL/N and SV40 virus-transformed (SVS-AL/N) cells grown in the presence of *N*-acetyl-[^{3}H] mannosamine. (From Brady and Mora[2], by courtesy of Elsevier Publishing Co.)

is a precursor of the sialic acid moieties of gangliosides. The entire series of gangliosides from G_{M3} through G_{D1a} are labelled in the control cells under these conditions but only ganglioside G_{M3}, however, is radioactive in SV40 virus-transformed cells (Figure 3.5). An identical labelling pattern of gangliosides is seen where *N*-acetyl-[^{3}H]glucosamine is used. These experiments strongly indicate that there is apparently a block in the synthesis of gangliosides larger than G_{M3} in the virus-transformed cells.

Since gangliosides are synthesised in a stepwise fashion, it therefore appears likely that the enzyme which catalyses the conversion of G_{M3} to G_{M2} may be deficient in the transformed cells. This enzyme is called hematoside: *N*-acetylgalactosaminyltransferase and it catalyses reaction (3.3):

$$\text{Cer-Glc-Gal-NeuNAc } (G_{M3}) + \text{UDP-GalNAc} \xrightarrow[\text{transferase}]{\text{amino sugar}}$$

$$\text{Cer-Glc-Gal-(NeuNAc)-GalNAc} + \text{UDP} \qquad (3.3)$$

where UDP = uridine diphosphate.

The activity of this amino sugar transferase is drastically reduced in SV40 and polyoma virus-transformed cell lines compared with that in the contact-inhibited parent cell lines (Table 3.2). The impairment of this reaction in the formation of gangliosides can account for the altered pattern in the transformed cells.

It then became of interest to determine whether other reactions involved in the synthesis of gangliosides were altered after tumourigenic virus transformation. These steps and the enzymes which catalyse the reactions are indicated in schematic form in Figure 3.6. The activity of these enzymes in

Table 3.2 Uridine diphosphate N-acetylgalactosamine hematoside: N-acetylgalactosaminyltransferase activity in mouse cell lines*

Cell line	*Glycolipid acceptor*† G_{M3}NAc‡	G_{M3}NGlyc§
N-AL/N‖	100	100
SVS-AL/N	0	10
Py-AL/N	0	0
Swiss 3T3¶	100	100
SV101	33	30
Py11	26	2

* From Ref. 16
† Activity as percent of control cells
‡ *N*-Acetylneuraminylgalactosylglycosylceramide (hematoside)
§ *N*-Glycolylneuraminylgalactosylglucosylceramide (hematoside)
‖ Transferase activity was determined in the microsomal fraction of the AL/N cell lines
¶ Transferase activity was determined in the whole homogenates of Swiss cell lines

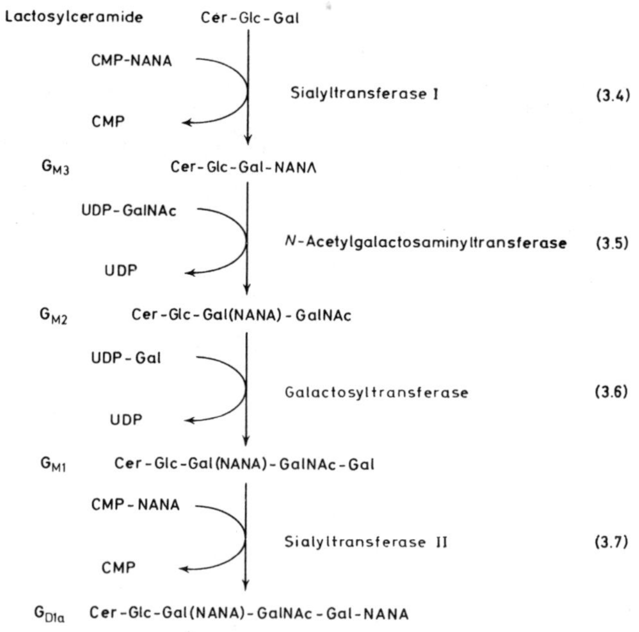

Figure 3.6 Pathway of the biosynthesis of gangliosides

Table 3.3 Glycosyltransferase activities in virus-transformed Swiss mouse cell lines*

Enzyme	*Cell line†*	
	SV101	Py11
Sialyltransferase I	105	110
N-Acetylgalactosaminyltransferase	13	17
Galactosayltransferase	655	168
Sialyltransferase II	36	82

* Calculated from Ref. 17
† Activity as percent of 3T3 control cells

Table 3.4 Glycosyltransferase activities in virus-transformed AL—N mouse cell lines*

Enzyme	*Cell line†*	
	SVS-AL/N	Py-AL/N
Sialyltransferase I	56	143
N-Acetylgalactosaminyltransferase	16	14
Galactosyltransferase	6	11
Sialyltransferase II	184	192

* Calculated from Ref. 17
† Activity as percent of contact-inhibited AL/N control cells

contact-inhibited and virus-transformed Swiss mouse cell lines is indicated in Table 3.3. The activity of these enzymes in various AL/N mouse cell lines is shown in Table 3.4. Sialyltransferase I activity is usually slightly increased in the transformed cells. Galactosyltransferase activity is increased in transformed Swiss mouse cells, but decreased in transformed AL/N cells from that observed in the respective parent cell lines. Sialyltransferase II activity is slightly decreased in transformed Swiss mouse cells, but its activity is markedly increased in the transformed AL/N cells. The consistent change in these various cell lines is the decrease of hematoside: *N*-acetylgalactosaminyltransferase activity. A similar diminution in the activity of this amino sugar transferase has been observed in SV40-transformed Balb cell lines[18].

3.3.2.2 Ganglioside synthesis in RNA virus-transformed cells

Again the chief alteration is a decrease in the activity of hematoside:*N*-acetylgalactosaminyltransferase[4]. The levels of other enzymes involved in ganglioside biosynthesis are not reduced when Swiss mouse 3T3 cells are transformed by Moloney sarcoma virus (MSV). In fact, there is a severalfold increase in galactosyltransferase (Figure 3.6, reaction (3.6)). Appropriate controls have been carried out by infecting the same line of cells with Moloney leukemia virus (MLuV) since the latter virus is required as a helper for the replication of MSV and is always present in sarcoma virus stocks. No change in ganglioside synthesis in the cells exposed to leukemia virus alone was found.

The kinetics of this metabolic alteration are interesting and may provide some insight into the nature of the enzymatic change. When the cells were treated with Moloney sarcoma virus at a high multiplicity of infection it was found that 95% of the cells appeared to be morphologically transformed by the second day after exposure to the virus. The decrease in activity of the hematoside:*N*-acetylgalactosaminyltransferase appeared to follow this phenotypic change. For example, by the fourth day after infection there was a 50% reduction in the activity of this enzyme compared with uninfected cells cultured at the same time. By the seventh day, there was an 80% reduction in amino sugar transferase activity. These results strongly suggest that critical studies must be carried out in detail in which the temporal sequence of changes in morphology, chemical composition of the membranes and alterations of enzymatic activity are carefully correlated.

3.3.2.3 Ganglioside synthesis in other cell lines

A reduction in sialyltransferase I activity has been reported in a polyoma virus-transformed clone of BHK cells[19]. This reduction may apparently be correlated with the lower amounts of G_{M3} observed in these cells[5]. However, our own studies on polyoma virus-transformed BHK cells indicate the presence of sialyltransferase I as expected from the presence of G_{M3} in these cells[6]. We have recently observed higher than normal levels of this enzyme activity in several clones of RSV-transformed chick embryo fibroblasts[10]. This finding correlates with our own ganglioside analysis of these cells as well as with that of Warren *et al.*[9] and contradicts the results of Hakimori *et al.*[5, 12]. Varying results from different laboratories point out the necessity of examining several separate transformed clones in order to more firmly relate glycolipid alterations with viral transformation.

3.3.2.4 Ganglioside galactosyltransferase in mouse cell lines

The increased levels of galactosyltransferase activity (Figure 3.6, reaction (3.6)) observed in DNA and RNA virus-transformed Swiss 3T3 cells (whether transformed by SV40, polyoma, or MSV) is quantitatively very similar and may be related to the mechanism of transformation or a subsequent perturbation of normal cellular regulation as discussed in Section 3.7. The decreased levels in this enzyme in AL/N cell lines is more readily explained. During the course of determining the ganglioside pattern of a highly malignant AL/N cell line that spontaneously transformed during sub-culturing of the normal N-AL/N parental line, we have observed a decrease in the G_{D1a} and G_{M1} content of this T-AL/N line after 200 passages in culture[20]. When the ganglioside glycosyltransferases were determined and compared to early passage and AL/N cells, the galactosyltransferase was reduced over 20-fold whereas the other three transferases (including amino sugar transferase) were similar in both passages. Since both SVS-AL/N and Py-AL/N cells have been extensively cultured, the reduced galactosyltransferase activity may be the result of a similar phenomenon.

3.3.3 Synthesis of neural glycolipids

It has been reported that the last step in the synthesis of ceramidetrihexoside (reaction (3.8)) is impaired in polyoma virus-transformed BHK cells and NIL fibroblasts[13].

$$\text{Cer-Glc-Gal} + \text{UDP-Gal} \xrightarrow[\text{transferase}]{\text{galactosyl-}} \text{Cer-Glc-Gal-Gal} + \text{UDP} \qquad (3.8)$$

Our results with the temperature-sensitive virus-transformed NRK cell line[11] are consistent with this finding. However, two aspects of this observation must be kept in mind. It has been stated that the activity of the galactosyltransferase which catalyses reaction (3.8) is dependent upon the density of cell growth in culture[13]. More will be said in this regard in Section 3.4.2. In addition, it is well known that hamster cells transformed with polyoma virus often show a positive reaction with antibody directed against the Forssman hapten (Gl–5, Figure 3.3) whereas they do not show this reaction prior to transformation[21-23].

It seems very difficult to reconcile this observation with the fact that the biosynthesis of glycosphingolipids is a stepwise process with each succeeding reaction depending upon the formation of the immediately preceding reactant[24]. Therefore, if the synthesis of Gl–3 is impaired in polyoma-transformed cells, it is hard to imagine how the Forssman hapten becomes apparent. We have not tested the transformed NRK cells for reactivity with Forssman antibody. However, we have no evidence from analytical thin-layer chromatography for the presence of Gl–5 in either the NRK or the temperature-sensitive transformant. A possible explanation of the transformed hamster cells becoming Forssman positive is that the antibody employed in the experiments cited in Refs. 21–23 is directed to a Forssman-reactive glycoprotein component of the membrane of the transformed hamster cells. The appearance of such an antigen might conceivably result from an impairment of synthesis of the normal oligosaccharide chain of a glycoprotein resulting in the exposure of a Forssman-reactive moiety rather than the synthesis of a neutral lipid component. Appropriate experiments can be designed and should be undertaken to resolve this dilemma.

3.4 PHYSIOLOGICAL CONSIDERATIONS

3.4.1 Productive infection of cells versus transformation

A number of cell lines infected by polyoma virus show an induction of cellular DNA synthesis and an increase in the activity of several enzymes related to DNA formation. The increase in enzymatic activity reaches its peak *ca.* 28 h after exposure to the virus. After 50 h, there is a considerable amount of virion protein and also completed but unreleased virus. Cell lysis

Table 3.5 Uridine diphosphate N-acetylgalactosamine hematoside: N-acetylgalactosaminyltransferase activity in lytically infected and in polyoma-transformed T-AL/N cell lines*

Harvest (h)	*Cell line*	*Glycolipid acceptor†* G_{M3}NAc	G_{M3}NGlyc
	T-AL/N (control)	100	100
28	T-AL/N (Py infected)	173	129
	Py-AL/N (transformed)	9	0
	T-AL/N (control)	100	100
50	T-AL/N (Py infected)	148	131
	Py-AL/N (transformed)	5	1

* From Ref. 29
† Percentage of radioactivity incorporated

and virus release occurs in 5–7 days[25-28]. These cells are not transformed and are said to be permissive for polyoma virus. The lytically-infected cells which produce polyoma virus do not show diminution of hematoside:*N*-acetylgalactosaminyltransferase activity (Table 3.5).

3.4.2 Culture conditions

It is extremely important to determine whether culture conditions cause an alteration in ganglioside composition and changes in the activity of enzymes involved in ganglioside formation. Different types of growth medium and changes in the pH have been shown to have no effect on ganglioside biosynthesis. The age of the culture in terms of the number of passages or subcultures is not a factor as far as hematoside:*N*-acetyltgalactosaminyltransferase activity is concerned. Virally-transformed mouse cells after more than 200 passages still have low levels of *N*-acetylgalactosaminyltransferase whereas control lines have high levels of this enzyme (comparable to early passages of the same cell line)[18]. The rate of cell growth and the level of saturation density in culture is also not a factor.

Spontaneously-transformed T-AL/N or Balb 3T12 cell lines which have good amino sugar transferase activity grow as fast and to the same level of saturation densities as comparable virus-transformed cells. In addition, neither the pattern of gangliosides nor the activity of enzymes involved in the biosynthesis of gangliosides are appreciably influenced by the density of cells in culture (Figure 3.7).

These findings are in contrast with the observation that the pattern and synthesis of neutral glycolipids appear to be influenced by culture conditions[14, 30]. Cell density-dependent changes in neutral glycolipids, and in some instances G_{M3}, have been observed in hamster cell lines. The effect appears to be abolished in both spontaneously- and virus-transformed cells[7].

However, even this generality has its exception; a highly malignant NIL cell sub-line maintains a density-dependent effect for Forssman hapten

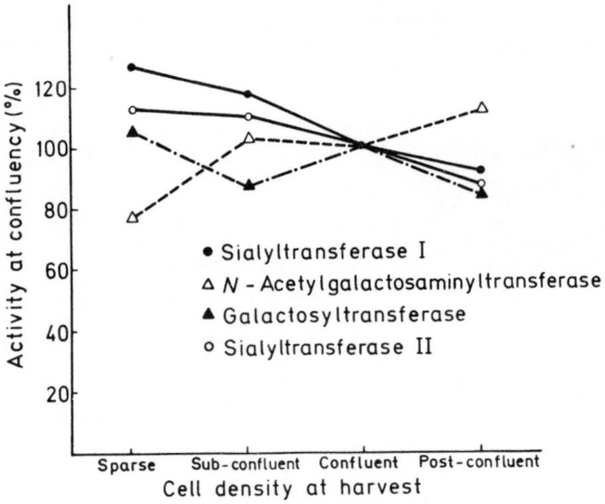

Figure 3.7 Activity of enzymes involved in ganglioside synthesis in Swiss mouse 3T3 cells at various cell densities in culture. (From Fishman *et al.*[17], by courtesy of Academic Press, Inc.)

(Gl–5, Figure 3.3)[31]. The significance of these density-dependent changes in glycolipids is not apparent at this time. These changes do not appear to be related to virus transformation, as has been demonstrated for the change in ganglioside pattern and synthesis.

3.4.3 Possible interactions between normal and virus-transformed cells

Since there is a decrease in the activity of hematoside: *N*-acetylgalactosaminyltransferase in cells transformed with tumourigenic DNA viruses as well as a tumourigenic RNA virus, it is conceivable that a common diffusible repressor substance may be elaborated in the cells which are transformed by these viruses. In order to investigate this possibility, contact-inhibited and virus transformed cells have been grown in the same plastic petri dish under a common medium but separated by a centre ridge. After harvesting, the activity of the amino sugar transferase was determined in the two types of cells No evidence was found for diffusion of an inhibitory factor from the transformed cells to the controls (Table 3.6). Furthermore, no correction of the decreased amino sugar transferase activity apparently occurs in the transformed cells under these conditions.

The possibility exists that a non-diffusible repressor material may be produced in the transformed cells which is passed from cell to cell if the cells. are sufficiently close to each other. In order to investigate this phenomenon, control and virus-transformed cells have been grown together in the same flask. The distribution of the two types of cells at harvesting was determined by the immunological identification of the transformed cells and the activity

Table 3.6 Uridine diphosphate N-acetylgalactosamine hematoside N-acetylgalactosaminyltransferase activity in cells cultivated separately or under common medium*

Cell line	*Radioactivity incorporated* (%)
T-AL/N (alone)	100
SVS-AL/N (alone)	18
T-AL/N (1]2 plate with SVS-AL/N)	95
SVS-AL/N (1/2 plate with T-AL/N)	10

* From Ref. 29

of the amino sugar transferase has been determined in homogenates of the mixed cells. A 36% decrease in the activity of the hematoside:*N*-acetylgalactosaminyltransferase was found based on the calculated value from the estimated number of the two different cells in the culture[29]. It is very difficult to decide whether this is a significant decrease in enzyme activity or whether there are flaws in the design and execution of this experiment. What is known from earlier work[16] and has been confirmed in subsequent experiments[29] is that there is no intracellular accumulation of an inhibitor of the enzyme in the virus-transformed cells. The level of hematoside:*N*-acetylgalactosaminyltransferase activity is exactly that predicted by theory in experiments in which the two different types of cells were homogenised separately and mixed in varying proportions in the enzyme assay.

3.5 STUDIES WITH 'FLAT-REVERTANT' CELL LINES

3.5.1 Ganglioside patterns

Another series of observations which we shall now discuss provides a potentially important insight into the relevance and mechanism of the change in ganglioside pattern and synthesis in tumourigenic virus-transformed cells. Phenotypically 'flat-revertant' cell lines have been obtained by treating SV40 and polyoma virus-transformed cells with fluorodeoxyuridine[32]. The 'flat-revertant' cloned cells obtained from the SV40 virus-transformed cells grow as a monolayer and are contact-inhibited in culture. Furthermore, they are less tumourigenic than the transformed cells from which they are derived. The ganglioside pattern in these 'flat-revertant' cells (F1–SV) is essentially the same as that in untransformed contact-inhibited parent Swiss mouse 3T3 cells (Figure 3.8 and Table 3.7). In a 'flat-revertant' cell line obtained from polyoma virus-transformed cells, the cells show some phenotypic and karyotypic differences from the control 3T3 cells and the

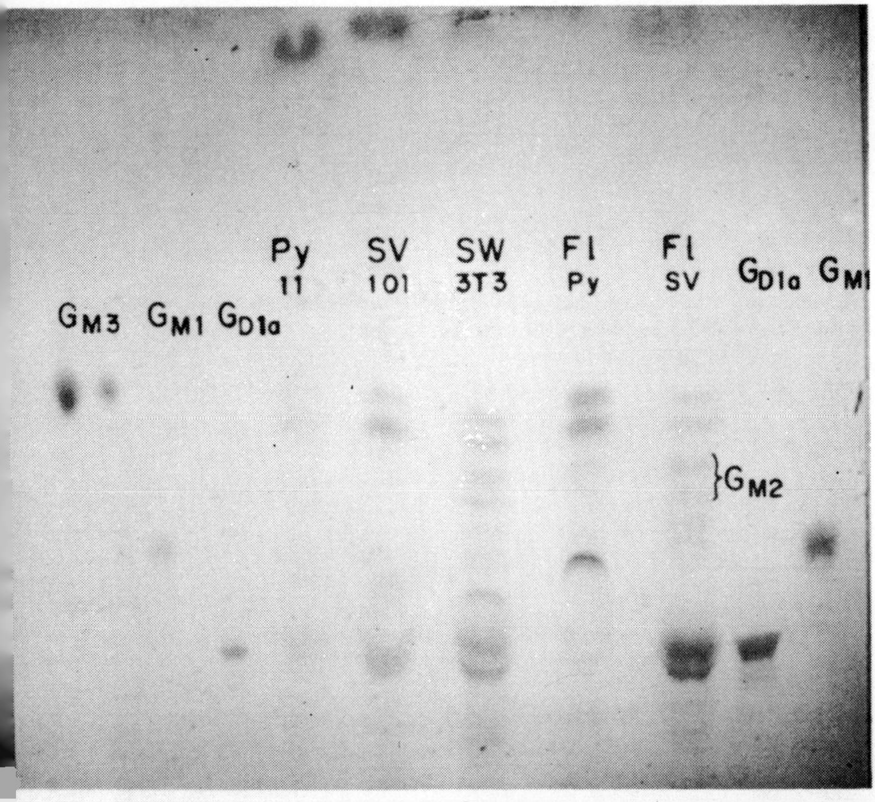

Figure 3.8 Thin-layer chromatogram of gangliosides in Swiss mouse cell lines. SW3T3 = contact-inhibited parent cell line; Py11 = polyoma virus-transformed derivative line; SV101 = Simian virus 40 transformed derivative; F1-Py = 'flat revertant' of Py11; F1-SV = 'flat revertant' of SV101; G_{M3}, G_{M1}, G_{D1a} and G_{M2} = authentic ganglioside standards. (From Mora *et al.*[29], by courtesy of Academic Press, Inc.)

Table 3.7 Distribution of gangliosides in Swiss 3T3 cell line, in SV40 and polyoma virus-transformed derivative lines, and in flat sub-lines derived from the virus-transformed lines*

Cell lines	*Gangliosides* (nmol mg^{-1} protein)				*Total*
	$G_{D}1_{a}$	G_{M1}	G_{M2}	G_{M3}	
Swiss 3T3	1.8	1.4	1.3	3.3	7.8
SV101	0.05	0.1	0.05	3.5	3.7
Py11	0.1	0.2	0.05	3.2	3.5
F1-SV	2.6	0.8	1.4	2.9	7.7
F1-Py	0.93	0.2	0.4	3.9	5.4

* From Ref. 29

'flat revertant' of the SV40-transformed cells. The polyoma 'flat-revertant' cells (F1–Py) are not as contact-inhibited in culture as the control 3T3 cells or the SV40 revertant cells. Also, the chromosome number does not increase to the sub-tetraploid state as in the F1–SV line. The ganglioside pattern in the F1–Py cells is intermediate between that of the parent cell line and that in the polyoma virus-transformed cells.

3.5.2 Ganglioside synthesis

The specific activity of the hematoside: *N*-acetylgalactosaminyltransferase in the F1–SV cell line is slightly increased over that in the original 3T3 cells (Table 3.8). The activity of this enzyme is only partially restored in the incompletely reverted polyoma cells. Thus, the ganglioside pattern and the activity of the hematoside: *N*-acetylgalactosaminyltransferase appear to be coordinately linked to the phenotypic growth properties of the cells in culture. Similar coordinate expression and repression of growth and hematoside: *N*-acetylgalactosaminyltransferase activity has been observed in SV40-transformed Balb/c 3T3 cells and 'flat' clonal lines selected out by a serum factor growth dependence (Table 3.9).

Table 3.8 Uridine diphosphate N-acetylgalactosamine hematoside: N-acetylgalactosaminyltransferase activity in Swiss mouse cell lines†

Cells	Glycolipid acceptor‡	
	Cer-Glc-Gal-NANA	Cer-Glc-Gal-NGNA*
3T3	100	100
SV101	33	30
F1-SV	202	128
Py11	26	2
F1-Py	46	22

* NGNA = *N*-glycolylneuraminic acid
† From Table III, Ref. 29
‡ Percentage radio activity incorporated relative to controls

Table 3.9 Hematoside: N-acetylgalactosaminyltransferase activity in Balb/c cell lines*

Cells	*Saturation density in culture* (10^{-5} cells cm^{-2}	*Enzymatic activity* (%)
Control Balb 3T3	0.5	100
SV40 transformant	4.2	38
'Flat revertant' of SV40 transformant	0.3	132

* From Ref. 18

3.6 GLYCOPROTEINS

3.6.1 Gross changes

Glycoproteins are recognised as important constituents of cell surfaces and membranes and differences between the glycoproteins of normal and tumour cells have been sought by many investigators. Due to the immense heterogeneity of this class of complex macromolecules in terms of both size and composition, the search has been most difficult. Indeed, the literature appears to be a morass of claims and counterclaims concerning the changes in glycoproteins and the corresponding glycosyltransferases when normal cells are compared to transformed cells. Many early studies, in fact, involved investigations of changes in the total carbohydrate content or enzyme levels.

Sialic acid is an important constituent of glycoproteins as well as glycolipids. The amount of this material has been reported to be lower in several SV40-transformed clones of Swiss mouse cell lines[33]. All of the membrane fractions of similar cell lines were found to have reduced amounts of amino sugars and sialic acid[34]. There were no significant differences in the sugar nucleotide pools in these cells. Thus, changes in the composition of membrane, associated complex carbohydrates probably reflect differences in the activity of glycosyltransferase enzymes, or perhaps glycosidic enzymes. Grimes has measured the levels of sialic acid and sialyltransferase activity with exogenous acceptors in SV40-transformed Swiss 3T3 and Balb/c mouse cells. There appears to be a close correlation between sialic acid content and sialyltransferase activity. The transformed cell lines possessed only 30–60% of that found in the corresponding control cells[35]. Fucosyltransferase activity is also reduced in the transformed lines. An inverse correlation exists between sialic acid content and the degree of contact inhibition of growth of various Swiss 3T3 lines[36]. Normal and 'flat-revertant' cells possess the highest amounts of sialic acid whereas spontaneously- and SV40 virus-transformed cells possess the lowest amounts. Surprisingly, the polyoma virus-transformed derivative does not fit into this pattern and possesses normal levels of sialyltransferase*.

In contrast, Bosmann and co-workers have reported that many glycosyltransferases including fucosyltransferase are more active in DNA virus- transformed Swiss 3T3 cells[37]. However, the glucosyltransferase involved in collagen biosynthesis decreases after transformation[38], a finding which agrees with the decreased collagen production observed in these cells[39]. These studies employed deglycosylated glycoproteins such as fetuin and bovine submaxillary mucin as exogenous acceptors for measuring transferase activity.

It is apparent that all of the above studies were seeking differences at very gross levels. Determination of various monosaccharide amounts or glycosyltransferase activities in crude cell extracts with non-specific substrates is not very selective. It is, therefore, apparent that more definitive methodology are necessary to detect alterations in glycoproteins.

3.6.2 Resolution of membrane glycoproteins

Several groups have attempted to resolve membrane glycoproteins of normal and transformed cells by chromatographic and electrophoretic techniques.

* Grimes, W. J., personal communication and Ref. 5.

The most acceptable of these techniques is to grow the cells in the presence or radioactive precursors such as L-fucose or D-glucosamine. Normal cells are labelled with one isotopic form of the sugar (i.e. L-fucose-[^{14}C]) and transformed cells with L-fucose-[^{3}H]. Preparations from the normal and transformed cells can then be co-chromatographed or co-electrophoresed and differences in radioactive glycoproteins observed directly by double-isotope counting techniques.

Using such a procedure, Meezan *et al.* have compared solubilised membrane fractions (surface, nuclei, mitochondrial and microsomal) from Swiss 3T3 and SV40 Swiss 3T3[40]. When comparable glucosamine-labelled fractions were co-chromatographed on Sephadex G-150 in sodium dodecyl sulphate solutions it was found that the labelled glycoprotein profiles of normal and transformed cells were significantly different. After pronase digestion of the various membrane fractions, the labelled glycopeptides were co-chromatographed on Sephadex G-50. Again, significant differences appeared which were typified by an increase in higher molecular weight glycopeptides from the transformed cell membrane fractions. However, when membrane fractions from the same cell lines were analysed by SDS–polyacrylamide gel electrophoresis, no great differences were observed[41]. Subsequent pronase digestion and chromatography on Bio Gel P-10 have also failed to confirm the previously observed changes. The main difference between the two studies was that the Swiss 3T3 cells had undergone 15 additional passages in tissue culture and were now less contact-inhibited.

Surface membranes have been isolated from Swiss 3T3, three clones of SV40 Swiss 3T3 and one clone of Py Swiss 3T3 and analysed by polyacrylamide gel electrophoresis[42]. On the basis of molecular size (in gels containing SDS), no differences were observed in the solubilised membrane proteins, whereas on the basis of charge (in gels containing urea) differences exist between the membrane proteins of normal and DNA virus-transformed cells. However, surface membrane proteins from each cell type have been resolved on individual gels and detected by protein staining. Subtle differences are hard to detect unless samples from different cell lines are co-electrophorised and glycoproteins were not specifically analysed. Apparently aware of these differences, the authors in a subsequent paper have analysed glucosamine-labelled plasma membranes by SDS–polyacrylamide gel electrophoresis[43]. In this study, double-isotope labelling was employed to directly compare plasma membrane glycoproteins from the normal and transformed lines. Substantial differences were observed between the glycoprotein patterns from normal cells and from transformed cells and there were also significant differences among the three SV40 transformed clones.

3.6.3 A consistent difference in glycopeptides

Warren and co-workers have observed a consistant change in glycopeptide composition when cells from various species transformed by a number of different viruses are compared with untransformed cells of the same species[44]. When either surface material removed by trypsin (trypsinates) or purified surface membranes isolated from normal baby hamster kidney cells (BHK) and cells transformed by two strains of Rous sarcoma virus (RSV-BHK) were compared by co-chromatography on Sephadex G-200 in SDS, there was a demonstrable difference in the fucose-labelled glycoproteins[45]. The major higher molecular weight glycoproteins from the transformed cell lines eluted from the column more rapidly than material from the normal cells. Subsequent co-chromatography of glycopeptides prepared by pronase digestion of the above material showed a similar difference (Figure 3.9). The same effect

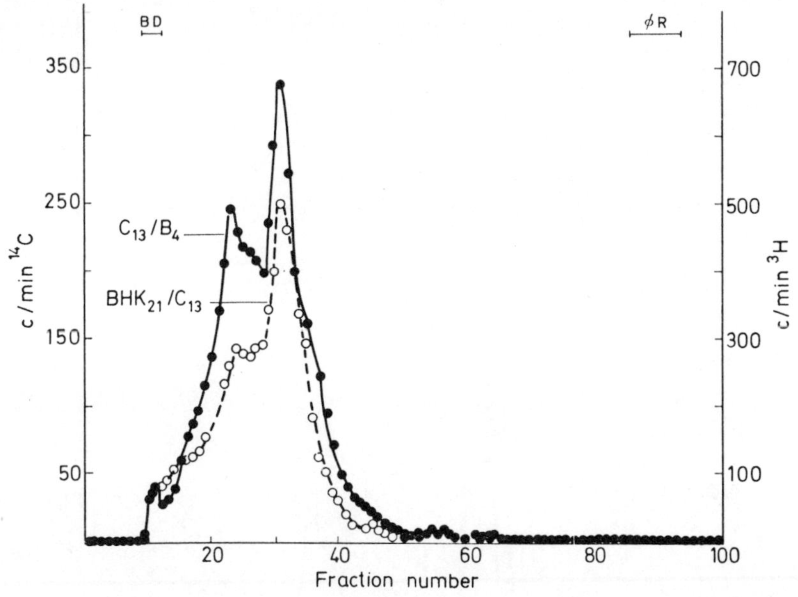

Figure 3.9 Co-chromatography on Sephadex G-50 of pronase-digested purified surface membranes from transformed (C_{13}/B_4) and control (BHK_{21}/C_{13}) hamster cells grown in the presence of radioactive L-fucose. Closed circles = ^{3}H c.p.m. from C_{13}/B_4 and open circles = ^{14}C c.p.m. from BHK_{21}/C_{13}. (From Buck *et al.*[45], by courtesy of the American Chemical Society)

has been observed with glucosamine labelling but not with leucine labelling and similar results have been obtained when fucose-labelled glycopeptides were compared between the following pairs of normal and transformed cells: Balb/c 3T3 and Moloney sarcoma virus-transformed Balb/c cells; BHK and polyoma virus-transformed BHK and chick embryo fibroblasts (CEF) and Rous sarcoma virus-transformed (RSV) CEF[46].

These particular cell-surface glycoproteins are strongly influenced by growth conditions[47]. When fucose-labelled glycopeptides isolated from normal BHK cells in exponential or stationary stages of growth are compared, there is an enrichment in higher molecular weight glycopeptides from the growing cells (Figure 3.10). Similar results have been obtained from RSV-BHK cells. Thus, differences in surface glycopeptides of normal and transformed cells are greatest when the cells are growing rapidly.

This observed difference in surface glycoprotein between normal and

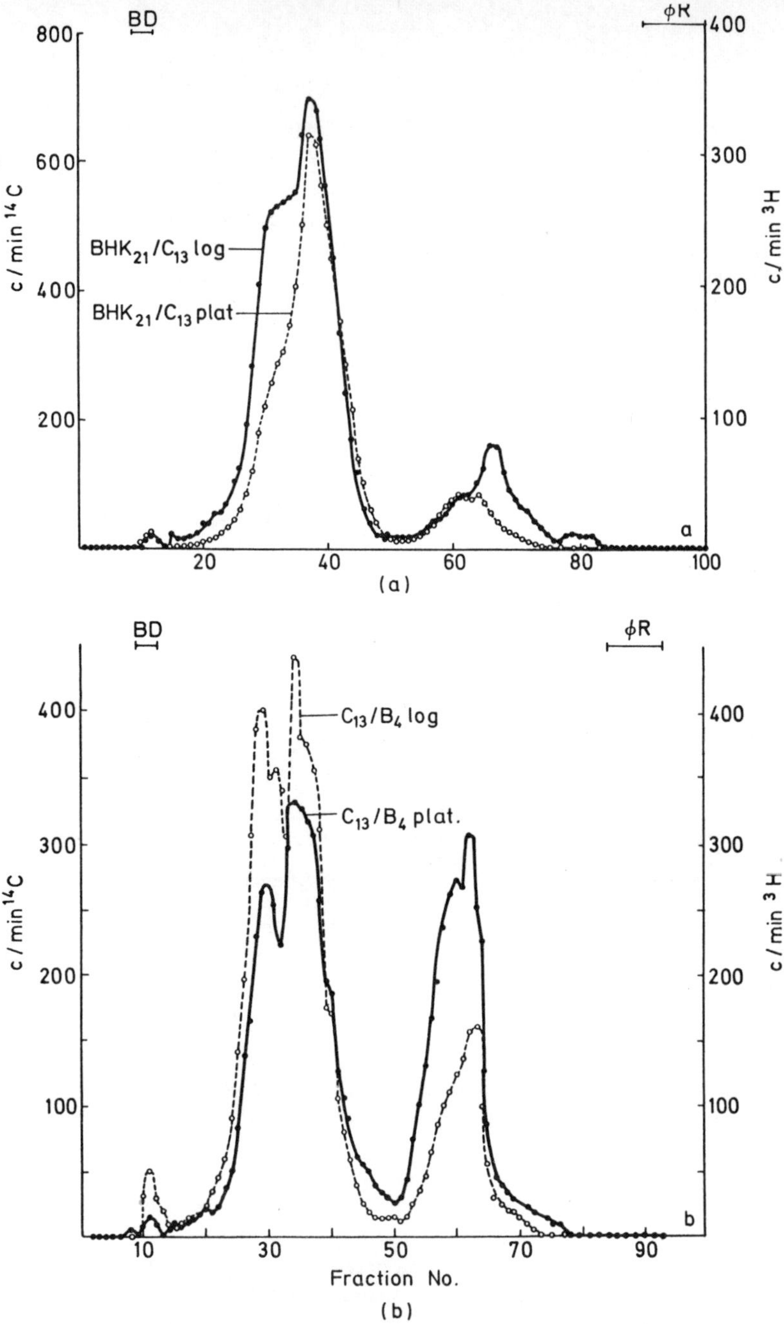

Figure 3.10 Co-chromatography on Sephadex G-50 of pronase-digested trypsinate from cells in logarithmic and plateau phases of growth. The cells in logarithmic phase were exposed to radioactive L-fucose during the first 3 days in culture and the plateau cells were exposed to radioactive L-fucose during the last 3 days in culture. (a) Closed circles = glycopeptides from BHK_{21}/C_{13} cells in log phase of growth in the presence of L-fucose-[^3H]; open circles = glycopeptides from BHK_{21}/C_{13} in plateau phase of growth in the presence of L-fucose-[^{14}C]. (b) Open cirlces = glycopeptides from C_{13} B_4 (transformed cells) in log phase of growth in the presence of L-fucose-[^{14}C]; closed circles = glycopeptides from C_{13}/B_4 cells in plateau phase of growth in the presence of L-fucose-[^3H]. (From Buck *et al.*[47], by courtesy of the American Chemical Society.)

virus-transformed cells appears to be associated with the transforming virus and only expressed when the cells are in a transformed state[9]. CEF cells infected with Rous associated virus do not exhibit a change in their surface glycoproteins.

More striking are the experiments with a temperature-sensitive (TS) mutant of RSV[48]. CEF cells transformed by TS behave as normal contact-inhibited cells at the non-permissive temperature (41°C) but acquire a transformed morphology when grown at 36°C. Fucose-labelled glycopeptides were prepared from CEF, RSV-CEF, and TS-CEF cells grown at both temperatures. Upon co-chromatography on Sephadex G-50, glycopeptides from RSV-CEF elute ahead of those from CEF irrespective of growth temperature; glycopeptides from TS-CEF are similar to those of RSV-CEF at 36°C and those of CEFat 41°C.

3.6.4 The role of sialic acid and sialytransferase

A biochemical and enzymatic explanation for the increase in higher molecular weight surface glycopeptides of transformed cells has recently been described[49]. When radioactive fucose-labelled glycopeptides derived from pronase-digested surface material of transformed cells are treated with neuraminidase, they appear upon Sephadex column chromatography as normal cell glycopeptides (Figure 3.11). Thus, the increased molecular weight appears to be due to an increase in the sialic acid content of these glycopeptides. When this desialylated material is used as a specific acceptor for sialyltransferase, more activity is found in extracts of transformed cells than normal cells (Table 3.10). This increased sialyltransferase activity is more pronounced in exponentially growing cells and is found on isolated surface membranes (Table 3.11). Finally, TS-CEF cells have five times as much of this specific sialyltransferase when grown at 36°C than at 41°C[44].

Several comments on the above experiments are in order. First, although the desialylated glycopeptides from transformed cells have the same apparent molecular weight as glycopeptides from normal cells, these two classes of glycopeptides are not identical. Thus, as seen in Table 3.10, the material from normal cells (designated as peak B) is a poor sialyltransferase acceptor compared to desialylated peak A material. Second, the similar sialyltransferase activities per mg protein of isolated surface membranes and crude homogenates (Table 3.11) suggest that this specific enzyme is found elsewhere in the cell. Indeed, the authors have indicated that glycopeptides similar to those of the surface membrane are found in all of the membrane fractions of the cell[44]. This observation is in agreement with those of others[40]. Third, the shift in glycopeptides and sialyltransferase activity is found only in rapidly growing cells, being more apparent in transformed cells than in normal cells. This observation, taken together with the second comment, would suggest these changes are not related to the phenomena of cell-to-cell contact and contact inhibition.

Bosmann employing a different approach has obtained similar results with mouse 3T3 cells[50]. Specific acceptors were prepared by treating intact cells first with neuraminidase then trypsin. The desialylated material released

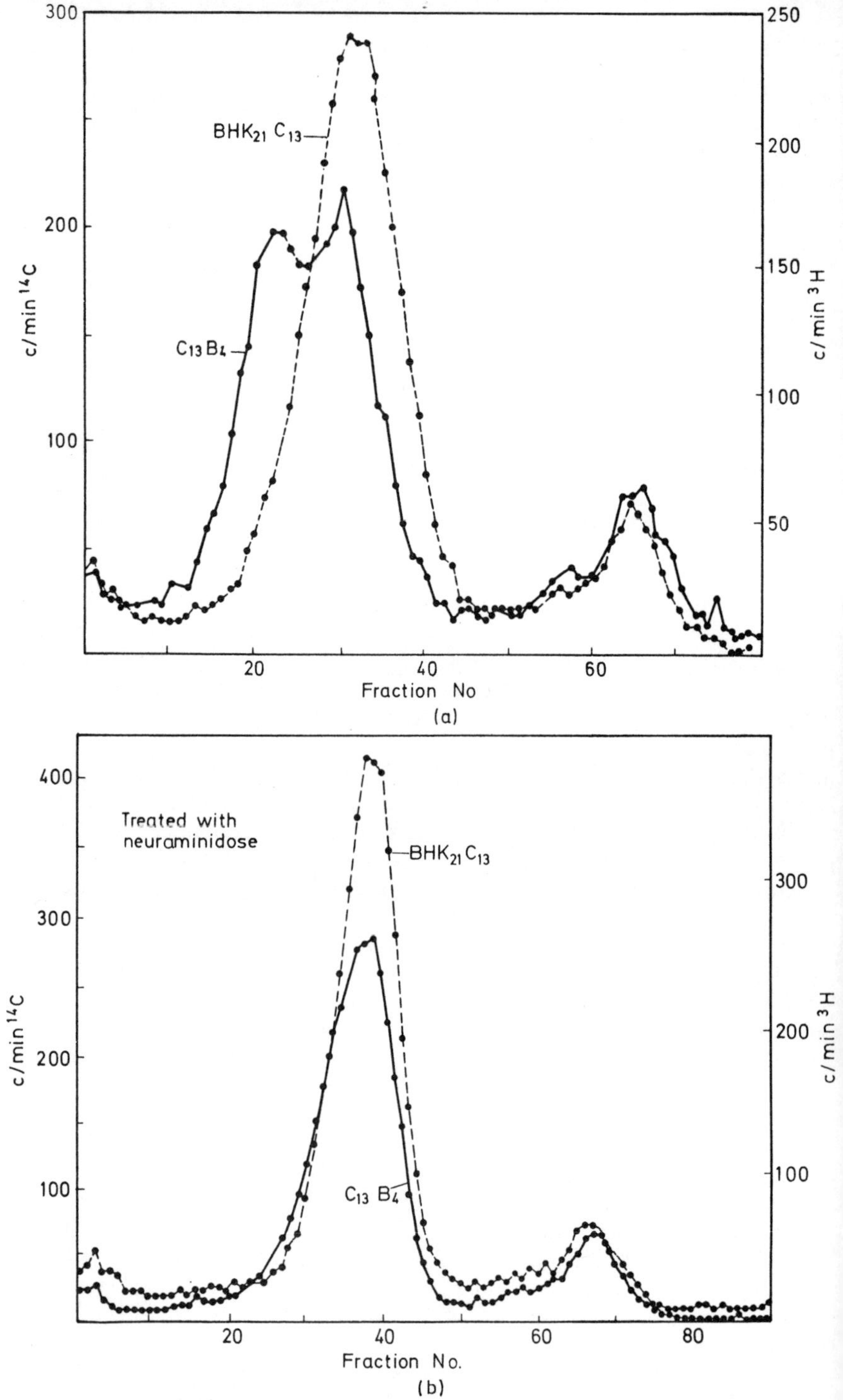

Figure 3.11 Elution patterns of pronase digests of trypsinates from Sephadex G-50. Material derived from BHK_{21}/C_{13} cells labelled in log phase for 3 days with L-fucose-[^{14}C] and from C_{13}/B_4 cells labelled with L-fucose-[^{3}H]. In (b), the pronase digests were treated with neuraminidase prior to application to the column. (From Warren *et al.*[44], by courtesy of the Federation of American Societies for Experimental Biology.)

Table 3.10 Transfer of ^{14}C-NAN to various acceptors by extracts from control and virus-transformed cells*

^{14}C-NAN acceptor	*Extract from BHK_{21}/C_{13} cells*†	*Extract from C_{13}/B_4 cells*†
None	0	0
Endogenous	1,690	2,170
Peak B	1,600	2,030
Peak A	0	0
Desialylated peak A	1,790	5,450
Desialylated fetuin	14,800	13,500
Desialylated bovine mucin	17,700	18,000

* From Ref. 44 by courtesy of the Federation of American Societies for Experimental Biology
† Measured as open (mg proteih)$^{-1}$ h^{-1}

from the surface of the cells was then used as an acceptor for sialyltransferase activity. The enzyme source was crude detergent extracts of the cells. By assaying acceptor from different cells with homogenates of different cells, it was found that virus-transformed 3T3 cells have both more acceptor and sialyltransferase activity than normal 3T3 cells in the order MSV-3T3, RSV-3T3, PY-3T3 and 3T3. When the acceptor was trypsin but not neuraminidase-treated cellular material, the activities were substantially reduced but the order of acceptor and enzyme content was the same for the various cell lines. When intact cells were used as the source of enzyme, 30% of the sialytransferase activity was observed. However, this value is of little use as it is well known that detergent treatment stimulates sialyltransferase activity[35].

Although both Bosman[50] and Warren *et al.*[49] have attempted to measure a similar phenomenon, the differences in their methodology may account for some of the differences in their results. Bosmann's acceptor represents a crude desialylated macromolecular fraction release from the surface of the cells

Table 3.11 Transfer of ^{14}C-NAN by fractions of cells in log and plateau*

Cell	*Growth*	*Fraction*	*^{14}C incorporated into desialylated peak A material*, c.p.m./(mg protein)$^{-1}$ h^{-1}]
(A) BHK_{21}/C_{13}	Log	Supernatant	2140
BHK_{21}/C_{13}	Plateau	Supernatant	546
C_{13}/B_4	Log	Supernatant	10850
C_{13}/B_4	Plateau	Supernatant	1900
(B) C_{13}/B_4	Log	Crude homogenate	9700
C_{13}/B_4	Plateau	Crude homogenate	2860
C_{13}/B_4	Log	Surface membrane	7400
C_{13}/B_4	Plateau	Surface membrane	2840

* From Ref. 44 by courtesy of the Federation of American Societies for Experimental Biology

by the action of trypsin. Warren *et al.*[49] digested further the surface material released by trypsin with pronase and isolated a glycopeptide fraction from transformed cells that is more abundant than that found in normal cells. This glycopeptide fraction was then treated with neuraminidase. Whereas Bosmann found up to 26% of the sialyltransferase activity without neuraminidase treatment of the acceptor, Warren *et al.* found none (Table 3.10)) Bosmann concludes that since there is more activity with such an acceptor from transformed cells rather than normal cells (26% *vs.* 14%), transformed cells contain higher amounts of incomplete surface glycoproteins. This conclusion is based on the rather small differences observed and seems to be at variance with the greater specific sialyltransferase activity found in the transformed cells, as well as with the greater amount of sialylated glycoproteins found in transformed cells[49]. However, the most disturbing aspect of Bosmann's report is the uncertainty of the origin of the various cell lines. Although referred to as mouse 3T3 cells (obtained from G. Todaro), it is apparent that some are of Balb/c origin and others of Swiss lineage.

3.6.5 Surface glycosyltransferases

A role for surface glycosyltransferases in intercellular recognition and contact inhibition has been suggested by Roseman[24]. His model envisages glycosyltransferases on the surface of one cell interacting with acceptors on another cell and vice versa. Depending on the number and type of acceptors and enzymes, and whether reactions (sugar transfers) are being initiated or terminated, intercellular recognition and adhesiveness may occur. Thus, normal cells with the appropriate surface glycosyltransferases and acceptors will strongly interact and become contact-inhibited. Transformed cells lacking specific surface gylcosyltransferases would be unable to synthesise and/or interact with the appropriate surface acceptors and hence lack contact-inhibition. Present experimental evidence as discussed below does not fully support this hypothesis as originally proposed.

The existence of such surface glycosyltransferases has been demonstrated recently. A surface galactosyltransferase has been found in chick embryo neural retinal cells[51]. The surface enzyme catalyses the transfer of galactose from UDP–galactose to either exogenous acceptors or endogenous acceptors on the surface of the cells. Exogenous acceptors inhibit intercellular adhesiveness of the cells. Galactosyltransferase activity has also been detected on the surface of contact-inhibited Balb/c 3T3 and spontaneously-transformed Balb/c 3T12 mouse cells[52]. More galactose is transferred to surface acceptors of 3T3 cells from sparse cultures than from confluent cultures. The effect of cell-to-cell contact has also been examined *in vitro* by agitating the cells during the transferase assay. The non-agitated cells incorporate up to five times as much galactose as agitated cells. In contrast, Balb/c 3T12 cells show no contact dependence in culture or *in vitro*. Galactose transfer to surface acceptors is similar in sparse or confluent cultures or whether the cells are agitated or not. The activities obtained were in all instances comparable to those obtained from sparse 3T3 cells.

The authors propose that 3T3 cells engage in *trans*-glycosylation, that is galactose is transferred by surface enzymes on one cell to surface acceptors on another cell. The transformed 3T12 cells undergo *cis*-glycosylation; cells transfer galactose to their own surface acceptors. Thus, surface galactosyltransferase activity is contact-dependent in 3T3 but not 3T12 cells. In addition 3T3 cells which have similar levels of surface galactosyltransferase activity irrespective of culture density appear to have less surface acceptors when confluent. As the cells come into contact, *trans*-glycosylation occurs and the number of surface acceptor sites available for the *in vitro* assay is reduced.

Bosmann has also reported that a number of different glycosyltransferases exist on the surface of mouse 3T3 cells which transfer glucose, galactose, mannose, sialic acid and *N*-acetylgalactosamine to surface acceptors[53]. When these activities were measured with cells from confluent cultures, cells transformed by MSV, RSV and Py had higher activities than untransformed 3T3 cells. Transferase activities were density-dependent in the untransformed 3T3 cells but not in the virally-transformed cells. Thus, sparse and confluent cultures of transformed cells have similar activities to one another and to sparse 3T3 cells. Using exogenous glycoprotein acceptors, it was also shown that the transformed cells also had greater sialyl and galactosyltransferase activities but less glucosyltransferase activity (with deglucosylated collagen as acceptor) than the normal cells.

There are several serious criticisms of these experiments of Bosmann. As mentioned previously, the lineage of these mouse 3T3 cells is unclear. The cells were harvested by scraping, whereas in the experiments by Roth and co-workers cells were harvested by trypsinisation. Cells harvested by mechanical means are found to exist as clumps, especially from confluent cultures, whereas trypsinisation results in single cell suspensions. Clumping is due to intercellular adhesion and to the extracellular coatings the cells excrete. It is well known that normal 3T3 cells are more adhesive[54] and synthesise and excrete more collagen than transformed cells[39]. Thus, the reduced incorporation observed by Bosmann in confluent 3T3 cells may represent a masking of surface glycosyltransferases and acceptor sites.

Scraping the cells was employed because Bosman found that prior treatment of the intact cells with trypsin reduced surface transferase activity by 85–100%. Studies in our own laboratory indicate that when surface sialyltransferase activity is measured in mouse cells, cells harvested by trypsin incorporate as much or more sialic acid than cells harvested by scraping or EDTA treatment.*

The results and the conclusions derived from these results by the different investigators imply different mechanisms and roles for these surface glycosyltransferases in the phenomena of contact inhibition and transformation. Roth and White[52] suggest that normal and transformed cells have similar levels of surface galactosyltransferase activity and these levels are not influenced by cell density in culture. Normal cells from confluent cultures have less surface acceptor sites than cells from sparse cultures because of *trans*-glycosylation, whereas transformed cells have the same amounts and can only undergo *cis*-glycosylation. Bosmann implies that transformed cells have more surface sialyl- and galactosyl-transferase activity than normal

* Fishman, P. and Schultz, A., unpublished observations.

cells when the cells are confluent, but similar activities when the cells are from sparse cultures. Bosmann has also presented data that transformed cells have more surface acceptor sites for sialic acid than normal cells (see Section 3.6.4). Finally, Warren *et al.* have measured sialyltransferase activity in isolated surface membranes of transformed cells and found a growth effect (Table 3.11). Warren *et al.* have also found less surface acceptors for a specific sialyltransferase in transformed cells (Table 3.10). Only after treatment with neuraminidase is this glycopeptide fraction an effective acceptor. Since there are more of these sialylglycopeptides in growing (sparse) cultures than in stationary (confluent) cultures, there must be more acceptor sites for sialyltransferase in confluent cultures of both normal and transformed cells.

Thus, Roth and White support Roseman's hypothesis only if the specific mechanism of *trans*- and *cis*-glycosylation is invoked. However, this implies that transformed cells have additional more extensive alterations on their surface membranes which prevent *trans*-glycosylation and the role of surface glycosyltransferases is secondary to transformation. The experiments of Warren and co-workers are completely antithecal to Roseman's hypothesis. Bosmann's data in general are not incompatible with Roseman's hypothesis. However, it is not clear why transformed cells have higher surface glycosyltransferase activities than normal cells as well as more surface acceptor sites (for sialyltransferase). More important is the point that the number of these sites in sub-confluent cultures is only a small percentage of the number of surface sites exposed by neuraminidase treatment. It can only be concluded that either the surfaces of both normal and transformed cells have undergone extensive *cis*-glycosylation or these surface enzymes and acceptors are not involved in cell-contact phenomena.

3.6.6 Glycoproteins and mitosis

Cells undergo mitosis in discrete phases. The four major phases of the cell cycle are S-phase during which nuclear events such as DNA synthesis occurs, G_2 which is the pre-mitotic phase, M-phase during which the cell actually divides and G_1, the phase between mitosis and DNA synthesis. It is now apparent that many of the alterations associated with transformed cells are expressed by normal mitotic cells. Thus, dividing cells are as readily agglutinated by plant lectins as transformed cells[55,56]. Transformed cells have lower levels of cyclic-AMP than normal cells[57]. However, cells in mitosis also have low levels of cyclic-AMP[58].

Do temporary alterations in membrane glycoproteins occur during mitosis which are similar to the permanent changes expressed in transformed cells? Bosmann and Winston have investigated glycoprotein synthesis in synchronised cultures of a mouse lymphoma line[59]. Glycoprotein synthesis as measured by the incorporation of radioactive fucose and glucosamine occurred throughout the cell cycle but reached a maximum value during S-phase. The levels of several glycoprotein: sugar nucleotide glycosyltransferases also fluctuated during the cell cycle and were highest during S-phase[60]. However, these studies measured total cellular glycoprotein synthesis and not surface membrane components.

Brown has examined the accumulation of surface glycoproteins during mitosis[61]. L cells were exponentially grown in the presence of radioactive glucosamine and their surface glycoproteins removed by trypsinisation. When chromatographed on a DEAE–cellulose column, four classes of glycoproteins were found. When glycoproteins were similarly analysed from a synchronised cell culture, one glycoprotein class accumulated preferentially during mitosis. Brown emphasised that his methodology only measures the accumulation of surface glycoproteins and not their synthesis. As glycosylation of proteins occurs on internal membranes[62], surface glycoproteins could be synthesised prior to mitosis and be incorporated into the surface membrane during cell division. Palade has suggested that membrane biogenesis occurs not *de novo* but by the fusion of pre-assembled membrane vesicles with existing plasma membrane[63].

Glycoproteins from the surface of metaphase BHK cells have been examined recently[56]. The cells were grown in the presence of radioactive fucose and arrested in metaphase by treatment with vinblastin sulphate. The surface glycoproteins were removed with trypsin, digested with pronase and the resulting glycopeptides chromatographed on Sephadex G-50. Metaphase cells when compared to non-metaphase cells were richer in the higher molecular weight glycoproteins found in RSV-transformed cells. It has been shown previously that the amount of these higher molecular weight glycoproteins is influenced by the growth state of the cells[47]. (See also Section 3.6.3). Thus, it appears that a glycoprotein alteration that is permanently expressed by virally-transformed cells is also temporarily expressed by normal cells when they are either rapidly growing or dividing.

3.6.7 Surface antigens and agglutination sites

The role that the carbohydrate moieties of glycoproteins (and glycolipids) play as determinant components of surface antigens and agglutination receptor sites is well established[64]. Examples are provided by the blood group substances, histocompatibility antigens, influenza virus receptor sites and phytohemagglutin receptor sites. On a molecular basis, the oligosaccharide chains of glycoprotein (and glycolipids) may be uniquely situated for an involvement in these surface-recognition phenomena. In a fluid membrane consisting of mobile molecules of protein floating in lipid[65], these carbohydrate components would appear as polar entities projecting from the membrane surface. Evidence that oligosaccharide chains are restricted to the external surface of plasma membranes has recently been presented[66]. The involvement of glycoproteins in two surface-recognition phenomena related to transformation will be considered, namely agglutination of transformed cells by plant lectins and tumour-specific transplantation antigens.

It was originally reported by Aub *et al.* that agglutinins from wheat germ strongly agglutinate neoplastic cells[67]. Since then Burger and co-workers and Sachs and co-workers have extensively investigated the effects of these plant lectins on tumour cells (for a recent review, cf. Ref. 68). The two most commonly used, concanavilin A and wheat germ agglutinin, are themselves both glycoproteins. Although these compounds preferentially bind to and agglutinate transformed cells at much lower concentrations than normal cells, normal cells will also be agglutinated at low concentrations of lectin under certain conditions. When normal cells are treated with trypsin, they are agglutinated in addition to transformed cells[69]. As mentioned earlier, mitotic cells are readily agglutinated by concanavalin[55]. Finally, cells undergoing lytic virus infection are readily agglutinated[70]. Thus, another surface alteration which is permanently expressed in many transformed cells can be temporarily invoked in normal cells. The relationship between this increased agglutination phenomenon and the neoplastic state is still unclear.

Evidence that the cell surface agglutination sites are glycoproteins is still presumptive. Various monosaccharides can act as haptens to inhibit the agglutination reaction or the binding of isotopically-labelled plant agglutinins. Thus, mannose blocks concanavalin A binding and *N*-acetylglucosamine interfers with wheat germ agglutinin. Crude glycoprotein fractions from cell membranes also inhibit agglutination[71]. The work of Kornfeld and coworkers has shown that glycoproteins are the receptor sites on erythrocyte membranes for a variety of phytohemagglutins[72, 73]. However, the final identity of the agglutinin receptor site await its isolation in the pure form.

When mice or hamsters are inoculated with polyoma virus, they are capable of rejecting polyoma virus-transformed cells. This transplantation rejection phenomenon is due to the presence of tumour-specific transplantation antigen (TSTA) on the surfaces of the transformed cells. Similarly, TSTA has been detected on SV40-transformed cells[75, 76]. TSTA is an antigen which is separate from the nuclear tumour (T-antigen) and viral capsio antigen. TSTA is viral-specific by *in vivo* immunogenicity tests[77]. In addition, use of *in vitro* assays has indicated that SV40-transformed cells of different species carry the same TSTA[78, 79]. TSTA, however, does appear during the lytic infection of monkey cells by SV40[80]. Thus, like the virus-induced agglutination sites, it is not restricted to the transformed state *per se*. The biochemical nature of TSTA is unknown. It appears to be separate from other surface changes associated with transformation. As it is virus-specific and absent in cryptic form, it is not part of the agglutination receptor site or the normal cell antigen. However, there may be some cross-reactivity between TSTA and embryonic antigen[81]. Whether TSTA is virally coded or represents derepression of host genes is also unknown. The fact that TSTA is virus-specific would suggest the former; however, different viruses may derepress different host cell genes or even cause rearrangement of the cell surface to form new specific antigenic groups.

There is no direct evidence that TSTA is a glycoprotein. Only by analogy to other surface antigens has it been included in this section. Recently, SV40-TSTA has been isolated in a soluble form[81] and it is hoped that its biochemical composition will soon be determined.

3.6.8 Conclusions

It would be enticing to speculate that alterations in the composition of membrane glycoproteins and/or levels of membrane-bound glycosyltransferases can explain the transformed state *in vitro* and the malignant state *in vivo*.

That such phenomena as loss of contact inhibition of growth and movement, altered morphology, reduced adhesiveness, increased immunological response and rejection, and increased invasiveness are the result of these changes is highly speculative. In fact, the relationship of these changes in glycoprotein metabolism to transformation and neoplasia is unclear. The changes in sialoglycoproteins that occur after transformation by many oncogenic viruses are related to the growth state of the cell and occur in untransformed cells during mitosis. In addition, these changes, including increased levels of a specific sialytransferase, are not limited to the plasma membrane but also occur in internal membranes. The increased sensitivity of tumour cells to certain plant agglutinins is also observed in normal cells during the cytolytic cycle and the presence of TSTA renders virus-transformed cells less malignant in immunocompetent syngenic animals.

The inter-relationship between these various phenomena is also unclear. Are the changes in surface glycoproteins related to the appearance of new surface antigens and agglutinin receptor sites? Are surface glycosyltransferases involved in their synthesis? Or is there no *de novo* synthesis but only a rearrangement of exisiting components or glycosylation and or deglycosylation of the existing components? Finally, what about cause and effect? Are these changes (in glycoproteins, glycosyltransferases, agglutinability) the mechanism by which the tumour virus maintains the transformed state or are they secondary consequences of an altered metabolic state of the transformed cell? Warren *et al.* argue that such a quantitative imbalance of glycoprotein components of various cellular membranes may interfere in the normal cell cycle and result in the cell undergoing another round of DNA synthesis and mitosis instead of becoming quiescent[44]. In a similar vein, Burger speculates that the surface alteration during mitosis may trigger the next chromosome replication which in turn induces mitosis[68]. It is just as easy to put the horse before the cart and argue that the loss of growth control in transformed cells results in continual expression of these temporary surface changes.

3.7 REGULATION OF ENZYME EXPRESSION

The consistency with which the changes in ganglioside pattern and the decrease of hematoside:*N*-acetylgalactosaminyltransferase activity occurs in tumourigenic virus-transformed cells argues that these biochemical alterations are the direct consequences of the integration of the virus genome into the genetic apparatus of the host cell. How then does the insertion of the virus DNA cause this diminution of amino sugar transferase activity? The solution of this question will provide an important insight into the regulation of cellular structure. There are many possible areas where the viral genome can exert its influence: at the level of nucleic acid synthesis, protein synthesis, or expression of enzyme activity. Unfortunately, the regulation of these events in mammalian cells is poorly understood compared with our knowledge in bacterial systems. However, based on available data, we can make a number of intelligent speculations. At the DNA level, one possibility is that the virus genome is integrated at or near the site of the gene coding for the amino sugar transferase in certain virus-transformed cells. Thus, rearrangement of the genetic code for this enzyme could cause a frame-shift mutation which may result in a block of transcription or formation of a partially active or even an inactive protein. Several inferences can be drawn from the studies with the 'flat-revertant' cells which have been shown in the case of F1–SV cells to continue to carry the viral genome; but they have a normal phenotype, and the pattern and synthesis of gangliosides are normal. Studies with virus-transformed hybrid cells indicate that the viral genome can be associated with more than one chromosome, possibly in a random fashion[82, 83]. Association of viral genome with one particular host chromosome would be required for the reduction of amino sugar transferase activity. Furthermore, if the cistrons for various other ganglioside-synthesising enzymes are chromosomal neighbours, one might expect occasional changes in the activities of other glycosyltransferases upon transformation. In fact, decreased activity of G_{M2}:galactosyltransferase activity (Figure 3.6, reaction (3.6)) and G_{M1}:*N*-acetylneuraminyltransferase (Figure 3.6, reaction (3.7)) are sometimes observed in certain virus-transformed cells. A model for this situation is the phenomenon of polarity observed in certain bacterial mutations. Here a mutation in one gene of a polycistron reduces the translation of subsequent genes[84].

Depending on the site and number of chromosomes involved, and the extent of diminution of amino sugar transferase activity attendant upon the insertion of a viral genome, one might expect a gene dose effect. Some support for this concept may be derived from the 'flat-revertant' Swiss cell lines. The SV40 revertants have an increased chromosome number to the sub-tetraploid level. The activity of hematoside:*N*-acetylgalactosaminyltransferase in this cell line was actually slightly increased over that in the control 3T3 cells. The 'flat-revertant' polyoma cell line which we examined did not shift entirely to the high chromosome number, and in this line it will be recalled that the activity of the hematoside:*N*-acetylgalactosaminyltransferase was only partially restored to the level in the contact-inhibited parent cell line.

Alternatively, the viral genome could assert its effect at the RNA level. It has been shown in eukaryotic cells that induction and repression of enzyme synthesis can occur at the level of messenger-RNA[85]. Although nothing is known about the regulation of ganglioside biosynthesis, the possibility exists that some substance is produced in virus-transformed cells which affects the synthesis of hematoside:*N*-acetylgalactosaminyltransferase and occasionally other glycolipid-synthesising enzymes as well. Thus, it may be surmised that the two tumourigenic DNA viruses SV40 and polyoma virus and the RNA Moloney sarcoma virus contain the genetic information for a common repressor molecule. Both of the DNA viruses code for 5–10 polypeptides[86], one of which might be a repressor of the amino sugar transferase.

Some recent experiments with a temperature-sensitive mutant of Moloney sarcoma virus in the NRK rat kidney cell line are instructive in this regard. The NRK cell line differs from the AL/N, Balb/c, and Swiss mouse cell lines in that the only ganglioside in the NRK cells is G_{M3}, a situation which is similar to that in BHK cells. When NRK cells are transformed with the temperature-sensitive Moloney sarcoma virus mutant, the cells assume the typical phenotypic properties of virus-transformed cells at the permissive temperature of 31 °C. However, if the cells are grown at the non-permissive temperature of 39 °C, they appear to have the phenotypic appearance of the non-transformed parent cell line. We have observed a temperature-dependent modulation of the quantity of G_{M3} and the neutral glycolipid G1–3 (Figure 3.3) in these cells[11]. There is considerably less G_{M3} in cells grown at 31 °C than at 39 °C. Thus, the activity of the enzymes which are involved in the formation of these compounds (Figure 3.6, reaction (3.4) and reaction (3.8), respectively) are subject to modification perhaps by some factor within the host cell. We are now attempting to develop a temperature-sensitive mutant virus which will transform the conventional mouse cell lines since the elucidation of this regulatory phenomenon is of paramount importance.

Another possibility is that the viral genome through one of its products directly regulates the activity of hematoside:*N*-acetylgalactosaminyltransferase. We have found no evidence for such a soluble cytoplasmic inhibitor; however, we have shown that this enzyme is bound to the membrane-rich fractions of cells[16]. It is conceivable that the activity and specificity of membrane-bound glycosyltransferases are determined by membrane-bound elements in close association with them. This would be analogous to the regulatory effect that α-lactalbumin has on galactosyltransferase[87]. The two proteins in a complex constitute the enzyme lactose synthetase. The galactosyltransferase by itself has a broad substrate specificity for monosaccharides other than glucose as well as for glycoproteins. The genome of the transforming virus might alter this hypothetical membrane-bound regulating element thus changing the activity and/or specificity of the amino sugar transferase. We are continuing our work on isolating this enzyme in a soluble, membrane-free form in order to test this hypothesis[88].

Another metabolic aspect which cannot be overlooked is the increase in the activity of certain glycosyltransferases which are occasionally seen in virus-transformed cells. Thus, we have observed a near doubling of sialyltransferase II activity (Figure 3.6, reaction (3.7)) in SV40 and polyoma virus-transformed AL/N cells and an increase of G_{M2}:galactosyltransferase activity (Figure 3.6, reaction (3.6)) in virus-transformed Swiss mouse cell lines. Warren and co-workers have reported an increase of glycopeptide: sialyltransferase activity in transformed BHK cells[49]. Since the increased glycolipid-synthesising enzymes occur in cells in which the activity of hematoside:*N*-acetylgalactosaminyltransferase is drastically reduced, we deduce that the increased activity of these glycosyltransferases is due to one or perhaps both of the following phenomena. There may be an impairment of end-product feed-back regulation of the activity of these transferases since the higher ganglioside homologues cannot be synthesised because of deficient glycosyltransferases. However, there may be a more subtle explanation whose mechanism is at best only vaguely conceptualised although it is possible that it also rests on an impaired feed-back mechanism. Patients with hereditary lipid-storage diseases due to decreased activity of specific lipid catabolising enzymes[89] usually have a several-fold increase in activity of auxiliary enzymes in their tissues[90]. One wonders then if the increased activity of the various glycosyltransferases in transformed cells is based on an apparently analogous attempt to overcome the metabolic disturbances.

3.8 SIGNIFICANCE

The discovery of the alteration in ganglioside pattern and synthesis as a consequence of transformation of cells by oncogenic viruses may be a useful and important contribution to cancer research. We know that gangliosides are highly concentrated in the plasma cell membranes[91,92], and therefore it is conceivable that they participate in the phenomenon of cell–cell recognition and contact-inhibition of growth. On the other hand, it is possible that gangliosides are not entirely exposed on the cell surface and may be overlayered by a glycoprotein coat; therefore, the magnitude of the contribution of gangliosides to surface-regulatory phenomena cannot be accurately evaluated at this time. Even though there is a dramatic decrease in the amount of higher ganglioside homologues in the tumourigenic virus-transformed cells, this alteration may not cause much change in the net negative charge on the surface of cells, since there is frequently an increase in the amount of G_{M3} in the transformed mouse cells[1] and most of the surface sialic acid is found in glycoproteins[93]. There seems to be a slight decrease in the net negative charge on the surface of most transformed cells and decreases in total sialic acid after virus-transformation have been reported[33,34]. What these findings mean with regard to the social behaviour of cells in culture is not clear. A primary consideration which must be kept in mind is the fact that a small percentage of all cells transform spontaneously after repeated passage in culture without added virus. These cells are tumourigenic and grow to a high density in culture. The ganglioside pattern of these cells closely resembles that in the contact-inhibited parental cell lines from which they were derived. Thus, at this moment, our conclusions must be restricted to the constant finding of the effect of insertion and expression of the tumourigenic virus genome on the altered pattern and synthesis of gangliosides. The principal correlation we have observed is that this effect is linked to the phenotypic properties of the cells and the expression of the transforming viral genome.

Many problems lie ahead and only a few will be mentioned here. We are obliged to obtain a clearer insight into the alterations of the glycoprotein patterns of virus-transformed cells. Obviously, it is imperative to carry out comparative investigations of the glycoprotein patterns of contact-inhibited cells and spontaneously-transformed derivative lines. Other investigations regarding ganglioside and glycoprotein metabolism in chemical carcinogenesis must be undertaken. Preliminary work along this line has indicated a difference in ganglioside metabolism in cultures of minimally-deviated chemically-induced hepatoma cells compared with hepatocytes[94]. Finally, much additional work is required to clarify the regulation of the activity of the hematoside: *N*-acetylgalactosaminyltransferase. This is a key enzyme for higher ganglioside synthesis and appears to be a pivotal reaction whose activity is under the influence of tumourigenic viruses. One of the first questions to be answered is whether the synthesis of gangliosides can be restored to normal if this amino sugar transferase is supplied exogenously. A likely candidate for this experiment is the polyoma virus transformant of Swiss mouse cells in which the activities of other enzymes involved in ganglioside synthesis appears to be normal. Experiments which clarify these aspects appear to offer immense reward for obtaining insight into the molecular events of viral carcinogenesis.

References

1. Mora, P. T., Brady, R. O., Bradley, R. M. and McFarland, V. W. (1969). *Proc. Nat. Acad. Sci. U.S.A.*, **63**, 1290
2. Brady, R. O. and Mora, P. T. (1970). *Biochim. Biophys. Acta*, **218**, 308
3. Dijong, I., Mora, P. T. and Brady, R. O. (1971). *Biochemistry*, **10**, 4039
4. Fishman, P. H., Bassin, R. and McFarland, V. (1973). *Fed. Proc.*, **32**, 462 Abstr.
5. Hakomori, S.-I. and Murakami, W. T. (1968). *Proc. Nat. Acad. Sci. U.S.A.*, **59**, 254
6. Fishman, P. H., Aaronson, S., Eckhart, W., Bradley, R. M. and Brady, R. O. (1973). In preparation
7. Sakiyama, H., Gross, S. K. and Robbins, P. W. (1972). *Proc. Nat. Acad. Sci. U.S.A.*, **69**, 872
8. Hakomori, S.-I., Saito, T. and Vogt, P. K. (1971). *Virology*, **44**, 609
9. Warren, L., Critchley, D. and Macpherson, I. (1972). *Nature (London)*, **235**, 275
10. Mora, P. T., Fishman, P. H., Basin, R. H., Brady, R. O. and McFarland, W. N. (1973). *Nature*, in press
11. Fishman, P. H., Aaronson, S. and Brady, R. O. (1973). In preparation
12. Hakomori, S.-I., Teather, C. and Andrews, H. (1968). *Biochem. Biophys. Res. Commun.*, **33**, 563
13. Kijimoto, S. and Hakomori, S.-I. (1971). *Biochem. Biophys. Res. Commun.*, **44**, 557
14. Hakomori, S.-I. (1970). *Proc. Nat. Acad. Sci. U.S.A.*, **67**, 1741
15. Kolodny, E. H., Brady, R. O., Quirk, J. M. and Kanfer, J. N. (1970). *J. Lipid Res.*, **11**, 144
16. Cumar, F. A., Brady, R. O., Kolodny, E. H., McFarland, V. W. and Mora, P. T. (1970). *Proc. Nat. Acad. Sci. U.S.A.*, **67**, 757
17. Fishman, P. H., McFarland, V. W., Mora, P. T. and Brady, R. O. (1972). *Biochem. Biophys. Res. Commun.*, **48**, 48
18. Fishman, P. H., Brady, R. O. and Mora, P. T. (1973). *J. Amer. Oil. Chem. Soc.*, in press
19. Den, H., Schultz, A. M., Basu, M. and Roseman, S. (1971). *J. Biol. Chem.*, **246**, 2721
20. Brady, R. O., Fishman, P. H. and Mora, P. T. (1973). *Fed. Proc.*, **32**, 102
21. Fogel, M. and Sachs, L. (1962). *J. Nat. Cancer Inst.*, **29**, 239
22. Fogel, M. and Sachs, L. (1964). *Exp. Cell. Res.*, **34**, 448
23. O'Neill, C. H. (1968). *J. Cell. Sci.*, **3**, 405
24. Roseman, S. (1970). *Chem. Phys. Lipids*, **5**, 270
25. Dulbecco R., Hartwell, L. H. and Vogt, M. (1965). *Proc. Nat. Acad. Sci. U.S.A.*, **53**, 403
26. Hartwell, L. H., Vogt, M. and Dulbecco, R. (1965). *Virology*, **27**, 262
27. Dulbecco, R. and Johnson, T. (1970). *Virology*, **42**, 368
28. Weil, R. and Kára, J. (1970). *Proc. Nat. Acad. Sci. U.S.A.*, **67**, 1011
29. Mora, P. T., Cumar, F. A. and Brady, R. O. (1971). *Virology*, **46**, 60
30. Robbins, P. W. and Macpherson, I. A. (1971). *Nature (London)*, **229**, 569
31. Sakiyama, H. and Robbins, P. W. (1973). *Fed. Proc.*, **32**, 86
32. Pollack, R., Wolman, S. and Vogel, A. (1970). *Nature (London)*, **228**, 938
33. Ohta, N., Pardee, A. B., McAuslan, B. R. and Burger, M. M. (1968). *Biochim. Biophys. Acta*, **158**, 98
34. Wu, H. C., Meezan, E., Black, P. H. and Robbins, P. W. (1969). *Biochemistry*, **8**, 2509
35. Grimes, W. J. (1970). *Biochemistry*, **9**, 5083
36. Culp, L. A., Grimes, W. J. and Black, P. H. (1971). *J. Cell. Biol.*, **50**, 682
37. Bosmann, H. B., Hagopian, A. and Eylar, E. H. (1968). *J. Cell Physiol.*, **72**, 81
38. Bosmann, H. B. and Eylar, E. H. (1968). *Nature (London)*, **218**, 582
39. Green, H., Todaro, G. J. and Goldberg, B. (1966). *Nature (London)*, **209**, 916
40. Meezan, E., Wu, H. C., Black, P. H. and Robbins, P. W. (1969). *Biochemistry*, **8**, 3518
41. Sakiyama, H. and Burge, B. W. (1972). *Biochemistry*, **11**, 1366
42. Sheinen, R., Onodera, K., Yogeeswaran, G. and Murray, R. K. (1971). *2nd LePetit Symposium, The Biology of Oncogenic Viruses*, 274
43. Sheinen, R. and Onodera, K. (1972). *Biochim. Biophys. Acta*, **274**, 49
44. Warren, L., Fuhrer, J. P. and Buck, C. A. (1973). *Fed. Proc.*, **32**, 80
45. Buck, C. A., Glick, M. C. and Warren, L. (1970). *Biochemistry*, **9**, 4567
46. Buck, C. A., Glick, M. C. and Warren, L. (1971). *Science*, **172**, 169
47. Buck, C. A., Glick, M. C. and Warren, L. (1971). *Biochemistry*, **10**, 2176

48. Martin, G. S. (1970). *Nature (London)*, **227**, 1021
49. Warren, L., Fuhrer, J. P. and Buck, C. A. (1972). *Proc. Nat. Acad. Sci. U.S.A.*, **69**, 1838
50. Bosmann, H. B. (1972). *Biochem. Biophys. Res. Commun.*, **49**, 1256
51. Roth, S., McGuire, E. J. and Roseman, S. (1971). *J. Cell Biol.*, **51**, 536
52. Roth, S. and White, D. (1972). *Proc. Nat. Acad. Sci. U.S.A.*, **69**, 485
53. Bosmann, H. B. (1972). *Biochem. Biophys. Res. Commun.*, **48**, 523
54. Abercrombie, M. and Ambrose, E. J. (1962). *Cancer Res.*, **22**, 525
55. Fox, T. O., Sheppard, J. R. and Burger, M. M. (1971). *Proc. Nat. Acad. Sci. U.S.A.*, **68**, 244
56. Glick, M. C. and Buck, C. A. (1973). *Biochemistry*, **12**, 85
57. Otten, J., Johnson, G. S. and Pastan, I. (1971). *Biochem. Biophys. Res. Commun.*, **44**, 1192
58. Burger, M. M., Bombik, B. M., Breckenridge, B. M. and Sheppard, J. R. (1972). *Nature* (*New Biol.*), **239**, 161
59. Bosmann, H. B. and Winston, R. A. (1970). *J. Cell. Biol.*, **45**, 23
60. Bosmann, H. B. (1971). *Arch. Biochem. Biophys.*, **145**, 310
61. Brown, J. C. (1972). *J. Supermolecular Struct.*, **1**, 1
62. Spiro, R. G. (1970). *Ann. Rev. Biochem.*, **39**, 599
63. Palade, G. (1959). In *Subcellular Particles*, (T. Hayashi, editor), 64, (Ronald Press: New York)
64. Winzler, R. J. (1970). *Int. Rev. Cytology*, **29**, 77
65. Singer, S. J. and Nicolson, G. L. (1972). *Science*, **175**, 720
66. Hirano, H., Parkhouse, B., Nicolson, G. L., Lennox, E. S. and Singer, S. J. (1972). *Proc. Nat. Acad. Sci. U.S.A.*, **69**, 2945
67. Aub, J. C., Sanford, B. A. and Wang, L. (1965). *Proc. Nat. Acad. Sci. U.S.A.*, **54**, 400
68. Burger, M. M. (1973). *Fed. Proc.*, **32**, 91
69. Burger, M. M. (1969). *Proc. Nat. Acad. Sci. U.S.A.*, **62**, 994
70. Benjamin, T. L. and Burger, M. M. (1971). *Proc. Nat. Acad. Sci. U.S.A.*, **67**, 929
71. Allan, D., Anger, J. and Crumpton, M. S. (1972). *Nature* (*London*), **236**, 133
72. Kornfeld, S., Rogers, J. and Gregory, W. (1971). *J. Biol. Chem.*, **246**, 6581
73. Presant, C. A. and Kornfeld, S. (1972). *J. Biol. Chem.*, **247**, 6937
74. Habel, K. (1961). *Proc. Soc. Exp. Biol. Med.*, **113**, 1
75. Khera, K. S., Ashkenazi, A., Rapp, F. and Melnick, J. L. (1963). *J. Immunol.*, **91**, 604
76. Butel, J. S., Tevethia, S. S. and Melnick, J. L. (1972). *Adv. Cancer Res.*,
77. Girardi, A. J. (1965). *Proc. Nat. Acad. Sci. U.S.A.*, **54**, 445
78. Tevethia, S. S., Grouch, H. A., Melnick, J. L. and Rapp. F. (1970). *Int. J. Cancer*, **5**, 176
79. Smith, R. W., Morganroth, J. and Mora, P. T. (1970). *Nature* (*London*), **227**, 141
80. Girardi, A. J. and Defendi, V. (1970). *Virology*, **42**, 688
81. Coggin, J. H., Ambrose, K. R. and Anderson, N. G. (1970). *J. Immunol.*, **105**, 524
82. Martin, G. and Littlefield, J. W. (1968). *J. Virology*, **2**, 69
83. Weiss, M., Ephrussi, B. and Scaletta, L. (1968). *Proc. Nat. Acad. Sci. U.S.A.*, **59**, 1132
84. Martin, R. G. (1967). *J. Mol. Biol.*, **26**, 311
85. Tomkins, G. M., Gelehrter, T. D., Granner, D., Martin, D. Jr., Samuels, H. H. and Thompson, E. B. (1969). *Science*, **166**, 1474
86. Echkart, W. (1969). *Nature* (*London*), **224**, 1069
87. Brew, K., Vanaman, T. C. and Hill, R. L. (1968). *Proc. Nat. Acad. Sci. U.S.A.*, **59**, 491
88. Cumar, F. A., Fishman, P. H. and Brady, R. O. (1971). *J. Biol. Chem.*, **246**, 5075
89. Brady, R. O. (1973). *Scientific American*, **229**, 88
90. Brady, R. O., O'Brien, J. S., Bradley, R. M. and Gal, A. E. (1970). *Biochim. Biophys Acta*, **210**, 193
91. Klenk, H.-D. and Choppin, P. W. (1970). *Proc. Nat. Acad. Sci. U.S.A.*, **66**, 7
92. Renkonen, O., Gahmberg, C. G., Simons, K. and Kaarianinen, L. (1970). *Acta Chem. Scand.*, **24**, 733
93. Parson, D. F. and Subjeck, J. R. (1972). *Biochim. Biophys. Acta*, **265**, 85
94. Brady, R. O., Borek, C. and Bradley, R. M. (1969). *J. Biol. Chem.*, **244**, 6552

4
Mobility of Membrane Components

F. PODO* and J. K. BLASIE
University of Pennsylvania

* On leave from the Instituto Superiore di Sanita, Laboratori di Fisica, Roma, Italy.

4.1 INTRODUCTION

A number of physical techniques may now be used to study the structure and dynamics of biological membranes at the molecular and even submolecular levels. Some of these techniques are direct in the sense that they probe the structure and dynamics with only a minimum perturbation of the membrane. Examples of such physical techniques are x-ray and neutron diffraction, nuclear magnetic resonance spectroscopy (n.m.r.) and Raman spectroscopy. Other techniques are somewhat indirect in the sense that they require the incorporation of an extrinsic probe molecule into the host membrane as an impurity and this may perturb the local environment in the host to varying extents depending on the physicochemical nature of the probe and its environment in the membrane. Such probes therefore 'see' some approximation to the natural membrane, the less perturbation the probe provides the better the approximation. Examples of such physical techniques are electron paramagnetic resonance spectroscopy (e.p.r.) of spin-labelled probe molecules, optical spectroscopy of chromophore-labelled probe molecules and fluorescence microscopy antibody-labelled membranes.

We shall concentrate on the application of 'direct' physical techniques to the problem of membrane structure and dynamics at the molecular level, utilising the more 'indirect' techniques wherever the approximation to the natural situation appears to be reasonable or whenever these methods are the only methods at present available for providing information regarding the mobility of some particular membrane component. We shall not consider here techniques which are even less direct in that they may involve even greater perturbation of the natural membrane under study, or those which are not directly sensitive to structure and dynamics of membranes at the molecular level or lower. Studies of model membrane systems will be used whenever they best exemplify the use or potentiality of the technique in similar studies of natural membranes. The work of other laboratories will be referred to appropriately.

Before turning to the main concern of this article, we should first briefly discuss some theoretical and practical aspects of the application of these physical techniques to the study of biological membrane systems.

4.1.1 X-Ray and neutron diffraction

X-Ray diffraction from orientated multilayers of natural membranes and model membranes can provide lamellar diffraction arising from the membrane's electron density profile out to $(2 \sin \theta)/\lambda \sim 0.2$ Å^{-1} where 2θ is the diffraction angle and $\lambda = 1.54$ Å for CuKα_1 x-radiation[1]. Such an electron-density profile is actually a projection of the electron-density distribution of the membrane in a direction parallel to the plane of the membrane on to an axis normal to that plane[2]. This lamellar diffraction may contain both Bragg reflections and paracrystalline reflections (which arise from disorder in the multilayer) and the phase problem must be solved unambiguously, including a proper treatment of the paracrystalline reflections (if these are present), in order to provide the membrane profile to an equivalent resolution of less than *ca.* 5 Å[3]. These profiles must then be interpreted in terms of the distribution of the membrane's molecular components in the membrane's cross-section (or profile), taking into account the various components' possible electron density structures, which is certainly not unambiguous and depends to a considerable extent on the resolution to which the profile was calculated. Diffraction which occurs at right angles to the lamellar diffraction arises from the packing of macromolecules and hydrocarbon chains in the plane of the membrane and is critically dependent on the nature of the packing, i.e. crystalline, paracrystalline, liquid or random, and the packing density of the various components[1a, 1b, 3b, 4]. This diffraction is of considerable use in interpreting the profile of the membrane, particularly if its variation with the electron density of the aqueous phase around the membrane is known[5]. In addition, comparison of the profiles calculated from x-ray diffraction data with those calculated from neutron diffraction data are particularly useful in this regard due to the large differences in relative atomic scattering factors for x-rays *vs.* neutrons[6]. The use of selectively protonated *vis-à-vis* deuterated components in natural and model membranes could lead to an unambiguous interpretation of the profile utilising the two diffraction techniques.

Elastic-scattering experiments involving x-ray and neutron diffraction cannot, in general, distinguish between the ensemble-average of a frozen statistically-disordered structure and a dynamical structure whose time-average is identical to the ensemble-average of the frozen structure[3b]. In order to gain information about molecular motion within a structure, it is necessary to correlate structure deduced by diffraction techniques with data from structure probes sensitive to molecular motion, such as the several forms of spectroscopy mentioned below. On the other hand, molecular motion may also be inferred from structural changes between the initial and final states (the structures of these states being determined by diffraction techniques) produced by natural or artificial perturbation of the structure.

4.1.2 Inelastic and quasi-elastic neutron scattering

Inelastic and quasi-elastic scattering techniques are potentially useful in detecting the discrete vibrational motions of molecules or a collection of molecules and the diffusive motions of molecules, respectively, in a natural membrane or a model membrane[7]. Because the energies of thermal neutrons are more closely related than x-rays of a comparable wavelength to those of the motions just described, inelastic and quasi-elastic scattering involving neutrons is more easily observable. The interpretation of inelastic neutron scattering is model dependent in the sense that the scattering data must be fitted with a calculation of the vibrational motion of the polyatomic molecule or collection of such molecules based upon a model for their structure (individually and collectively) and their interactions (intra-molecular and inter-molecular). The interpretation of quasi-elastic neutron scattering is somewhat model dependent in that it also relies on one of the several models available for diffusive motion. Nevertheless, qualitative indications of the kinds of motion available to these molecules can be obtained.

Since quasi-elastic neutron scattering has to date only been applied to the study of diffusive motion in lipid bilayer model membranes[8], we shall not consider it further except to point out its potential as a direct probe of membrane component dynamics.

4.1.3 Raman spectroscopy

Both resonance and non-resonance Raman spectroscopy are essentially inelastic light scattering arising from the discrete vibrational structure of polyatomic molecules in the non-resonance case and of portions of molecules in the vicinity of, and including, a chromophore in the resonance-enhanced case[9a]. Both cases are model-dependent in that they eventually involve a calculation of the vibrational structure of the entire molecule, or portions thereof, in the resonance-enhanced case, based on the relative atomic co-ordinates and model interatomic potential energy functions. In spite of this, qualitative information relating to the intramolecular motion available to these molecules may be obtained through careful correlative studies using model compounds[9b].

4.1.4 Magnetic resonance: n.m.r. and e.p.r.

A magnetic resonance experiment consists of inducing and detecting (by means of radiation of an appropriate frequency) transitions between the Zeeman levels of particles possessing a non-zero magnetic moment and angular momentum, the splitting of the Zeeman levels depending on an external magnetic field H_0. These requirements are satisfied by electrons as well as by several natural nuclei. Although based on strictly analogous physical principles, the n.m.r. and e.p.r. methods differ strongly not only in the instrumental details (the former occurring in the radio frequency, the latter in the microwave frequency range) but also, and essentially, in the requirements imposed on the system in order for it to be properly studied. Biological and model membranes, which possess natural nuclei such as ^{1}H, ^{13}C, ^{31}P, ^{15}N, ^{17}O, are in fact suitable for study by n.m.r. spectroscopy without any further difficulty other than that arising from the complexity of the system and the low natural abundance of some nuclei. On the contrary, these systems are not directly suited, in general, for e.p.r. experiments, because of the diamagnetic character of their natural components resulting from paired selection spins. Under such circumstances, studies using e.p.r. spectroscopy are only possible by incorporation of radicals or atoms with unpaired spins into the system[10]. The difficulties which could arise from such an intrinsically perturbative method are obvious, although they can sometimes be minimised by testing the biological functions of the membrane after the introduction of the spin label, by testing the conclusions through the use of several labels having different structures, or by the use of independent physical techniques.

4.1.4.1 N.M.R. spectroscopy

The present article is not intended to provide an exhaustive treatment of n.m.r. theories and experimental methods[11-13]. Only some basic principles will be summarised, in order to indicate briefly the type of information the technique is able to provide about the organisation and dynamic structure of a molecular system.

A nucleus situated in a molecular system will be magnetically screened by the extranuclear electron orbitals, in which orbital currents will be induced by the application of a magnetic field H_0. If the rate of rotation of the system is sufficiently rapid to neglect the static dipole–dipole interactions, the nuclear spin Hamiltonian is composed of two terms expressing respectively the interaction with the local field and the indirect nuclear spin–spin interaction or J-coupling. The local field differs from the external field H_0 because of the electronic screening σ. The latter reflects the variations in the local electronic structure in the vicinity of the nucleus and for this reason the nucleus can be considered as a probe, intrinsic to the molecular system, for assessing information on molecular structure and conformation, as well as local solvent effects[14]. This information can be obtained from an analysis of the absorption n.m.r. spectrum, in the frequency domain of a molecular system, the various resonances being due to nuclei occupying various chemical positions (chemical-shift determination). For example, in the proton spectrum obtained from sonicated dispersions of diacyl lecithins in water it is possible to resolve and assign several signals which may be attributed to the methyl and methylene groups in the choline and along the fatty-acid chains, as well as in the glycerol moiety[15, 16]. A significant example of solvent effects studied by n.m.r. spectroscopy is the possibility of obtaining information about the degree of penetration of water into aggregates of amphipathic molecules dispersed in aqueous solutions[17]. Variations in the ^1H chemical shifts obtained for the various proton-containing groups along a phospholipid molecule when the latter is transferred from a non-polar solvent to water in which it exists as a bilayer structure, suggest that water might penetrate the structure as far as the glycerol backbone, the fatty-acid chains being exposed to a non-polar environment[15, 16].

Intermolecular shielding effects are also produced by molecules possessing a strong diamagnetic anisotropy such as those containing aromatic rings. These effects can be used to locate such ring structures to within a few ångstroms in natural and model membrane systems[14] due to the short-range nature and anisotropy of this effect.

The Boltzmann distribution of the spins over the nuclear Zeeman levels creates a net magnetisation $\vec{M}$ in the system in the direction of the field. Following disruption of this distribution by any process, the spin system returns to equilibrium with its surroundings by means of first-order relaxation mechanisms. The $\vec{M}$ component along the field relaxes with a characteristic time T_1 through radiationless transitions in which the magnetic energy is exchanged with the lattice as rotational or translational energy. The $\vec{M}$ component in a plane perpendicular to the field relaxes with a characteristic time T_2 through spin–energy exchange between the nuclei themselves and the T_1 relaxation process. The best way of measuring both T_1 and T_2 is by the use of pulse Fourier-transform n.m.r. methods[18]. The experiment used involves the detection and Fourier transformation of the free induction decay signal obtained after an appropriate sequence of radiofrequency pulses. The times T_1 and T_2, which are usually referred to as the spin–lattice and spin–spin relaxation times, respectively, are related to the molecular arrangement and dynamic structure of a spin system. In fact, local magnetic fields are created at a given nucleus depending on the relative geometric arrangement of the neighbouring nuclear spins; the fluctuations of these microscopic fields, generated as the molecules move about in space, create the conditions for inducing among the magnetic levels the transitions responsible for relaxation. The fluctuations which occur in the microscopic magnetic fields follow the frequency spectrum corresponding to the molecular motion[11]. The latter can be described mathematically in terms of the correlation functions which express the probability that a dynamic system maintains, at a given instant, a memory of its past motional history. The correlation functions in the time domain characterise spectral density functions in the frequency domain, the two functions being related through a Fourier transformation. The relaxation times depend on the intensity of the component of the spectral density functions at the frequency of the nuclear resonance as well as on the strength of the several interaction mechanisms which are able to provide a coupling between the nuclear precession and the molecular motions. The two relaxation times show a different dependence on the spectral density functions, the spin–lattice relaxation in particular being only affected by high-frequency fluctuations and the spin–spin relaxation by both low- and high-frequency processes. There are several mechanisms which are able to provide magnetic relaxation. In many cases the dominant mechanism is provided by the magnetic dipole–dipole interactions. The static contribution to the local magnetic field due to dipole–dipole interactions is averaged out as soon as the molecular system is moving in space at a re-orientation rate faster than the width of the spectrum. In deriving the spin–lattice relaxation rates resulting from dipolar interactions in a dynamic system, with the static interactions already averaged out, it is useful to distinguish between the intra- and inter-molecular contributions, the former being determined only by intramolecular rotations and the latter by relative intermolecular motions, resulting from a combination of relative translations and rotations of the molecules.

In the case of isotropic random rotational motion, the Stokes–Einstein approach of a rigid sphere rotating in a continuous medium can be used to express the correlation functions of the molecule; in particular, the dipole–dipole intramolecular relaxation rates may therefore be expressed in terms of only one correlation time τ_c, which describes the isotropic rotational diffusion of the internuclear vectors. Using these assumptions, when $\tau_c \ll \frac{1}{\nu_0}$, both the spin–lattice and the spin–spin relaxation rates are identical, frequency-independent and proportional to τ_c ('extreme-narrowing' condition). The T_1 rate, plotted as a function of the correlation time, shows a maximum, the coupling between the magnetic and the lattice energy levels being most effective when $\tau_c \simeq \frac{1}{\nu_o}$. In the region of the long correlation times ($\tau_c > \frac{1}{\nu_o}$), as τ_c increases with decreasing temperature or increasing viscosity of the medium, the decrease in T_1 rate follows a $\frac{1}{\tau_c}$ law. The T_2 rate increases continuously; in particular, in the region of the long correlation times the T_2 rate is proportional to τ_c. The temperature dependence of τ_c is usually expressed by the equation

$$\tau_c = \tau_o \exp(\Delta E/RT)$$

where ΔE is the activation energy for the particular motional relaxation mechanism considered and τ_o is a constant. ΔE may be determined directly, for example, under 'extreme-narrowing' conditions.

If a molecule undergoes anisotropic rotational diffusion, different correlation times must be introduced to account for the rotations of the internuclear vector about the different axes. Nuclear magnetic relaxation has been derived for like spins in ellipsoids of revolution undergoing rotational Brownian motions and obeying the Stokes–Einstein approximation for motion in a continuous medium[19]. Anisotropic considerations become important when the molecules are non-spherical (even in an isotropic medium) and when the geometry of the system imposes anisotropic restrictions on molecular re-orientation (e.g. long-axis molecules aligned in a layer and molecules moving on a surface) and also when independent stochastic rotations of the internuclear vectors can occur within a molecule due to internal motions[20]. The treatment carried out by Woessner[20] for the spin-relaxation processes of a spin pair re-orientating randomly about an axis, which in turn tumbles randomly, is of particular interest for applications to lipid bilayers in membrane systems. It has been shown that although the two types of motion are independent, their contribution to the relaxation are not. The correlation times are a combination of those describing the re-orientation of the axis and the internal rotation about that axis. The weighting factors of the contributions to the relaxation corresponding to the various correlation times depend on the angle of the internuclear vector with the axis of re-orientation; T_2 is dominated by the slower re-orientational motion while T_1 contains contributions from both motions whose combination depends on the ratio of the two correlation times.

Many of the attempts made to date of quantitatively interpreting the n.m.r. relaxation in a model membrane system show the basic limitation of only considering random and isotropic models of motion for fitting the data. This will be discussed in a latter section.

4.1.4.2 E.P.R. spectroscopy

In an e.p.r. spectroscopic experiment, transitions between the Zeeman energy levels of an electron system containing one or more unpaired electrons are induced and detected by means of microwave radiation. These Zeman levels are non-degenerate in a static magnetic field, the unpaired electron(s) being distributed at thermal equilibrium among these levels according to the Boltzmann distribution[12, 13]. The basic electron-spin Hamiltonian (neglecting nuclear Zeeman interactions) for the system is written in terms of the electron Zeeman interaction $\beta\vec{\mathbf{H}}\cdot\vec{\vec{g}}\cdot\vec{\mathbf{S}}$, nuclear hyperfine interactions, $\vec{\mathbf{S}}\cdot\vec{\vec{\mathbf{T}}}\cdot\vec{\mathbf{I}}$ and, for spin multiplicities greater than 1/2, an electron spin interaction $\vec{\mathbf{S}}\cdot\vec{\vec{\mathbf{D}}}\cdot\vec{\mathbf{S}}$. The $\vec{\vec{g}}$, $\vec{\vec{\mathbf{T}}}$ and $\vec{\vec{\mathbf{D}}}$ tensors are characteristic of the local environment of the unpaired electron system including solvent interactions. The $\vec{\vec{\mathbf{T}}}$ tensor contains both the isotropic hyperfine and dipolar hyperfine interactions with neighbouring nuclei. Hence, the resonance energy and the hyperfine structure of an e.p.r. signal provide information for identifying a paramagnetic species and assessing the nature of its environment, including the effects of solvent. It is essential to determine independently the components of the generally anisotropic $\vec{\vec{g}}$ and $\vec{\vec{\mathbf{T}}}$ tensors of the paramagnetic species in a host single crystal. The motional modulation of these $\vec{\vec{g}}$, $\vec{\vec{\mathbf{T}}}$ $\vec{\vec{\mathbf{D}}}$ tensors is mainly responsible for the spin–lattice relaxation of the electron spin system. Knowledge of the anisotropy of these tensors then allows a determination of the motion of that species in its local environment within a natural or model membrane to be made through a comparison of the experimentally-observed e.p.r. spectra of the probe in the membrane with simulated spectra assuming various models for the ensemble and motional averaging of these tensors (including the orientational nature of the spin system in the membrane as detected by the anisotropy in its motion).

4.1.5 Optical spectroscopy

Polarisation absorption and fluorescence spectroscopy of intrinsic membrane chromophores and extrinsic chromophores incorporated into the host membrane structure may provide information on the anisotropic rotational mobility of the chromophore in the membrane[21]. Whilst, by definition, there is no perturbation of the local environment of an intrinsic membrane chromophore, some perturbation of the local environment of an extrinsic probe necessarily occurs. Whilst information on the anisotropic rotational mobility of an intrinsic probe relates directly to the motion of the chromophore itself in the membrane, it relates only indirectly to the structural and dynamic nature of its local environment in the membrane. The latter situation is even more indirect in the case of incorporated extrinsic chromophores since in this case the local environment is only an approximation to the actual dynamic structure of the membrane in the region occupied by the probe before the latter was incorporated. For a detailed review, the reader is referred to the recent article by Radda and Vanderkooi[22].

We shall now discuss, after this brief introduction, the intra- and intermolecular motions of lipid and protein components of biological membranes as indicated through singular or collective use of these physical techniques.

4.2 INTRAMOLECULAR MOTION OF MEMBRANE COMPONENT

4.2.1 Lipid molecules

Physical techniques have provided considerable evidence for the occurrence of relatively pure lipid bilayer regions within the natural mosaic structure of the natural membrane[23]. This evidence comes from (a) the interpretation of electron-density profiles for several natural membranes[24] in relation to those for lipid model membranes, (b) the interpretation of x-ray diffraction data arising from lipid fatty-acid chain packing in the plane of several membranes with the chains in a melted or frozen state in relation to that for lipid model membranes and the result that thermal phase transitions and phase separations occur over similar temperature regions for some natural membranes and the model membrane composed of only its lipid components[25], and (c) that the lipid fatty-acid chain mobility in some natural membranes and lipid bilayer membranes are similar as determined by a variety of spectroscopic techniques including n.m.r. studies of fatty-acid chain nuclear relaxation times, e.p.r. studies of incorporated spin labels and fluorescence polarisation studies of incorporated chromophore labels, as will be discussed in detail later in this section.

This evidence suggests that it is reasonable to consider the nature of intramolecular motion of lipid molecules in a lipid bilayer model membrane as being approximately related to the motion in similar regions in the mosaic structure of a natural membrane. Above the thermal transition for chain melting, the dipalmitoyl lecithin bilayer structure is similar to that of natural lipids[1a, 24a] as determined by x-ray diffraction. The natural lipid structures are usually inherently lower-resolution structures owing to the natural mixture of fatty-acid chains and polar head-groups.

The electron-density profiles for the dipalmitoyl lecithin bilayer at 25 °C and 49 °C at similar water contents for the multilayers show that the thickness of the bilayer is decreased at the higher temperature, while both the polar head-group region and hydrocarbon core of the bilayer are considerably modified. Defining the axis of the lamellar reflections as the meridian in the diffraction pattern from the lecithin multilayers, a sharp Bragg reflection at 4.15 Å occurs on the equator with relatively little off-equatorial spreading for the lower temperature bilayer, whilst a broad maximum at 4.5 Å with considerable off-equatorial spreading occurs at 49 °C. The degree of orientation of the lamellar reflections is similar at the two temperatures. The electron-density profiles for dipalmitoyl lecithin bilayers at 25 °C and two different water contents show some alterations in the polar head-group region while the hydrocarbon core is relatively invariant. With increasing water content, the 4.15 Å equatorial Bragg reflection becomes relatively less intense with an increase in off-equatorial spreading of this reflection into a considerably broader line-profile possessing a maximum at nearly the same diffraction angle at 25 °C as judged by microdensitometry. The electron-density profiles for dipalmitoyl lecithin and 1:1 dipalmitoyl lecithin:palmitate bilayers show large differences in the polar headgroup regions while the hydrocarbon cores are nearly identical, as are the near-equatorial Bragg reflections at *ca.* 4.15 Å at 25 °C. The electron-density profile in the hydrocarbon core for dipalmitoyl lecithin below 40 °C, together with the nature of the near-equatorial diffraction, indicate that the fatty-acid chains occur predominately in an extended from normal to the plane of the bilayer and are probably arranged in a hexagonal lattice in the plane of the bilayer with an interchain separation of 4.8 Å. The Bragg reflection at 4.15 Å for this hexagonal lattice may be a higher-order reflection of a 'super-lattice' of twice the dimensions since several lower-angle equatorial reflections including the 4.15 Å reflection from such a lattice appear at very low water contents. This 'super-lattice' may represent the lecithin molecular lattice. The interior methylene groups of the chains form the region of constant electron density in the core region of the profile while the terminal methyl group and possibly 1–2 adjacent methylene groups lie in a trough of relatively lower electron density at the centre of the bilayer. Higher-order lamellar reflections (>10) which could narrow this feature of the profile are not observed except at very low water contents when more than 22 orders appear. The 2–3 carbon atoms at the terminal methyl end of the chains may then be statistically or motionally disordered in the unit cell structure in order to account for the width of this central trough. Above 45 °C, the hydrocarbon core of the bilayer thins dramatically and the fatty-acid chains are highly disordered*, existing in a state similar to that of liquid long-chain normal hydrocarbons as indicated by the nature of the broad intensity maximum at *ca.* 4.5 Å as described. The terminal methyl ends of the chains appear to remain predominantly localised near the centre of the bilayer as indicated by the fact that a somewhat broader trough of lower relative electron density is still distinguishable in the centre of the bilayer. The general features of the high-temperature profile for the hydrocarbon core of the bilayer may be indicative of the increased statistical or motional disorder along the chains towards their terminal methyl groups, particularly the last 3–4 carbon atoms.

The outer peak of the two peaks in the polar head-group region of the dipalmitoyl lecithin electron-density profile (<40 °C) may be associated with the phosphate and trimethylammonium groups while the inner peak is associated with the ester groups, with one water molecule coordinated to each ester group as indicated by model calculations. This assignment is supported by the invariance of the hydrocarbon cores and the differences in the polar head-group regions of the dipalmitoyl lecithin profile in comparison with the 1:1 dipalmitoyl lecithin: palmitate profile, as indicated by model calculations.

Increasing the water content in the lecithin multilayers makes the ester section of the polar head-group region of the profile less structured. In addition, a slightly greater proportion of the fatty-acid chains segments are packed in a statistically-disordered amorphous state* and a smaller proportion of segments occur in an extended chain hexagonal packing, as indicated by changes in the profile in the hydrocarbon core and near-equatorial diffraction.

The detailed dynamic structure of both synthetic diacyl lecithin bilayers and

* The statistically-disordered 'amorphous' packing refers to chain packing in the plane of the bilayer membrane. It results for a combination of lattice disorder *coupled with* substitution disorder, i.e. although the chains are relatively extended, the conformation of neighbouring chains differ. Hence, diffraction occurs from the most probable interchain vectors instead of from a Bragg plane as for a well-ordered lattice.

natural lipid bilayers must, as indicated earlier, be determined by other, primarily spectroscopic, physical techniques.

Broad n.m.r. spectra are observed in unsonicated lecithin bilayers[26] above the melting point of the fatty-acid chains, T_m, at least in the absence of a considerable degree of chain unsaturation. The broad proton signals observed above T_m have been attributed to slow molecular motions in the structure, leading to an incomplete averaging of dipole–dipole interactions[26, 27]. Both the *N*-methyl and the terminal methyl groups appeared more mobile than the rest of the molecule. ^{13}C spectra show rather narrower peaks[28] due to the reduced spin–spin intermolecular interactions. However, it has been suggested that only the outer layers of the unsonicated multi-bilayer liposomes are able to produce the high-resolution spectrum[29]; furthermore the broadening which is still visible in the fatty-acid signals has been attributed to chemical shift differences within the multilayer vesicles rather than to residual ^{13}C–proton dipolar broadening. While the non-exponential free-induction decay indicates a distribution of transverse relaxation times for the various protons in the lecithin bilayers, the longitudinal relaxation time T_1 in unsonicated vesicles is apparently described by only one characteristic time[26]. This observation was at first interpreted on the basis that only a small number of protons in the hydrocarbon chain are significantly coupled to the lattice with a spin-diffusion mechanism probably permitting the magnetisation to diffuse to an appropriate heat sink[30]. However, it has since been demonstrated that although the presence of such a mechanism cannot be completely ruled out, the apparent uniformity of spin–lattice relaxation times along the lecithin molecules is essentially a consequence of the poor spectral resolution obtained with unsonicated systems[29, 31].

The line-width narrowing observed in the spectra of sonicated aqueous suspensions of lecithin liposomes[32] above T_m has been used as the basis of most recent n.m.r. applications to the problems of organisation and dynamic structure of model membrane systems. Static magnetic dipole–dipole interactions are apparently averaged out in sonicated phospholipid dispersions due to the formation of rapidly tumbling single bilayer vesicles (diameter about 200–300 Å)[33]. Many physicochemical studies on the mechanisms and kinetics of sonication as well as on the nature of the vesicles have indicated that the characteristics of molecular packing are retained on going from coarse to sonicated phospholipid dispersions[34-36]. This evidence has confirmed the validity of using phospholipid vesicles as a reasonably correct system for studying structural and dynamic properties of lipid bilayers.

The different ^{13}C and ^{1}H T_1 relaxation times exhibited by various chemical groups in lecithin vesicles offer, in principle, a powerful tool for assessing local information on the dynamic structure of these systems, provided that the mechanisms responsible for the temperature and frequency dependence of relaxation can be correctly recognised and quantitatively tested. From this point of view, such studies are at present at a rather primitive stage, even in simplified systems like pure lecithin vesicles, in spite of the availability, in some cases, of sufficient sets of data. The main relaxation mechanism for ^{13}C resonances has been demonstrated as being due essentially to dipolar interactions between the nucleus and directly-bonded protons. Carbon nuclei having uniform mobility should therefore show relaxation times inversely proportional to the number of directly-bonded protons. In the case of proton relaxation, in addition to the contribution resulting from modulation of dipolar coupling by the geminal and (to a lesser extent) vicinal protons, intermolecular dipole–dipole interactions are also expected to influence spin–spin and spin–lattice relaxation rates. Attempts which have been made to isolate the intermolecular contribution to the proton relaxation will be discussed in a later section.

At this point, some mainly qualitative results suggested by studies of the mobility of various chemical groups along the lecithin molecules in a bilayer will be summarised and discussed. Both ^{13}C and ^{1}H T_1 values at a fixed temperature above T_m indicate, qualitatively, that a progressive increase occurs in the molecular motion in the lecithin bilayer from the glycerol group towards both the final methyl group in the chains and the *N*-methyl group in the polar head-group; such motion will be discussed in detail later in this section[29-31]. The region with the least mobility in the model membrane is thus situated in the vicinity of the glycerol groups and may be identified with the main permeability barrier in the structure[31]. A study of the differential changes in mobility exhibited by various chemical groups as the temperature is increased through T_m has been achieved via determinations of T_1 values. Below T_m, the chain signals first disappear, followed by the resonance attributable to the *N*-methyl group as the temperature is further decreased[31]. Studies of T_1 values indicated that a conformational change takes place in the vicinity of the *N*-methyl group, such a change being linked with the immobilisation of the fatty-acid chains. A mechanism which has been suggested for this transition implies that below T_m the choline phosphate dipoles lie in the plane of the vesicle surface, thus favouring the observed reversible aggregation of the vesicles (presumably by lateral electric dipole–dipole interactions). Above T_m, the dipoles extend to a greater extent from the esterified glycerol group towards the solvent, the positively charged head-groups dominating at the surface[29]. A gradual crystallisation or melting of the chains within the vesicles with coexisting fluid and solid-like regions are observed following the reversible intensity changes in the spectra of the fatty acid chains through the thermal transition. This evidence has been suggested as having a direct relevance for an evaluation of the percentage of lipids organised in a bilayer structure in a natural membrane, by determining the percentage of those exhibiting a well defined transition (e.g. mycoplasma membrane). The presence of unsaturated bonds in the chains of dioleoyl lecithin leads to considerable increase in the T_1 values of the chain carbons relative to those exhibited by saturated chains (as well as a large increment in T_1 for the $N^+CH_3)_3$ group)[29]. This result also suggests that such measurements may be used to gain insight into the role of unsaturated bonds at specific points along the lipid chains in the bilayer regions contained in natural membranes. Another interesting result, in agreement with the lower transition temperature and the greater permeability at the glycerol level found in dimyristoyl lecithin relative to dipalmitoyl lecithin, is the observation that, on the basis of ^{13}C T_1 measurements, the whole bilayer becomes more fluid as the length of the hydrocarbon chain decreases[38].

Different degrees of mobility arise in the vicinity of the methyl and various methylene units along the fatty-acid chains as the temperature is increased above T_m. Resolved ^{13}C resonances are observed in dipalmitoyl lecithin fatty-acid chains for the carbonyl, the terminal methyl and the carbon atoms at positions 2, 3, 14 and 15. The methylene peaks which include the carbons atoms at positions 4–13 occur in an unresolved band. The diamagnetic anisotropy of the carbonyl group allows in the proton spectra the resolution of the methylene peaks in positions 2 and 3 from the main methylene envelope. Different T_1 values, which increase in proceeding from the carbonyl to the terminal methyl group, are found in both the ^{13}C and ^{1}H spectra[29, 37]. If the main relaxation mechanism were attributable to the tumbling of the molecule as whole, a uniform T_1 value would be observed for all the methylene carbons. The fact that the molecular motion increases steeply toward the terminal methyl group is suggestive of intramolecular motions, which may be attributed qualitatively to rotational oscillations about the C–C bonds within a given conformation[38] or to conformational changes along the chain (*trans/gauche* interconversions[39]).

That the hydrophobic core of the bilayer exhibits a liquid-like behaviour is suggested by the fact that the proton spin–lattice relaxation times increase with temperature for the various groups of lecithin bilayers in both sonicated and unsonicated dispersions, indicating that even at the working frequency of $\nu_o = 220$ MHz the proton groups are in the short correlation time regime $(\tau < \frac{1}{2\pi\nu_o} \simeq 10^{-9}\text{–}10^{-10}\text{s})$. This is in agreement with studies carried out by Luzzati *et al.*[40] emphasising the liquid-like nature of the hydrophobic cores of phospholipid lamellar structures as suggested by x-ray diffraction studies. It is difficult to reconcile the observed temperature behaviour of the various proton groups along the chains with the observed values of T_2 or of the line widths $\Delta\nu_{1/2}$ on the basis of only one correlation time mechanism of random and isotropic motion[41-43]. In fact, T_2 values evaluated from line widths, or more precisely from pulsed n.m.r. techniques[39, 44], are generally found to be much shorter than T_1 values in these systems. On the other hand, it is known that the condition for the short correlation time region is that $T_1/T_2 \simeq 1$ according to the model mentioned above.

Only qualitative interpretations of the T_1 and T_2 behaviours have to date been generally outlined. However, a quantitative attempt to treat the two correlation times for the relaxation mechanisms has been put forward by Chan *et al.*[45] in order to reconcile T_1 and line-width values of the choline and acyl chain methyl groups in unsonicated locithin bilayers. The motional model adopted was based mainly on the treatment developed by Woessner[46] for magnetic dipole–dipole relaxation of spins attached to groups undergoing anisotropic motion. The correlation times introduced, which express the re-orientational motion of the rotor axis and the rotation about the axis, were estimated to be $\sim 10^{-6.5}$ and $\sim 10^{-10}$s, respectively. The relaxation of the two groups was attributed completely to intramolecular magnetic dipole–dipole interactions.

Three types of motion were considered in principle by Metcalfe[38] in an attempt to interpret the ^{13}C T_1 values of the fatty-acid chains in dipalmitoyl lecithin vesicles. These were:

(a) Isotropic tumbling of the whole vesicle, for which, through the use of the Stokes–Einstein equation, a correlation time of about 10^{-6}s was obtained.

(b) Motion of the lecithin molecule *as a whole* about its long axis for which a correlation time of 10^{-8}–10^{-9}s was assumed in the basis of e.p.r. data arising from spin-label experiments[48].

(c) Motion about individual carbon–carbon bonds ($\tau_{cc} \leq 10^{-10}$ s).

The tumbling rates of the vesicles were shown to have a negligible effect on the T_1 values of all carbon atoms (except C_2) in the range of 10^{-5}–10^{-8} s. The second term also appears to be negligible, so that the treatment was finally based on the third motional mechanism only. The latter was interpreted as consisting completely of oscillatory motions rather than rotational motion involving conformational changes and leading to its treatment as a one-correlation time mechanism. On this basis τ_{cc} maybe expressed as the product of the true correlation time and a factor describing the amplitude of the oscillatory motion[49]. Under these conditions it appeared necessary to postulate a model involving little or no decrease of the correlation time in the first part of the chain, followed by a sharp decrease toward the terminal methyl group.

In all the treatments mentioned above, anisotropic contributions to the T_1 temperature behaviour have been considered to be negligible. Activation energies for thermal relaxation processes have therefore usually been calculated[31, 39] from Arrhenius plots of $\ln T_1{}^{-1}$ *vs.* the reciprocal temperature. The values obtained were generally of the order of the potential barriers to internal rotation in alkanes (*trans–gauche* transitions[50]). This agreement has been considered as an indication that the major source of thermal relaxation is modulation of the intramolecular dipolar interactions by internal rotations with a correlation time $\tau < 10^{-9}$–10^{-10} s[39].

Recent experimental evidence, however, does not confirm this hypothesis as shown by studies carried out[37] on the temperature and frequency dependence of T_1 and T_2 for dipalmitoyl lecithin (DPL) bilayers. A treatment involving at least two correlation times is required to explain the frequency dependence of the spin–lattice behaviour with temperature for the *N*-methyl, *N*-methylene, glycerol β-hydrogen, terminal methyl and internal methylene groups along the chain. Depending on the proton group considered, differences of 1–3 orders of magnitude between (at least) two correlation times offer possible solutions to the T_1 and T_2 behaviour. An explanation for these two different correlation times in terms of the intramolecular motion of the various proton groups along the molecule may be obtained by assuming that a rapid rotational motion of these groups about an axis normal to the plane of the bilayer contributes to T_1 and that a slower re-orientational motion about an axis parallel to the plane of the bilayer also contributes to T_1 and dominates T_2. This explanation was tested by incorporating fatty acids whose chain length was equal to or less than that of DPL where, in the latter case, x-ray diffraction studies[51] predicted an increased reorientation motion to be expected near the terminal methyl ends of the chains based on the interpretation of electron density profiles for the various mixed bilayers. T_1 and T_2 measurements on these mixed bilayers[52] verified this explanation for these motions, the latter being generally consistent with the *trans–gauche* interconversion along the chain described by Horwitz and Klein[39].

However, this evidence also points out that calculations of activation energies using the above mentioned Arrhenius plots are not generally justified, the slope of the spin–lattice relaxation rates *vs.* temperature depending on the activation energies of at least two mechanisms and on the ratio of their correlation times. It is interesting to note at this point that non-resonance Raman spectroscopy of dipalmitoyl lecithin[53] and egg lecithin bilayers[54] in the frequency region corresponding to the skeletal vibrations of hydrocarbon chain indicates that a significant number of *gauche* conformations occur in an irregular fashion along the chain above T_m.

Collectively, these studies indicate that while the fatty-acid chains within the hydrocarbon cores of these pure and mixed lipid bilayers are in a generally fluid state, the rate and anisotropy of intramolecular motion varies considerably within different regions of the lipid molecule.

Several results obtained by n.m.r. spectroscopy on the structure and dynamics of phospholipid bilayers are in general qualitative and sometimes quantitative agreement with the conclusions of e.p.r. studies performed on paramagnetic probes incorporated in lipid vesicles and some natural membranes[55]. The solubility and e.p.r. spectra exhibited by some spin labels (such as TEMPO[56]) in some natural membranes have offered a useful tool for assessing the characteristics of the lipid chain packing and comparing the latter with that of suitable model systems[57]. These paramagnetic probes show very similar e.p.r. spectra in both lipid vesicles and several natural membranes, consisting of two partially overlapping signals originating from the probe fractions dissolved in the non-polar and aqueous solution respectively. The probe appears to undergo a rapid and effectively isotropic motion in the non-polar region with a correlation time of about 10^{-10} s in agreement with the liquid-like nature of this region suggested by n.m.r. spectroscopy. The structure and mobility of the hydrocarbon region of the phospholipid bilayers in model and natural membranes have also been studied by an analysis of the e.p.r. spectra of incorporated phospholipids and fatty acids, spin-labelled at selected positions[58,59]. Quantitative treatments of the motional freedom of the various spin labels, carried out using the computational approaches referred to in Section 4.1, indicate that rapid anisotropic motions occur in the lipid bilayers and that such motions may be accounted for in terms of rapid *trans–gauche* interconversions of the polymethylene chains. The motional freedom of the spin label appears to depend on the position of the probe along the chain, and also on the length of the chain, the freedom of motion increasing as the probe is moved away from the polar end of the molecule. The anisotropic rotational diffusion of the spin labels was, in general, found to be at least as rapid as 10^{-7}–10^{-8} s, suggesting a relatively high degree of motion of the label, again in agreement with the liquid nature of the hydrophobic region in some membranes and in phospholipid dispersions.

It should perhaps be noted that although the data obtained from fluorescence polarisation spectroscopy of chromophore-labelled fatty-acid molecules incorporated into lipid bilayers and membranes is generally consistent with the lipid intramolecular motions in the systems described above, the size of the chromophore itself appears to severely perturb its local environment in the lipid bilayer as judged by x-ray diffraction and n.m.r. spectroscopic studies of these systems[60]. For this reason, further quantitative description of either intramolecular or intermolecular motion of lipids in natural or model membranes using such probes is probably not warranted at this time.

Lipids whose intramolecular motion is modified by the heteromolecular interaction(s) with other components of the membrane will be referred to in the section on intermolecular motion.

4.2.2 Protein molecules

Evidence obtained from physical techniques concerning the vibrational and rotational motions of either 'membrane-associated' or 'integral-membrane' proteins[23] is rather small. Perhaps the best example available concerning the rotational mobility of an 'integral-membrane' protein is that of the visual pigment rhodopsin in the disc membranes of retinal rods. Lamellar x-ray diffraction information from intact retinal rods[61] and orientated pellets of isolated disc membranes[62] indicates that the membrane probably contains a predominately lipid fatty-acid chain hydrocarbon core. The reduction in the scaled total diffracted intensity from the planar arrangement of rhodopsin molecules (in the case of orientated isolated disc membranes) on increasing the average electron density of the sedimentation medium surrounding the disc membranes indicates that the rhodopsin molecule is embedded *ca.* 14 Å into the lipid hydrocarbon core of the membrane with *ca.* 28 Å of the molecule protruding into the aqueous surface layer of the membrane[5]. The average nearest-neighbour separation for the planar arrangement of the rhodopsin molecules in the isolated disc membranes varies with the pH values of the sedimentation medium indicating that the surface of that portion of the molecule protruding into the aqueous phase consists primarily of polar residues possessing a net negative electric charge at physiological pH values[63].

Absorption polarisation studies of the 11-*cis* retinal chromophore of rhodopsin indicate that rotational motion of the chromophore, which is thought to accurately reflect the motion of the entire rhodopsin molecule, indicates that rhodopsin has a very limited rotational mobility about an axis lying in the plane of the membrane since the plane of the chromophore is restricted to lie within $\sim$20 degrees of the plane of the membrane[64]. Transient photodichroism studies of the chromophore indicate that rhodopsin has a high degree of rotational mobility about an axis normal to the plane of the membrane with a rotational relaxation time of *ca.* 20 μs[65]. The viscosity of the medium in which rhodopsin rotates appears to be about 2 poise from these measurements. These results are in agreement with those obtained from x-ray diffraction studies which indicate that the tumbling of the rhodopsin molecule about an axis in the plane of the disc membrane would be highly unfavourable energetically while rotation about an axis normal to the plane could occur, the viscosity 'seen' for this motion being dominated by that of the lipid hydrocarbon core of the membrane.

Similar studies may be undertaken on other 'membrane-associated' and 'integral-membrane' proteins possessing an intrinsic rigidly-bound chromophore, especially if the membranes can be orientated. Extrinsic chromophores covalently bound to and immobilised on the protein may also be used if the presence of the chromophore does not alter the protein structure or the location of the protein and its interaction with its environment in the membrane. In addition, resonance-enhanced Raman spectroscopy of such an intrinsic chromophore could provide information on the vibrational structure of the chromophore and its local environment in the protein in the natural membrane (or in model membranes) because of the selectivity of the resonance-enhancement effect. Such studies would be similar to those already performed on the 'membrane-associated' protein, cytochrome *c*, and other hemoproteins in solution[66].

4.3 INTERMOLECULAR MOTION OF MEMBRANE COMPONENTS

4.3.1 Translational motion in the plane of the membrane

4.3.1.1 Lipid–lipid interactions

Considerable evidence now exists that phase separations occur among the various lipids in natural membranes and mixed lipid bilayer model membranes as the temperature of the system is lowered below that at which all the chains are melted. Such phase separations have been demonstrated by x-ray diffraction studies of the packing of fatty-acid chains as a function of the temperature in both natural[25] and mixed lipid bilayer membranes[67], and through the co-existence of regions of frozen and melted lipid fatty-acid chains within the same lamellae, the relative amounts of the different phases depending directly on whether the temperature is closer to that at which all the chains are melted or that at which all chains are frozen. Similar results have been obtained by e.p.r. studies of fatty acids and phospholipids spin-labelled in various positions, in which the spin labels appear to be sequestered only into regions of melted chains and excluded from regions of frozen chains at temperatures where the two regions co-exist in the same lamellae, at least in mixed lipid model membranes[68]. Only in a few special cases have co-lattices of hydrocarbon chains in mixed lipid bilayer model membranes been observed when all the chains are frozen. One such example is the co-lattice formed between the isoprenoid chains of reduced ubiquinone UQ_3 and the fatty-acid chains of dipalmitoyl lecithin in their mixed bilayer as determined from x-ray diffraction studies of chain packing and high-resolution electron-density profiles[1a]. Thus, it appears that lipid–lipid interactions are dominated by homomolecular interactions when all the hydrocarbon chains are frozen, except for certain special cases.

At those temperatures at which all the lipid hydrocarbon chains have melted, evidence from both n.m.r. and e.p.r. studies of natural and model membranes suggest ideal mixing of the lipid and lipid-like components in the membrane* as well as the formation of specific complexes. Studies carried out on membrane model systems have suggested the possibility of using high-resolution n.m.r. relaxation studies for assessing the existence and dynamic characteristics of intermolecular magnetic dipole–dipole interactions among

* An example of ideal mixing in mixed lipid bilayers on the n.m.r. time scale of 10^{-9} s is provided by the experimental evidence for lipid translational motion in the plane of the bilayer to be discussed later in this section.

lipids and between lipids and other components of biological membranes. Proton spin–lattice relaxation times, measured alternately in ^{1}H-DPL vesicles and in ^{1}H-DPL vesicles diluted (1:10) in ^{2}H-DPL have shown that at a temperature *ca.* 10 °C above T_m proton–proton intermolecular interactions provide a considerable contribution to the dipolar relaxation of the chains, especially in the region of the terminal methyl group[69]. A study of the temperature dependence of intra- and inter-molecular contributions to the spin–lattice relaxation mechanisms in ^{1}H-DPL vesicles containing various concentrations of ^{2}H- and ^{1}H-palmitic acid chains, the palmitate chains being incorporated into the DPL bilayer in a relatively extended form as judged by x-ray diffraction studies[1a], has proved interesting in that it has suggested that the intermolecular interactions become increasingly less effective as the temperature is further increased with respect to the melting point[70]. Furthermore, the average correlation time for intermolecular motion appears to be less than the rotational intramolecular correlation time, which is not unexpected since the former would arise from chain–chain relative rotational and vibrational motions.

With regard to the translational motion of phospholipids in the plane of lipid bilayer model membrane and at least one natural membrane, both n.m.r. studies on the host phospholipids and e.p.r. studies of the incorporated spin-labelled phospholipids indicate that such a translation occurs with a diffusion constant on the order of 10^{-8} cm^2 s^{-} [68].

Studies of molecular organisation and dynamic modifications induced in phospholipid bilayers by the incorporation of certain natural membrane components such as chlorophylls and steroids could be utilised as the first steps in an attempt to assess the characteristic arrangements and mobilities of the more complex multicomponent natural membranes.

In agreement with the conclusions of x-ray diffraction studies[71] carried out at temperatures below T_m, proton chemical shifts and spin–lattice relaxation times indicate that above T_m chlorophyll molecules are incorporated on both sides of a lecithin bilayer, with the porphyrin rings located in the region of the polar head-groups; the phytyl chains extend into the hydrophobic core thus modifying its dynamic structure and molecular packing to differing extents depending on whether they are introduced into the vesicle as phytol molecules or as the esterified alcohols of chlorophyll[72]. Furthermore, in agreement with the conclusions already described regarding the anisotropic characteristics of the intramolecular motions in pure lecithin bilayers, nuclear magnetic relaxation results obtained in the systems DPL–chlorophyll and DPL-phytol require an analysis involving at least two correlation times, taking into account the anisotropic combination of rotational motions about the bonds and the segmental re-orientation of the rotor axes. The latter motion appears considerably retarded by the introduction of phytyl chains into the lipid bilayers, phytol being more effective than chlorophyll. The rotational mobility of the fatty-acid chains could be isolated from that of the chlorophyll phytyl chains by the incorporation of ^{2}H-chlorophyll into ^{1}H-lecithin vesicles. Under the effect of the intermolecular lecithin–chlorophyll interactions, the intramolecular rotational mobility of the lecithin chains appears to be strongly reduced ($\tau_c > 10^{-9}$ s) whilst, in contrast, the phytyl chains retain a high degree of intramolecular rotational mobility in the bilayer ($\tau_c < 10^{-9}$–10^{-10} s) which is an ideally mixed bilayer on a 10^{-9} s time scale.

Different degrees of anisotropy have been suggested in an attempt to provide a consistent picture of the dynamic structure of lecithin–cholesterol model membranes reconciling the T_1 values[73] with the line-width data[74]. N.m.r. studies on sonicated and unsonicated lecithin–cholesterol co-dispersions have shown a differential broadening in the lecithin spectrum, the greatest broadening corresponding to the signals from the alkyl chains. This broadening has been attributed to a motional restriction imposed on especially the first 10 methylene units due to the effect of apolar interactions involving the steroid molecules[74]. Similar conclusions were deduced from e.p.r. studies on positionally spin-labelled polymethylene chains in cholesterol-containing lecithin bilayers[59] and from x-ray studies through the interpretation of high-resolution electron-density profiles of cholesterol-containing lipid bilayers[75] as well as modified fatty-acid chain packing induced by cholesterol incorporation[24]. Furthermore, the n.m.r. evidence suggested the formation of an equimolar complex (with a lifetime greater than 30 ms) in cholesterol: lecithin co-dispersions with the phosphate and hydroxyl groups probably hydrogen bonded in such a complex. The complex, which becomes less rigid at *ca.* 20 °C above the normal transition temperature for lecithin chains, is practically insensitive to temperature changes below this temperature (in particular, the phase transition is itself removed in 1 : 1 dispersions of cholesterol: DPL)[76]. A cluster-like distribution of such complexes has also been suggested as being formed in those co-dispersions where cholesterol is present in less than equimolar amounts[74]. The greatly reduced value of the proton spin–lattice relaxation times in cholesterol–lecithin relative to that for lecithin bilayers has also been interpreted qualitatively[73] on the basis of a considerable reduction in the mobility especially along the chains. However, the changes in T_1 appear inadequate to explain the large change (mentioned above) in the line widths caused by the introduction of cholesterol, as long as a single relaxation mechanism is taken into account. It is suggested that the reduction in the proton T_1 values is mainly related to a slowing down of the rotational motions of several of the methylene groups around the C–C bonds due to complex formation; on the other hand, the presence of such rigid complexes in the bilayer also reduces the ease of segmental re-orientation along the chains, thus strongly affecting the lecithin line widths in particular.

In view of the results obtained from these n.m.r. studies on model membranes, careful analysis of n.m.r. relaxation data may allow an accurate determination of certain lipid–lipid and lipid–lipid-like component interactions in natural membranes as well.

4.3.1.2 Lipid–protein interactions

A comparison of the n.m.r. relaxation times in ^{13}C-enriched intact membranes and in the vesicles formed with their extracted lipid appears to be a useful approach for assessing the dynamic structure of lipid regions in natural membranes, even in relation to lipid–protein intermolecular interactions and relative mobility. The use of specific ^{13}C enrichment in n.m.r. techniques appears promising as an attempt to extend n.m.r. methods from model systems to biological membranes, the main advantages consisting of the enhanced signal-to-noise ratios in enriched *vs.* naturally-abundant membranes, and the possibility of resolving otherwise overlapping resonances[47, 77].

The resonance from ^{13}C-enriched lipids at the $>C=O$ position may be observed only above the lipid thermal transition in *Acholeplasma laidlawii* membranes[47]. Below the transition, the signal broadens and gradually disappears in a similar fashion to the situation with the vesicles of extracted membrane lipids. The fact that the signal line width does not show any transition corresponding to the thermal denaturation of the membrane proteins has been taken as an indication that the latter have only a minor effect on lipid–lipid interactions.

Well-defined resonances have also been observed from membrane lipids in sarcoplasmic reticular membranes[77]. ^{13}C T_1 values measured for the chains of extracted lipids organised in vesicles are very close to those found in the intact membranes, indicating a strictly similar organisation and packing of lipids in the two systems at the level of the hydrophobic core. Comparison of ^{13}C and ^{1}H spectra arising from such chains leads to the interesting observation that about three-quarters of the lipid chains may be detected in the former spectra but only about one quarter in the latter. The appearance of only 25% of the lipid chain protons has been tentatively attributed either to the fraction of lipids in a single bilayer structure which are unperturbed by the presence of membrane proteins, or to the fact that a fraction of the lipids exists as a 'superfluid' which is directly related to the calcium transport system. The resonance broadening of the other protons has been attributed to nuclear dipole–dipole intermolecular interaction involving membrane proteins which is more effectively exhibited in ^{1}H rather than ^{13}C spectra. A comparison of the ^{13}C *vs.* ^{1}H spectra arising from polar head-groups has showed that the $N^{+}(CH_3)_3$ signals, which are fully developed in both the resonances in the natural as well as in the synthetic lecithin mixtures, can, however, only be detected in the ^{13}C spectra arising from the intact membranes. This fact, together with the enhanced ^{13}C T_1 rate of this group in intact membranes relative to that in extracted lipid bilayers, has suggested the existence of intimate interactions between the phospholipid head-groups and the membrane proteins.

Finally, the possibility of relating relaxation measurements to the characteristic organisation of lipids (such as the existence of different lipids clustering in mosaic regions within the bilayer) has been suggested by the observation of a multiple-relaxation time behaviour for the chemical groups of various lipids from both intact membranes and extracted lipid vesicles. However, more detailed analysis is required on model systems before arriving at firm conclusions. Moreover, although promising, these techniques have not yet provided any direct or quantitative information on the reciprocal motions of lipids and proteins in membranes.

More direct evidence on lipid–protein interactions has been recently obtained from e.p.r. studies on membrane model systems formed by cytochrome oxidase (from beef heart mitochondria) and phospholipid organised in vesicular structures[78]. The dynamic properties of the phospholipid bilayers, especialy in relation to the characteristics of the lipid–protein interfacial region, have been explored with the use of lipid spin labels. If it is assumed that cytochrome oxidase extends into the lipid bilayer, analysis of the e.p.r. spectra suggests the presence of two distinct phospholipid regions with considerably different fluidity characteristics: (a) a boundary lipid region, consisting essentially of the first lipid layer in contact with the protein, which is almost completely immobile (the principal value of the electron–nuclear hyperfine interaction measured along the z-axis of the radical being very close to that found for a crystalline state) and (b) a much more fluid region composed of the remaining lipids, whose dynamic structure is very similar to that exhibited by the liposomes of egg lecithin. Lipid exchange may occur between the two regions, the immobilised lipid: protein ratio remaining approximately constant while the mobile lipid: protein ratio is variable being greater than or equal to zero. From a consideration of the 'intrinsic' character of this and many other membrane proteins, it has been suggested that the existence of immobilised boundary lipids in exchange with a more mobile lipid region may be a rather general phenomenon occurring in biological membranes.

4.3.1.3 Protein–protein interactions

Translational mobility of the 'integral-membrane' protein rhodopsin relative to other rhodopsin molecules has been inferred from x-ray diffraction studies[4a, 63]. A Fourier analysis of x-ray diffraction data arising from the planar arrangement of rhodopsin molecules in orientated isolated disc membranes from retinal rods has provided a description of the local planar arrangement of rhodopsin molecules in terms of their average number and separation of nearest-neighbours. The fact that temperature, pH and variation of ionic strength are all capable of altering the particular local planar arrangement or clustering of the rhodopsin molecules without any detectable alteration in the average lipid fatty-acid chain packing in the plane of the disc membrane implies a lateral mobility for these protein molecules in the plane of the membrane. Such a lateral mobility would be expected to depend primarily on the viscosity of the lipid hydrocarbon core of the disc membrane arising from the substantial embedding of the rhodopsin molecule in the core as mentioned earlier and on the net electric charge occurring on that portion of the molecule protruding into the aqueous surface layer of the membrane, as well as on the shielding of this electric charge by counter-ions in this layer, due to nearest-neighbour interactions.

In these isolated disc membranes, particularly in view of the temperature dependence of the local planar arrangement of rhodopsin, it would appear that the molecules occur in a planar liquid-like state. The planar arrangement of rhodopsin molecules in disk membranes of retinal rods in the intact retina based on x-ray diffraction results may have even shorter range order more similar to that of a planar real-gas arrangement with no well-defined nearest neighbours, but only a distance of closest approach[1b, 36]. In contrast to this, a similar 'intrinsic-membrane' protein of the *Halobacterium halobium* membrane appears to occur in a planar crystalline lattice in the membrane surface with long-range statistical and thermal order[4b].

Other physical evidence for the translational mobility of 'intrinsic-membrane' proteins is provided by the observation from fluorescence microscopic studies of the diffusion of fluorescent antibody-labelled membrane antigens over the surface of the membrane[79]. These observations must remain somewhat qualitative since even under the best conditions what is observed is the diffusion of an antigen–antibody complex in which the antigen and its environment in the membrane may be severely perturbed relative to the natural membrane through the formation of such a complex.

Quasi-elastic neutron scattering dominated by 'intrinsic-membrane' protein, e.g. rhodopsin in orientated isolated disc membranes or model disc membranes, may provide direct evidence regarding such translation mobilities in the plane of the membrane.

4.3.2 Translational motion normal to the plane of the membrane

4.3.2.1 Lipids

The transverse motion of lipids across the membrane has been investigated by the use of e.p.r. spin-label experiments. The half-time value for the transition of phospholipids from the inside to the outside of vesicles has been measured by following the decay of the asymmetry created by sodium ascorbate reduction in the distribution of paramagnetic phospholipids (spin-labelled at the head-group) between the two apposed monolayers[80] ($t_{1/2} = 6.5$ h at 30 °C). The probability for the occurrence of a transverse transition from the inner to the external monolayer has been assessed as 0.07 h^{-1} and 0.04 h^{-1} in the opposite direction. The rate of such transitions has also been shown to depend on the composition of the lipid bilayer 'through which' the transition must occur.

4.3.2.2 Proteins

This mobility of an 'intrinsic-membrane' protein has been demonstrated by x-ray diffraction experiments on rhodopsin molecules in isolated disc membranes from retinal rods. Using the method described earlier for determining the extent to which rhodopsin protrudes into the aqueous surface layer of the membrane, it was shown that rhodopsin bleaching results in a decreased net electric charge on the molecule and a model-independent sinking of the rhodopsin molecule further into the lipid hydrocarbon core of the membrane occurs, i.e. rhodopsin protrudes to a lesser extent into the aqueous surface layer after bleaching[5, 63]. A similar result should also be obtained by the use of higher-resolution electron-density profiles for the disc membrane and rapid x-ray detection devices[81]. Forster energy transfer between a chromophore on the surface of rhodopsin exposed to the aqueous phase at the membrane surface and an incorporated extrinsic fluorescent probe molecule could also provide a method for the rapid detection of such motion provided that the problems created by the perturbation of the membrane by probe labelling and incorporation and the mutual orientation factor of the transition dipoles can be resolved.

4.4 SUMMARY

In this article we have attempted to demonstrate that both intramolecular and intermolecular mobilities of lipid and protein components of biological membranes may be detected in certain special systems particularly suitable to the physical techniques utilised. General extensions to other lipid, lipid-like, 'intrinsic' and 'associated' protein membrane components, as well as to more complex forms of these motions which are easily imaginable, may not be warranted at this time. However, there seems to be evidence based on reasonably 'direct' physical techniques that both lipid and protein components of biological membranes may possess varying degrees of intra- and inter-molecular mobility (or immobility) depending on their collective intermolecular interactions in the membrane. In terms of these intermolecular interactions, both homo- and hetero-molecular, one should consider the possibility of both 'fast' (10^{-8} s) and 'slow' (10^{+4} s) lipid–lipid, lipid–protein and protein–protein interactions, considering both 'membrane-associated' and 'integral-membrane' proteins, within the total dynamic structure of the membrane.

References

1. (a) Cain, J., Santillan, G. and Blasie, J. K. (1972). *Membrane Research* (C. F. Fox, editor), (New York: Academic Press); (b) Blaurock, A. E. and Wilkins, M. H, F. (1969). *Nature (London)*, **223,** 906; (c) Cain, J., Santillan, G. and Blasie, J. K., unpublished results on orientated multilayers of isolated disk membranes from frog retinal rods and chromatophore membrane from *Chromatium*
2. Lesslauer, W. and Blasie, J. K. (1971). *Acta Crystallogr.*, **A27,** 456
3. (a) Lesslauer, W. and Blasie, J. K. (1972). *Biophys. J.*, **12,** 175; (b) Hosemann, R. and Bagchi, D. N. (1962). *Direct Analysis of Diffraction by Matter* (Amsterdam: North Holland Publishing Co.)
4. (a) Blasie, J. K. and Worthington, C. F. (1969). *J. Mol. Biol.*, **39,** 417; (b) Blaurock, A. E. and Stoeckenius, W. (1971). *Nature (New Biol.)*, **233,** 152
5. Blasie, J. K. (1972). *Biophys. J.*, **12,** 191
6. Bacon, G. E. (1962). *Neutron Diffraction* (London: Oxford University Press)
7. (a) Egelstaff, P. A. (1965). *Thermal Neutron Scattering* (London: Academic Press); (b) Marshall, W. and Lovesey, S. W. (1971). *Theory of Thermal Neutron Scattering* (London: Oxford University Press)
8. Wilkins, M. H. F. *et al.*, unpublished work
9. (a) Albrecht, A. C. (1961). *J. Chem. Phys.*, **34,** 5, 1476; (b) Lippert, J. L. and Peticolas, W. L. (1972). *Biochim. Biophys. Acta*, **282,** 8
10. Jost, P. and Griffith, O. H. (1972). *Methods Pharmacol.*, **2,** 223.
11. Abragam, A. (1961). *The Principles of Nuclear Magnetism* (London, New York: Oxford University Press)
12. Slichter, C. P. (1963). *Principles of Magnetic Resonance* (New York: Harper and Row)
13. Carrington, A. and McLachlan, A. D. (1967). *Introduction to Magnetic Resonance* (New York: Harper and Row)
14. Jackman, L. M. and Sternell, S. (1969). *Applications of Nuclear Magnetic Resonance Spectroscopy in Organic Chemistry* (Oxford: Pergamon)
15. Finer, E. G., Flook, A. G. and Hauser, H. (1972). *Biochim. Biophys. Acta*, **260,** 49
16. McLaughlin, A., Podo, F. and Blasie, J. K. (1973). *Biophys. J.*, **13,** 251, in preparation

17. Podo, F., Ray, A and Nemethy, G. (1972). *164th ACS National Meeting, New York*
18. Farrar, T. C. and Becker, E. D. (1971). *Pulse and Fourier Transform NMR* (New York: Academic Press)
19. Woessner, D. E. (1962). *J. Chem. Phys.*, **37,** 647
20. Woessner, D. E. (1962). *J. Chem. Phys.*, **36,** 1
21. Bradley, R. A., Martin, W. G. and Schneider, H. (1973). *Biochemistry*, **12,** 2, 273
22. Radda, G. K. and Vanderkooi, J. (1972). *Biochim. Biophys. Acta*, **265,** 509
23. Singer, S. G. and Nicolson, G. L. (1972). *Science*, **175,** 720
24. (a) Levine, Y. K. and Wilkins, M. H. F. (1971). *Nature* (*New Biol.*), **230,** 69; (b) Wilkins, M. H. F., Blaurock, A. E. and Engelman, D. M. (1971). *Nature* (*New Biol.*), **230,** 72
25. Engelman, D. M. (1972). *J. Mol. Biol.*, **58,** 153
26. Chan, S. I., Feigeuson, G. W. and Seiter, C. H. A. (1971). *Nature* (*London*), **231,** 110
27. Veksli, Z., Salsbury, N. J. and Chapman, D. (1969). *Biochim. Biophys. Acta*, **183,** 434
28. Metcalfe, J. C., Birdsall, N. J. M., Feeney, J., Lee, A. G., Levine, Y. K. and Partington, P. (1971). *Nature* (*London*), **233,** 199
29. Levine, Y. K., Birdsall, N. J. M., Lee, A. G. and Metcalfe, J. C. (1972). *Biochemistry*, **11,** 1416
30. Bloembergen, N. (1949). *Physica*, **15,** 386
31. Lee, A. G., Birdsall, N. J. M., Levine, Y. K. and Metcalfe, J. C. (1972). *Biochim. Biophys. Acta*, **255,** 43
32. Sheard, B. (1969). *Nature* (*London*), **223,** 1057
33. Finer, E. G., Flook, A. G. and Hauser, H. (1971). *FEBS Lett.*, **18,** 331
34. Huang, C. (1969). *Biochemistry*, **8,** 344
35. Finer, E. G., Flook, A. G. and Hauser, H. (1972). *Biochim. Biophys. Acta*, **260,** 49 and 59
36. Penkett, S. A., Flook, A. G. and Chapman D. (1968). *Chem. Phys. Lipids*, **2,** 273
37. McLaughlin, A., Podo, F. and Blasie, J. K. (1973). *Biophys. J.*, **13,** 251a, in preparation
38. Levine, Y. K., Partington, P., Roberts, G. C. K., Birdsall, N. J. M., Lee, A. G. and Metcalfe J. C. (1972). *FEBS Lett.*, **23,** 203
39. Horwitz, A., Horsley, W. J. and Klein, M. P. (1972). *Proc. Nat. Acad. Sci. USA*, **69,** 590
40. Luzzati, V., Reiss-Husson, F., Rivas, E. and Gulek-Krywickii, T. (1966). *Ann. N.Y. Acad. Sci.*, **137,** 409
41. Bloembergen, N., Purcell, G. M. and Pound, R. V. (1948). *Phys. Rev.*, **73,** 679
42. Kubo, R. and Tomita, K. (1959). *J. Phys. Sci.* (*Japan*), **9,** 888
43. Solomon, I. (1955). *Phys. Rev.*, **99,** 559
44. Carr, H. Y. and Purcell, E. M. (1954). *Phys. Rev.*, **94,** 630
45. Chan, S. I., Seiter, C. H. A. and Feigeuson, G. W. (1972). *Biochem. Biophys. Res. Commun.*, **46,** 1488
46. Woessner, D. E. (1962). *J. Chem. Phys.*, **36,** 1
47. Levine, Y. K., Partington, P., Roberts, G. C. K., Birdsall, N. J. M., Lee, A. G. and Metcalfe, J. C. (1972). *FEBS Lett.*, **23,** 335
48. Hubbell, W. L. and McConnell, H. M. (1971). *J. Amer. Chem. Soc.*, **93,** 314
49. Van Putte, K. (1971). *J. Magn. Resonance*, **2,** 23
50. Abe, A., Ternigan, R. L. and Flory, P. J. (1966). *J. Amer. Chem. Soc.*, **88,** 631
51. Santillan, G. and Blasie, J. K., unpublished results
52. Podo, F. and Blasie, J. K., unpublished results
53. Lippert, J. L. and Peticolas, W. L. (1971). *Proc. Nat. Acad. Sci., USA*, **68,** 7
54. Mendelsohn, R. (1972). *Biochim. Biophys. Acta*, **290,** 15
55. McConnell, H. M. and McFarland, B. G. (1970). *Quart. Res. Biophys.*, **3,** 91
56. Rozantzev, E. G. and Neiman, M. B. (1966). *Tetrahedron*, **20,** 131
57. Hubbell, W. L. and McConnell, H. M. (1968). *Proc. Nat. Acad. Sci. USA*, **61,** 12
58. Hubbell, W. L. and McConnell, H. M. (1969). *Proc. Nat. Acad. Sci. USA*, **64,** 20
59. Hubbell, W. L. and McConnell, H. M. (1971). *J. Amer. Chem. Soc.*, **93,** 314
60. Podo, F. and Blasie, J. K., unpublished results
61. Worthington, C. R. and Gros, W. J. (1972). *Biophys. J.*, **12,** 255a
62. Santillan, G. and Blasie, J. K., unpublished results
63. Blasie, J. K. (1972). *Biophys. J.*, **12,** 205
64. (a) Liebman, P. A. (1962). *Biophys. J.*, **2,** 161; (b) Liebman, P. A. (1972). *Handbook of Sensory Physiology*, Vol. VIII/1, *Photochemistry of Vision* (H. J. A. Dartnall, editor), (Berlin: Springer-Verlag)

65. (a) Cone, R. A. (1972). *Nature* (*New Biol.*), **236,** 39; (b) Brown, P. K. (1972). *Nature* (*New Biol.*), **236,** 35
66. Spiro, T. G. and Strekas, T. C. (1972). *Proc. Nat. Acad. Sci. USA*, **69,** 9, 2622
67. Santillan, G. and Blasie, J. K., unpublished results on mixed lipid bilayers of egg lecithin or dioleoylecithin containing incorporated saturated fatty acid molecules of varying chain length
68. McConnell, H. M., Devaux, P. and Scandella, C. (1972). *Membrane Biology* (C. F. Fox, editor), (New York: Academic Press)
69. Metcalfe, G. C., Birdsall, N. J. M. and Lee, A. G. (1972). *N.Y. Acad. Sci. Meeting, New York*
70. Podo, F., Santillan, G. and Blasie, J. K., unpublished results
71. Cain, J. E. and Blasie, J. K. (1972). *Biophys. J.*, **12,** 45a, in preparation
72. Podo, F. and Blasie, J. K. (1973). *Biophys. J.*, **13,** 251a, in preparation
73. Lee, A. G., Birdsall, H. J. M., Levine, Y. K. and Metcalfe, J. C. (1972). *Biochim. Biophys. Acta*, **255,** 43
74. Darke, A., Finer, E. G., Flook, A. G. and Phillips, M. C. (1972). *J. Mol. Biol.*, **63,** 265
75. Blasie, J. K., Goldman, D. E., Chacko, G. and Davey, M. M. (1972). *Biophys. J.*, **12,** 253a, in preparation
76. Ladbrook, B. D., Williams, R. M. and Chapman, D. (1968). *Biochim. Biophys. Acta*, **150,** 311
77. Robinson, J. D., Birdsall, N. J. M., Lee, A. G. and Metcalfe, J. C. (1973). *Biochemistry*, **11,** 1416
78. Jost, P. C., Griffith, O. H., Copaldi, R. A. and Vanderkooi, G. (1973). *Proc. Nat. Acad. Sci. USA*, **70,** 480
79. Edidin, M. (1972). *Membrane Research* (C. F. Fox, editor), (New York: Academic Press)
80. Kornberg, R. D. and McConnell, H. M. (1971). *Biochemistry*, **10,** 111
81. Dupont, Y., Gabriel, A., Chabre, M., Gulik-Krywickii, T. and Scheckter, E. (1972). *Nature* (*London*), **238,** 331

5
Molecular Orientation of Proteins in Membranes

V. T. MARCHESI
Yale University

5.1 INTRODUCTION

Proteins make up approximately one-half of the mass of most animal surface membranes. These molecules, which seem to fall into two sub-classes as described below, are associated in some as yet undefined way with the lipids of the membrane, and together both form a stable–selective barrier at the cell periphery. Our ideas as to how such lipid–protein complexes are held together in the membrane are still largely speculative, and this state of uncertainty will remain until we learn more about membrane proteins and how they might be arranged in different cells. Among the many unanswered questions is whether membrane proteins have a unique chemical composition which makes them especially suitable to form complexes with lipids. It has been suggested, for example, that such proteins might have a high content of non-polar or hydrophobic amino acids which might account for their apparent insolubility in aqueous solutions and their capacity to stick to lipids. This idea initially gained some acceptance in the past on the basis of preliminary attempts to isolate so-called 'structural proteins' from membranes, which ostensibly had these very features, but more recent studies have failed to confirm these findings.

This review is an attempt to analyse some of the recent experimental data concerning the number and types of polypeptide chains which form the cell membranes of human red blood cells and to review the available evidence concerning their orientation in the membrane. The red cell membrane has been chosen as a model system for studies on the mammalian cell membrane because of its availability and apparent simplicity. This particular membrane may not be representative of all or even most surface membranes, but at the present time it still represents the most heavily investigated system and most of our ideas about membrane organisation are based largely on these results.

5.2 THE LIPID BILAYER AS A BASIC FRAMEWORK

Most of the phospholipids bound to red cell membranes appear to be arranged in the form of a bilayer as depicted schematically in Figure 5.1. This idea was based originally on an ingenious experiment performed by Gorter and Grendel[1], in which it was observed that the surface area of lipids extracted from red cell membranes was approximately twice that of the surface areas of the original cells from which they were derived. Although many interpretations could account for these findings these investigators proposed that the lipids formed a double layer around the cell essentially as depicted in this diagram. Many arguments have been advanced both in favour and against this interpretation[2-4].

It has been argued that a bilayer arrangement of lipids would be thermodynamically stable under physiological conditions and would also impart to the membranes the electrical and permeability properties which are observed in living cells[5]. Further support for this model comes from x-ray diffraction data[6-8], the results of freeze-etching and electron microscopy[9,10], and many recent attempts to analyse natural and artificial membranes using spin resonance and other types of physical probes[11-15]. The sum of these results strongly support the idea that the *bulk* of the membrane lipids are in this general configuration. It is important, however, to consider the possibility that a small proportion of the membrane lipids may have other configurations (e.g. globular micelle type[16] which cannot be detected by these techniques, but which might significantly affect the functional properties of the membrane. Local variations in the arrangement of lipids might exert a significant effect on transport across the lipid bilayer or might even be involved in the formation of the often-postulated aqueous channels across membranes.

If such sites represent a small fraction of the membrane barrier, which would be consistent with physiological data, the bilayer arrangement of lipids could represent the impermeable or 'insulating' region of the membrane, by analogy with the myelin sheath, while the 'functional' areas could have other, as yet unidentified, specific lipid arrangements. Some support for the idea that regional variations in the arrangement of membrane lipids may exist

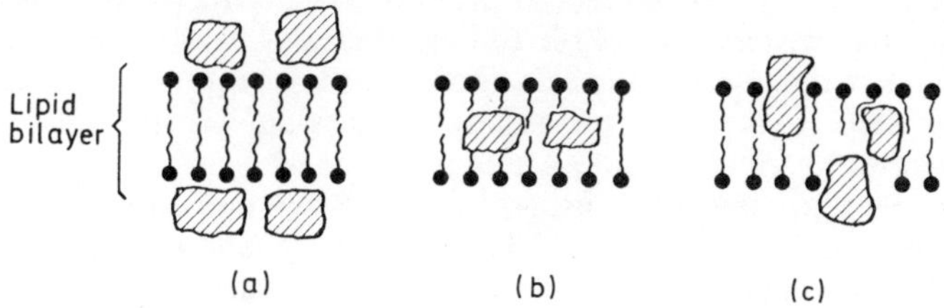

Figure 5.1 Schematic diagrams of possible ways which membrane proteins (cross-hatched) might associate with a phospholipid bilayer

is discussed below in terms of how certain specific membrane proteins might exert a modulating influence on their neighbouring lipids.

Many recent studies on the behaviour of electron spin resonance probes in artificial membrane systems also indicate that lipids of membranes are in a more dynamic state than previously suspected. The hydrocarbon chains of phospholipids appear to flex at rapid rates, especially at sites removed from the polar head groups, and individual molecules can shift laterally within the plane of the membrane at moderately fast rates[17]. Interestingly, both the flexing movements of hydrocarbon chains and the lateral shifts of individual molecules appear to take place only within the plane of individual leaflets, since other studies with e.s.r. probes suggest that phospholipid molecules in one half of the bilayer probably do not shift over to the other side (so-called flip–flop motion) at significant rates[18]. The lack of significant movement of lipids from one leaflet of a bilayer to the other is consistent with the recent suggestion by Bretscher[19] that particular types of phospholipids are confined to the inner leaflets of the red cell membrane.

If we accept the idea that the cell membranes have a lipid bilayer as a foundation, it is possible to conceive of at least three general ways in which proteins might be inserted into this substructure. These are depicted, very schematically, in Figure 5.1.

Modes A and B represent the two most divergent possibilities in which all of the proteins are either entirely outside or entirely inside the lipid matrix. The protein–lipid interactions in A would be primarily electrostatic, or by hydrogen bonding and one might predict that the proteins could be dissociated (and solubilised) from the lipid by manipulating the ionic strength and pH of the medium; this need not necessarily be so, however, if some of these proteins have special affinities for the bases of different phospholipids. Proteins bound to the polar groups of lipids would not necessarily have an unusual amount of non-polar amino acids or any other unique features.

Protein–lipid interactions in Mode B would presumably involve associations between non-polar side chains of amino acids and the hydrocarbon chains of fatty acids. These 'hydrophobic' associations would not be readily disrupted by manipulating the ionic strength or pH of the medium, but if the proteins were separated from the lipids, they would tend to form insoluble aggregates in aqueous solutions unless appropriate dissociating agents or detergents were present. Such proteins might also have a high proportion of non-polar amino acids.

Both modes A and B have been proposed in the recent past as possible models for the organisation of the cell membrane. A was considered a likely possibility at a time when little was known about membrane proteins, and it was suggested originally as an attempt to account for the relatively low surface tension of cells[20]. Mode B was proposed as a result of studies which suggested that a substantial proportion of 'structural protein' existed in membranes which were composed largely of hydrophobic amino acids[21-23]. This model became less attractive when more detailed studies of membrane proteins failed to confirm the earlier results[24]. Now it appears that many of the studies on so-called 'structural proteins' were carried out on denatured enzymes which were insoluble as a result of denaturation rather than due to an especially high content of hydrophobic amino acids.

Mode C has emerged as a compromise candidate which incorporates aspects of both A and B and, as we shall see below, is more consistent with most of the more recent experimental findings. This general model has been described in some detail by Singer and Nicolson[25] as the *Fluid–Mosaic Model* and will be discussed in more detail below.

5.3 EXPERIMENTAL APPROACHES TO THE STUDY OF PROTEIN ORIENTATION IN MEMBRANES

Table 5.1 lists some of the approaches which have been used to study the orientation of membrane proteins.

Table 5.1

(a)	*Physical analysis of protein conformation*
	x-Ray diffraction
	Calorimetry
	Spectroscopy (Ord, cd, ir)
	Probe analysis (fluorescence, electron spin resonance)
	Nuclear magnetic resonance
(b)	*Localisation by electron microscopy*
	Specific labelling (antibody, lectins, histochemical) after differential extraction or chemical/enzymatic modifications
(c)	*Chemical modification on* in situ *proteins*
	Radio-labelling of 'accessible' polypeptide chains
	Enzymatic digestion of 'accessible' polypeptide chains
	Cross-linking with bifunctional reagents

5.3.1 Physical studies

Many attempts have (and are still being) made to adapt some of the standard techniques of protein analysis to intact and partially disrupted membranes. Some of these are listed under Group (a) Table 5.1. The results of these approaches have been described in many recent monographs and review articles[26-28], and only the findings which bear on the discussion below will be cited here.

x-Ray diffraction analysis has been a useful way to determine the probable order of the lipids in a variety of biological membranes, which include those of the myelin sheath[7], the retinal rod outer segment[29] and mycoplasma[8]. So far, however, x-ray diffraction studies have not provided very much information about the location and molecular orientation of membrane proteins. Diffraction patterns of myelin are consistent with the idea that protein molecules are associated in some way with the polar groups of lipids but more detailed information is lacking. Some investigators feel that x-ray analysis can be used to detect changes in the lipid-rhodopsin interactions in light-activated retinal rod membranes[30], but other investigators are sceptical of these claims. Studies on the orientation of proteins in red cell membranes have not been rewarding so far.

Spectroscopic studies were initially thought to offer great promise in the study of membrane protein conformation since it appeared that conformational analysis could be carried out on membrane proteins while they were still *in situ* rather than after isolation. It was reasoned that studies of intact or near-intact membranes would be much more informative since it seemed likely that protein conformations would be modified by the methods commonly used to isolate and purify membrane components.

O.R.D. and C.D. spectra of intact red cell ghosts suggested that a substantial amount (perhaps 40 %) of the protein was in the α-helix conformation[31-33]. Certain anomalies were also noted in these spectra which were interpreted as evidence in favour of the idea that most of the polypeptides in the α-helical conformation were in close proximity to the hydrocarbon segments of phospholipids. On the basis of this interpretation, it was suggested that much of the protein was situated within the lipid bilayer, essentially as depicted in Figure 5.1b. More recent work suggests that the anomalous spectra are probably due to light scattering effects of the membranes themselves[34] rather than to some special environment of the helical peptides. Some investigators also question whether conclusions concerning the degree of helicity of membrane proteins can be accurately estimated by these spectroscopic techniques. This point is still open.

Spectroscopic studies also suggest that membrane proteins do not exist in the β or extended conformation to any appreciable extent[35]. This finding, if valid, rules out the earlier suggestion that membrane proteins might exist as macromolecular sheets which 'coated' the polar groups of phospholipid molecules. Studies using enzymatic probes and labelling reagents (described below) also make this idea extremely unlikely. A new technique employing energy-transfer measurements between fluorescent probes has recently been described by Stryer, and preliminary results on the application of this approach to the orientation of rhodopsin molecules in rod membranes appear promising[36].

5.3.2 Electron microscopy

The electron microscope has great potential as a tool to study the distribution and orientation of proteins in membranes for a variety of reasons. Most instruments have the resolving capability (5–10 Å) to examine individual protein molecules or even peptide segments, and they have, in addition, the great advantage of being able to study individual cells or specific parts of their membranes. In spite of these assets, it is surprising how little information about membrane proteins has been obtained with this instrument. This lack of success is probably due to our inability to devise appropriate ways to prepare and study biological material so that specific proteins can be identified and studied.

At the present time, we have only a few ways to identify individual proteins in thin sections of membranes or on freeze-etched preparations, and these are almost all based on the use of ferritin-conjugates of different reagents. Antibodies directed against specific components can be visualised by preparing ferritin or peroxidase conjugates and these can be used to identify the general location of the antigenic groups in or on the membrane. The main obstacle to this technique is the lack of specific antisera for different membrane proteins. Conjugates of lectins, specific for a variety of carbohydrate groups, have also been prepared, and these have been used to localise glycoproteins with some encouraging success. Histochemical methods for identifying certain chemical groups (such as carboxyls with low pK values) or certain enzymes are also available[37,38], but these methods suffer from their relative lack of specificity.

In some cases, ultrastructural studies have been carried out on membranes which have been modified by specific enzymes or differential extraction procedures to find out whether the selective removal of particular proteins results in a change in the ultrastructural appearance of the membrane. This approach has been used with great success in the past when dealing with the relationships between cytoplasmic organelles, but interpretation becomes more difficult when applied to membrane structure since it is likely that subtle changes in molecular orientation or arrangement of proteins would not be detectable with our present technology.

Recently, two new ways to study the topographical distribution of receptor sites on membranes by electron microscopy have been developed which have certain advantages over the older methods. Nicolson and Singer[39] have devised an ingeniously simple way to study the location of antigenic and other receptor activities on the surfaces of red cells based on the fact that when red cells are dropped on a hypotonic medium some of the cells lyse and flatten into pancake-like structures. These can be collected on collodion-coated electron microscope grids and then treated with different ferritin-conjugated labelling reagents (antibodies or lectins) which bind to the appropriate sites on the exposed membrane. The distribution of the ferritin label is detected on the basis of its iron core with a standard transmission electron

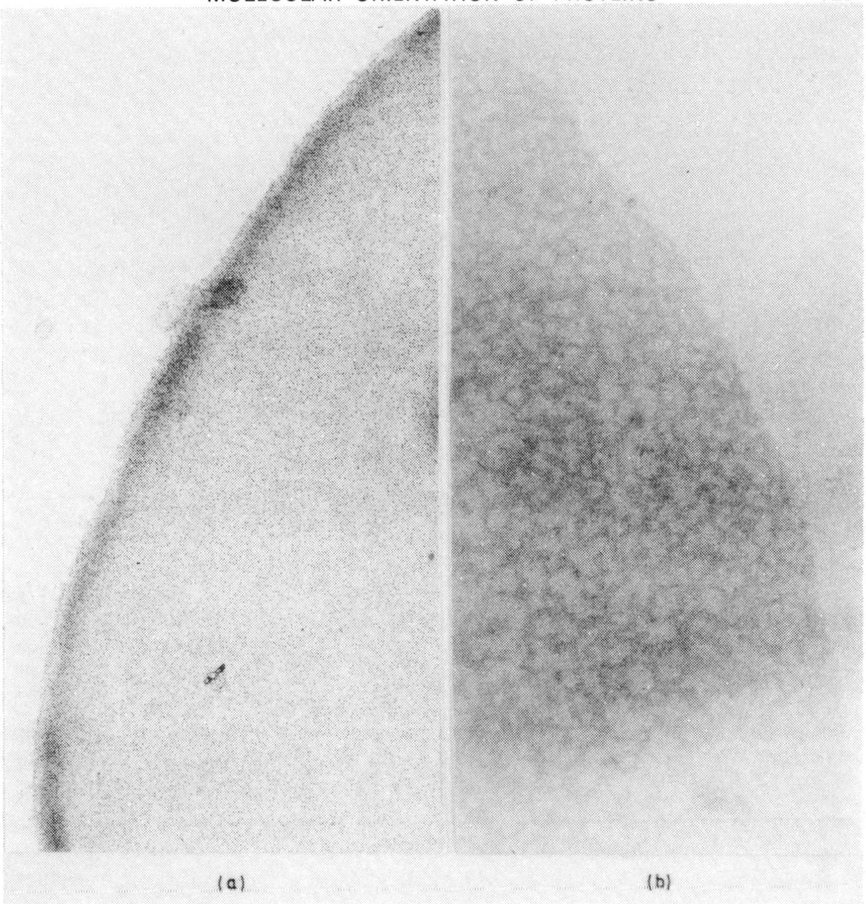

Figure 5.2 Electron micrographs of red cell ghost membranes prepared according to the technique of Nicolson and Singer[39]. The electron-dense iron cores of ferritin are distributed uniformly over the membrane surface when cells are treated with conjugates of ferritin and purified wheat germ agglutinin (a); this is readily distinguishable from unstained membranes prepared by the same technique (b)

microscope. Examples of red cell membranes labelled by this procedure are shown in Figure 5.2. Since large areas of membrane surface can be examined it is much easier to determine patterns of receptors than would be possible by a reconstruction based on multiple electron micrographs taken of serially-examined thin sections. The main disadvantage of this technique is that it depends upon the capacity of cells to lyse and form flattened regions of membrane which are thin enough to be examined without embedding and sectioning. Many cells do not behave in this way.

Since the labelling of receptor sites on the cell surfaces takes place after the cells have been spread on hypotonic medium and then partially dried on coated grids, the question has been raised as to whether rearrangements might take place in the membrane during one of these steps. Although

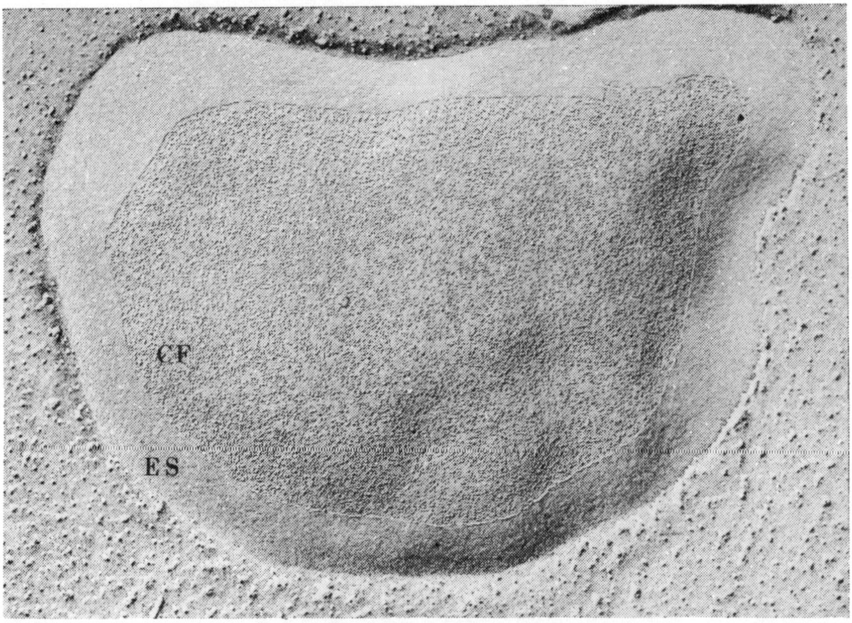

Figure 5.3 Electron micrograph of a red cell ghost membrane prepared by the freeze-etching technique. Cleavage of the membrane results in the removal of part of the outer half of the membrane and exposes an internal region, called the cleavage face (CF), which contains globular structures known as intramembraneous particles. Etching exposes the true external surface of the cell (ES) which appears relatively smooth, *ca.* × 50 000 (reduced $\frac{8}{10}$ths on reproduction). (From Tillack and Marchesi[41], by courtesy of Little, Brown and Co.)

there is no clear evidence that artifactual shifts of receptor sites do occur, this possibility has not been rigorously excluded.

Freeze-etching is another way to prepare cells so that large areas of surface membrane can be examined with the conventional electron microscope. Although this technique has been available for some years, its great utility for membrane studies has only recently become evident. Cells or membrane fragments can be frozen rapidly, either with or without chemical fixation, and then subjected to a fracturing step, usually with a microtome knife inside a temperature-controlled high vacuum chamber, which splits individual membranes along a plane more or less down the middle of the membrane[40,41]. As a result of this splitting, large regions of membranes *within* t he lipid bilayer are exposed and these can be replicated by plantinum–carbon shadowing. After internal regions of the membrane are exposed by this fracturing step a process called 'deep-etching' can be carried out on the same specimen which exposes either the true external surfaces of the cells, or the true cytoplasmic surfaces of the membranes, depending upon the position of the cell relative to the fracture plane.

As a result of combining both processes, *freeze-cleavage* and *deep-etching*, on the same cell a replica can be obtained which provides a remarkably clear picture of both the external surfaces of cells and regions of the membrane within the bilayer. An example of such a replica is shown in Figure 5.3.

Several new features of membrane structure were uncovered by this technique, the principal one being the discovery that globular units of *ca.* 75 Å in diameter are present within the bilayer region of the membrane (Figure 5.3). The nature and significance of these structures will be discussed in more detail below. Large areas of the external surface of cells can be exposed by etching, and macromolecular labels, previously applied to specific receptors on the cell

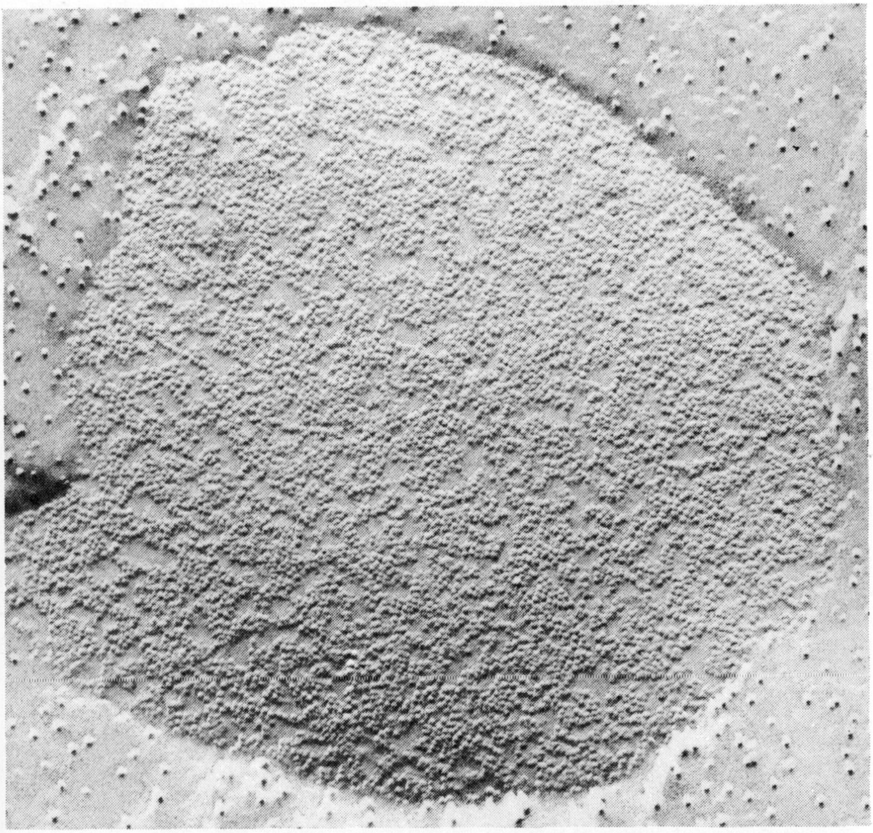

Figure 5.4 Electron micrograph of a red cell ghost incubated with ferritin–phytohaemagglutinin conjugate and prepared by freeze-etching. The conjugates appear as globular bodies distributed uniformly over the external surface of the cell. This surface of the membrane was exposed during the deep-etching step

surface, can be visualised solely on the basis of their molecular mass. Conjugates of ferritin and specific reagents, such as kidney bean phytohaemagglutinin (PHA) can be readily identified in replicas (Figure 5.4), but it is also possible to identify native, unconjugated PHA as well[42].

The principal advantage of freeze-etching over conventional shadowing and cell-spreading methods is that cell membranes need not be subjected to drying or other stabilising procedures as all steps are carried out on frozen material maintained in the temperature range of −150 to −100° C. Replicas prepared of freeze-etched material seem to be capable of resolving macromolecules in the 30–50 Å range; this is less than the resolving power of thin-sectioning techniques but is still more than adequate for most studies on receptor localisation. In addition, freeze-etching offers a unique opportunity to examine structural features in the *interior* of the membrane, and this has been used to correlate receptor sites on the external surface of cells with structures inside the membrane[42], and this is illustrated later.

5.3.3 Chemical modifications

Many attempts to modify membrane proteins *in situ* with enzymes or radioactive reagents have been carried out in recent years, and some of these have provided some interesting new information about protein orientation in membranes. The basic strategy of these experiments has been to incubate membranes with a specific reagent, such as a proteolytic enzyme or a radioactive alkylating compound, which is able to modify a particular membrane protein, either by removing a segment of the molecule or by adding an isotope, only if the modifying agent has direct access to the affected polypeptide chain.

As an example of the simplest case, it is possible to determine which membrane proteins are located on the external surfaces of cell by exposing the cells to agents which label or otherwise modify the exposed proteins without entering the cells. The lack of a reaction with haemoglobin, in the case of red cells, is usually taken as evidence that the reagent was not able to cross the cell membrane. This approach has been carried out by incubating intact cells, permeable membranes, and 'inside-out' membranes with a diverse group of reagents and enzymes, and the results provide us with a reasonably consistent picture of the relative positions of many of the major proteins of the red blood cell membrane.

Progress using this approach has been accelerated by the introduction of a variety of different chemical and enzymatic probes for different membrane components and also by the introduction and development of the SDS acrylamide gel-electrophoresis technique as a convenient and reproducible way to analyse membrane proteins. Using the latter procedure, it is possible to identify specific membrane-associated polypeptides and determine which are labelled or otherwise modified by the perturbing agents.

Since this approach seems promising, some of the findings obtained so far will be described in detail later.

5.4 ANALSIS OF MEMBRANE PROTEINS

Although red cell membranes have been the subject of active study for a number of years, it is still not clear how many different proteins contribute to their structure and functional properties. When the proteins of such membranes are analysed by acrylamide gel electrophoresis in the presence of sodium dodecyl sulphate at least eight major classes of polypeptide chains can be identified by appropriate staining techniques; these are illustrated in

Figure. 5.5. These eight bands may represent the minimum number of proteins. Since this technique depends on the capacity of SDS to dissociate proteins into polypeptide subunits which are then separated on the basis of their molecular size, it is conceivable that each of these eight bands contains more than one polypeptide chain, although it is also possible that several of the individual chains are derived from the same protein.

The basic procedure for analysing membrane proteins using the SDS–gel electrophoresis system has been modified by different investigators[43-46], and, as a result, some reports suggest that human red cell membranes contain a few more (or less) polypeptide chains than are illustrated in Figure 5.5. Unfortunately, there is no way at present for us to decide which system is providing the most accurate representation of the normal membrane complement, since both the artifactual loss of membrane proteins (by proteolysis) or the production of new bands (by aggregation) can occur with this technique. The results of the system devised by Fairbanks *et al.*[46], illustrated in Figure 5.5, embody the salient findings of most of the other methods, and their nomenclature will be used as a reference in the discussion to follow.

The proteins which make up four of the eight major bands can be isolated from red cell membranes by what are generally considered 'mild' extraction procedures. The term 'mild' is strictly operational and refers to the fact that the extraction conditions do not involve subjecting the membranes to extremes of temperature of pH or to detergents or other denaturing agents. The term 'mild' also implies that the proteins which are extracted from the membranes under these conditions are not denatured or otherwise altered by the procedure, but this is an unproven assumption.

Polypeptides which correspond to bands I, II, V and VI can be extracted from membranes with EDTA in low ionic strength buffer followed by an extraction in a concentrated salt solution. Bands I, II and V are soluble in the low ionic strength media, while band VI, now known to be glyceraldehyde 3-P04 dehydrogenase[47], requires an elevation in the ionic strength of the extracting media before it is released from the membrane. When the extracted, water soluble proteins are dissolved in SDS and electrophoresed, they have the same pattern as was obtained with the intact membranes (Figure 5b,c), and only the extracted bands are missing from the residual membrane fragments.

The major component of bands I and II has been isolated by a variety of techniques and corresponds to the protein called spectrin[46,48,49]. This material was originally found to share certain solubility properties with muscle actin[50,51], and it was suggested that this material might form a fibrous network along the inner surfaces of the red cell membrane and perhaps act in some way to stabilise the membrane. Subsequent work has shown that this material is confined to the inner surface of the membrane (as described later), and more recently it has been suggested that this material might be altered in the membranes of osmotically fragile red cells[52]. These results seem promising, but more work is needed.

The relative ease with which these proteins could be extracted from membranes prompted the suggestion that they be grouped into a sub-class and designated 'peripheral' membrane proteins[53]. The remaining proteins of the membrane which could not be solubilised by these treatments were considered to form another class more tightly bound and hence more

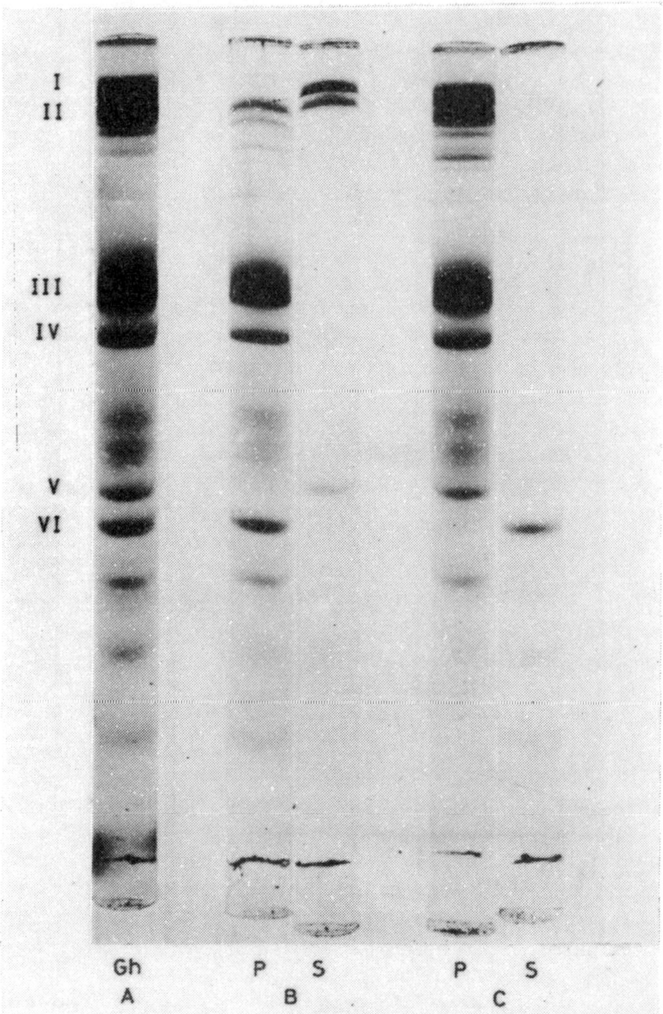

Figure 5.5 Analysis of RBC membrane proteins by the SDS-acrylamide gel technique. Eight major bands can be distinguished (a) in addition to other minor components. Four of these bands can be extracted from membranes by treatment with EDTA (b-s) in combination with high salt treatment (c-s). The membrane fragments obtained after these extractions (b-p) and c-p still retain bands III and IV. These probably represent the so-called integral proteins. (From Fairbanks, Steck and Wallach[46] by courtesy of *Biochemistry*)

'integral' to the membrane. The term 'peripheral' was originally meant to denote 'loosely bound' rather than to suggest a particular location relative to the lipid bilayer, but subsequent work, described below, indicates that these 'peripheral' proteins are probably also peripheral or external to the lipid bilayer. Some investigators have also suggested that peripheral membrane proteins should be considered 'membrane-associated' rather than true membrane components, but since we have no way of knowing where the membrane ends and the rest of the cell begins, this argument seems academic at present.

5.5 MOST MEMBRANE PROTEINS ARE SITUATED INTERNAL TO THE LIPID BARRIER

It is possible to determine which polypeptide chains of the membrane are exposed to the external environment of the cell by treating intact cells with enzymes or reagents which cannot enter the cells nor react with proteins which are located internal to the lipid barrier of the membrane. The results of such experiments have provided the somewhat surprising conclusion that only a small number (perhaps two or three) of the major polypeptide chains are located on the external surfaces of red blood cells.

Tryptic digestion of intact red blood cells causes the release of glycopeptide fragments from the membrane[54,55] but does not produce any detectable changes in the major protein components[56], Figure 5.6a. If, however, the membrane is modified by osmotic lysis, and the ghosts are not resealed, almost all of the major polypeptides of the membrane become susceptible to tryptic degradation (Figure 5.6b).

Essentially the same result is obtained when cells are incubated with chemical reagents capable of reacting only with groups on proteins which are accessible to the aqueous medium surrounding the cells. Radioactive compounds such as the diazonium salt of sulphanilic acid[57] or formyl methionyl sulphone methyl phosphate (FMMP)[58] react with amino groups of proteins (and probably at other sites as well) to form stable covalent bonds, thereby labelling the exposed portions of such proteins. After the labelling reaction is completed, the membranes can be isolated and the polypeptide chains containing the radioactive reagents can be analysed by autoradiography or by scintillation counting of SDS–gel fractions. Figure 5.7 shows that when intact red cells are incubated with the sulphanilic acid reagent only one or two of the eight major bands contains significant quantities of this label. Essentially the same results are obtained when other labelling methods are used which have different sites of action.

It is difficult to determine whether the proteins which are labelled by these techniques correspond to any of the major bands illustrated in Figure 5.5. One of the proteins on the external surface of the red cell is a sialoglycoprotein (to be described in detail below) which cannot be identified in Figure 5.5, since its capacity to bind the Coomassie Blue protein stain is less than other polypeptides which migrate to the same place on the gels. Most of the labelled material corresponds to the mobility of band III, and since this is also sensitive to chymotrypsin cleavage[56], it is likely that both the sialoglycoprotein and the

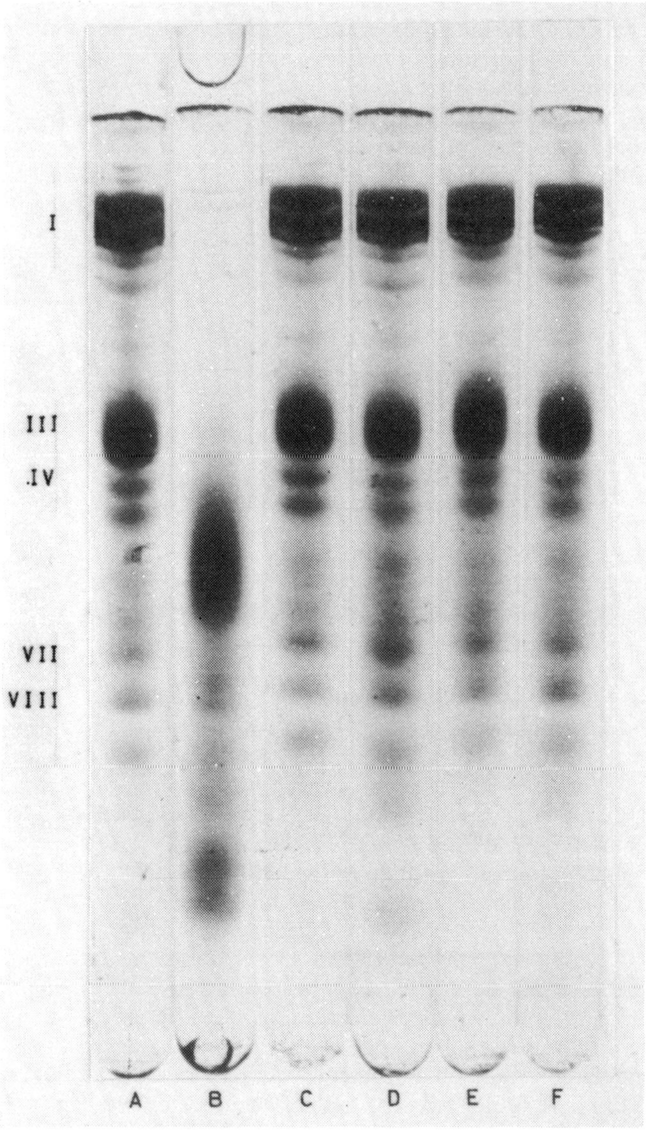

Figure 5.6 Incubation of intact red blood cells with trypsin does not result in any detectable change in the major polypeptides of the membrane (a), but if 'leaky' ghosts are incubated under the same conditions, most of the polypeptides are digested (b). Ghost membranes which are 'resealed' by the addition of salt solutions (d) or divalent cations (f) are no longer susceptible to tryptic cleavage. Analysed by SDS–gel electrophoresis. (From Triplett and Carraway[56], by courtesy of *Biochemistry*)

components of band III are at least partially exposed to the outside of the cell.

Recent studies by Tanner and Boxer[59] indicate that band III contains two other glycoproteins in addition to the major sialoglycoprotein, and it is likely that the oligosaccharide-containing segments of all three proteins extend outside the cell.

Berg and co-workers combined both the sulphanic acid labelling and proteolytic digestion techniques and were able to identify the polypeptide fragments of band III which are not susceptible to proteolytic digestion[60], Figure 5.7b. This experiment also showed that proteolytic digestion of intact cells removed only the exposed segment of band III but did not expose or otherwise affect the 'labellability' of other membrane proteins.

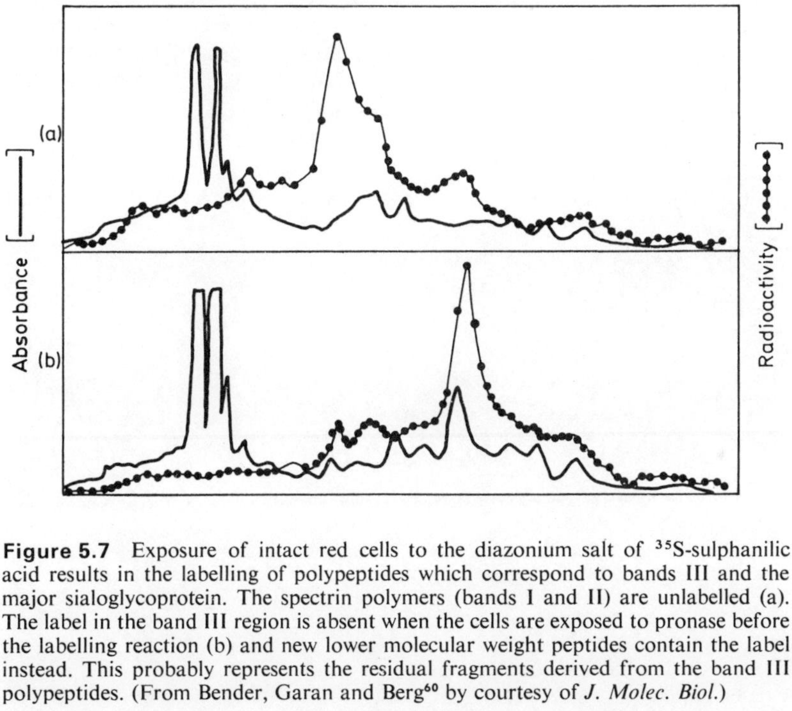

Figure 5.7 Exposure of intact red cells to the diazonium salt of ^{35}S-sulphanilic acid results in the labelling of polypeptides which correspond to bands III and the major sialoglycoprotein. The spectrin polymers (bands I and II) are unlabelled (a). The label in the band III region is absent when the cells are exposed to pronase before the labelling reaction (b) and new lower molecular weight peptides contain the label instead. This probably represents the residual fragments derived from the band III polypeptides. (From Bender, Garan and Berg[60] by courtesy of *J. Molec. Biol.*)

A marked difference was noted in the accessibility of membrane proteins to labelling reagents (or enzymatic digestion) when the incubations were carried out with intact cells as opposed to isolated membranes of the same preparations. While only a few membrane proteins could be labelled in intact cells, essentially *all* the membrane proteins could be labelled or digested when red cell ghosts were incubated under the same conditions as described earlier. The explanation usually offered for this result is that the osmotically lysed ghost membranes are permeable to the labelling reagents and proteolytic enzymes while the intact cells are not. According to this interpretation, the proteins which are only accessible when the ghost forms are incubated should be located on the inner surface of the membrane. However, another interpretation of these findings which is favoured by some investigators is that osmotic lysis might cause significant changes in the organisation of the membrane so that the protein elements are reoriented with respect to the external environment of the cell and thereby rendered accessible to labelling or proteolytic cleavage. Since we know little about the changes which accompany osmotic lysis or even the mechanism by which materials cross the lipid barrier during lysis, it is not possible to rule out this alternative explanation. Nevertheless, most of the information we do have concerning haemolysis and the properties of ghost membranes does not support this suggestion.

The structure of a red cell ghost membrane is indistinguishable from that of an intact cell when both are examined by freeze-etching and electron microscopy. This technique is capable of detecting rearrangements in the membrane which are caused by salt extractions, EDTA treatment and even pH shifts[61]. Furthermore, if red cell ghosts are deliberately modified so as to create 'inside out' vesicles, the freeze-etch image of such distorted vesicles is 'true' in the sense that the structure of the membrane is completely reversed and is not a morphological mosaic[62].

Additional evidence against the idea that osmotic lysis causes a gross rearrangement of membrane proteins has come from studies on 'resealed'

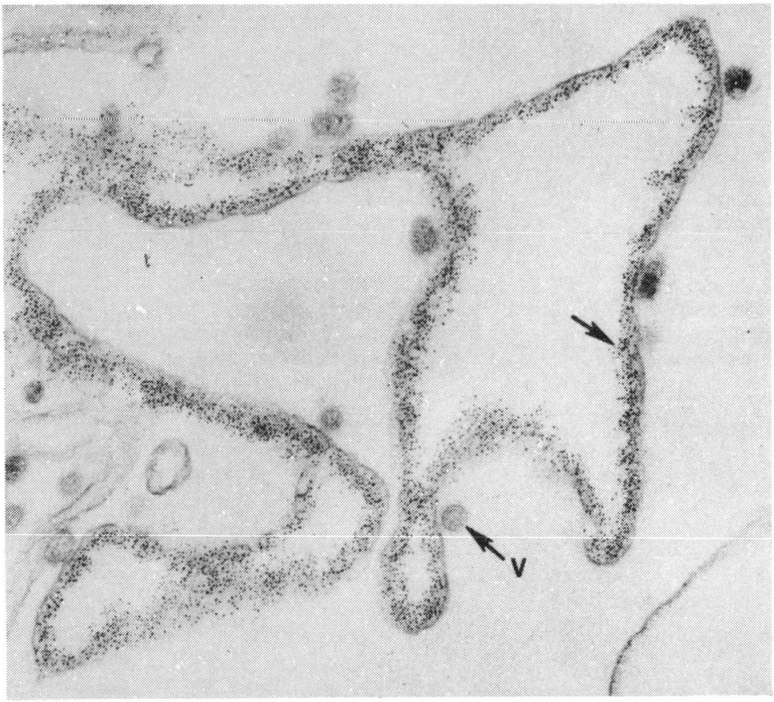

Figure 5.8 Electron micrograph of a red cell ghost labelled with ferritin-conjugated anti-spectrin antibody. Influenza viruses (V) were also added to serve as markers for the external surfaces of the cells

ghost membranes[56]. Red cells lysed rapidly and then treated with divalent cations and isotonic salt solutions seem to have permeability properties which are more similar to intact red cells than to the usual 'leaky' ghosts described above. Ghost membranes prepared in this way are not as susceptible to proteolytic digestion as are the leaky ghosts, in fact, the membrane proteins of resealed ghosts are just as inaccessible to trypsin as those in the intact cells (Figure 5.6d,f). The most plausible explanation for these findings is that 'resealing' prevents the proteases from gaining access to the inside surface of the membrane. Proteins which are normally exposed to the outer surface of intact red cells are, of course, still exposed on the surfaces of 'resealed' ghosts. Sceptics might argue that 'resealing' the ghost membranes causes a

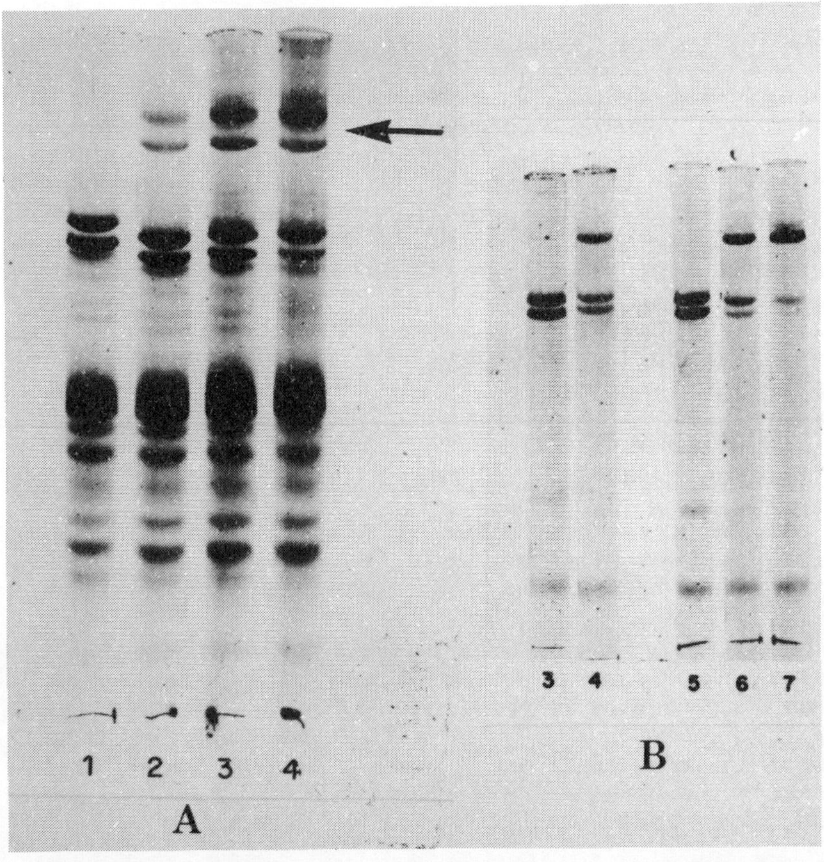

Figure 5.9 High molecular weight polypeptides (arrow) are formed when red cell membranes are exposed to dilute glutaraldehyde for short periods (a). Bands I and II seem to decrease correspondingly. I-control, 2, 3, 4 treated with 1, 2 and 3 mm glutaraldehyde for 20 min. (From Steck[66], by courtesy of *J. Molec. Biol.*) A single large molecular weight polypeptide is also formed when solubilised spectrin is treated with mild oxidising agents; bands I and II decrease proportionately (b) (unpublished obserbations of T. Steck)

rearrangement of membrane proteins either back to their original state or to another inaccessible arrangement, but this seems unlikely.

The results of both the external labelling and the proteolytic digestion experiments indicate that bands I and II which correspond to the protein called spectrin are located at sites which are internal to the lipid barrier of the membrane. More direct evidence in support of this has been provided by studies which show that ferritin-conjugated anti-spectrin antibodies react along the inner surfaces of the red cell membrane[63], Figure 5.8. The region of the membrane which binds anti-spectrin antibodies contains a network of filamentous structures which are evident when ghost membranes are examined both by conventional thin sections[64] and by freeze-etching techniques[65]. This fibrous material is not seen after ghosts are extracted with EDTA (to remove spectrin) or after tryptic digestion of the ghost membranes which achieves the same result.

Electrophoresis of spectrin in SDS-gels results in the production of two prominent bands (labelled I and II) which differ slightly in molecular weight, but appear to be present in the membrane in roughly equal amounts. Some indication that these two large (*ca.* 225 000 dalton) polypeptide chains may be in close proximity in the intact membrane has come from some recent attempts to analyse neighbouring membrane proteins by determining their reactivity to various cross-linking reagents[66]. Treatment of red cell ghosts with low concentrations of glutaraldehyde for short periods results in the production of large polypeptide aggregates which appear to be derived from band I and II (Figure 5.9a). The approximate size of the polypetides formed as a result of the cross-linking is consistent with their being polymerised forms of spectrin, and the bands of non-cross linked spectrin are correspondingly reduced in the same preparations. Steck has also shown that treatment of isolated spectrin with oxidising agents results in the formation of macromolecular aggregates presumably as a result of disulphide interchanges which are also large multiples of the native material. Both sets of experiments are consistent with the idea that the two polypeptide chains of spectrin are normally associated together as a macromolecular complex in the membrane, and this persists when spectrin is isolated under non-denaturing conditions.

Taken together, these results indicate that the bulk of the proteins of the red cell membranes are located internal to the lipid barrier of the membrane. Several polypeptides are exposed to the cell exterior, but it is interesting that the so-called *peripheral* proteins are all associated in some way with the internal surface of the membrane. Some of the tightly-bound or *integral* proteins are accessible to the outside of the cell, but these appear to be tightly bound to membrane lipids, and in some cases, they may extend completely across the membrane. The basis for their association with lipids and the evidence that they do span the bilayer is discussed later.

5.6 MOLECULAR PROPERTIES OF MEMBRANE GLYCOPROTEINS

Between 5 and 10% of the dry weight of the red cell membrane is composed of carbohydrates which are attached covalently to either lipid or protein elements[67]. In the human red blood cell membrane a relatively small fraction of the sugar (< 5%[68]) is in the form of glycolipids, and the rest is on glycoprotein, but this ratio fluctuates in red cells of other species, and is also markedly different in the membranes of other cell types.

It has been assumed for some time that the glycoproteins of the red cell membranes were probably situated on or close to the external surface of the membrane. This view was based on the finding that most of the sugar could be released from the membrane by enzymes which were presumably capable of acting only on the external surface of the cell. Neuraminidase added to suspensions of intact red cells is capable of releasing essentially all of the membrane-bound sialic acid of the cell, without, apparently, crossing the cell membrane[69]. Further indications that membrane-bound carbohydrates reside on the cell exterior were provided by numerous serological studies which showed that antigens and receptor sites composed of carbohydrates were accessible to specific antibodies (or lectins) added to the surrounding medium.

The early attempts to isolate intact glycoproteins from red cell membranes achieved only limited success, since, as we now know, these molecules are bound tenaciously to the lipids of the membranes and cannot be extracted by the usual protein solvents[70]. Recently, several new methods have been developed for the isolation of these molecules which seem to be effective probably because of their capacity to dissociate lipid–protein interactions. Solvents suitable for this purpose include phenol[71], pyridine[72], chloroform–methanol[73] and low concentrations of lithium di-iodosalicyclate solutions[74]. The latter method has been developed in our laboratory and has proved to be a simple and rapid way to isolate and purify the major sialoglycoprotein of the red cell membrane in a homogeneous water-soluble form. This component, called glycophorin[75], represents *ca.* 10% of the total protein of the red cell ghost and has blood group antigens and other receptors as well as most of the membrane-bound sialic acid.

5.6.1 Properties of glycophorin

Glycophorin appears to be composed of a single polypeptide chain to which are attached multiple oligosaccharides of varying composition and length. Two basic types of oligosaccharides exist which are attached to the polypeptide backbone through either threonine/serine residues (Type I via *N*-acetyl galactosamine linkage) or to asparagine residues (Type II with *N*-acetyl glucosamine as the linking sugar). The chemistry of these oligosaccharides has been studied by Winzler and co-workers[76,77] and Kornfeld and co-workers[78,79], and the results are described in detail in several recent reviews[67,80-83].

One glycophorin molecule may have as many as 20–30 separate oligosaccharide units attached to the polypeptide chain, the exact number is not known, and most of these seem to be attached to the *N*-terminal half of the molecule.

The idea that glycophorin is a single polypeptide chain is based on the results obtained after cyanogen bromide cleavage experiments and preliminary amino acid sequence data described below. However, these results do not exclude the possibility that two or more polypeptides of similar composition exist which have essentially identical properties. Most glycoproteins have microheterogeneity in their oligosaccharide portions[84], and it is also conceivable that minor differences in amino acid sequence exist as well.

The size of this glycoprotein molecule is not known with certainty as its high content of carbohydrate (60% by weight) and sialic acid (25% by weight) complicate attempts to study it by the standard analytical methods. Initial attempts to determine the size of the glycoprotein monomer in sodium dodecyl sulphate produced molecular weights in the 100 000 dalton range[85,86], but later studies showed that glycoproteins behave anomalously when analysed by this technique, and a corrected value of *ca.* 55 000 was obtained which agrees with more recent determinations[87]. However, this value must still be considered provisional.

Glycophorin is readily soluble in water and in all aqueous buffers we have tried, but under these conditions it appears to be a macromolecular aggregate with an approximate size in the 400 000 dalton range. This value is based on gel filtration and equilibrium centrifugation and is also consistent with measurements made by electron microscopy. These macromolecular aggregates, which contain a small amount of tightly bound lipid, carry a variety of blood group antigens (A, B, MN, 1) and also contain the receptors for influenza viruses and for different plant lectins. This capacity of glycophorin to bind specific plant lectins and viruses has been used to determine its distribution on the surfaces of intact red cells and red cell membranes (see later).

5.6.2 Linear arrangement of glycopeptide components

Many investigators have shown that sialoglycopeptides can be isolated from intact red cell membranes by treatment with proteolytic enzymes[54,55,88]. A number of different glycopeptides have been isolated from red cell membranes by tryptic digestion, but their relationship to each other and to the original membrane-bound glycoproteins have been unclear.

We have found that at least four unique glycopeptides are generated when isolated glycophorin preparations are digested with trypsin[89]. These peptides which account for more than 90% of the total sugar of the molecule, have been purified and partially characterised in terms of amino acid and carbohydrate content and receptor activities, and it appears that they are all derived from the same molecule. If this glycoprotein is trypsinised while it is still bound to the intact membrane, however, only three of the four glycopeptides can be released from the membrane. The fourth glycopeptide (labelled β), *ca.* 25 amino acids long, is only accessible to tryptic digestion after the molecule is isolated from the membrane, and we have taken advantage of this property to determine how the glycopeptides are arranged in the molecule. Intact red cells were trypsinised and the released glycopeptides isolated, then the remaining glycoprotein fragments were extracted from the membrane and subjected to a second trypsin digestion. By determining which glycopeptides are released at the different steps, we were able to show that the α-glycopeptides (α-1, α-2 and probably α-3) are located distal to (i.e. on the

N-terminal side) of the β-glycopeptide as shown in Figure 5.10. Since the β-glycopeptide is not released from the cells by trypsin, we assume that its *C*-terminal linking residue (arginine) is shielded from the tryptic activity possibly because of its close proximity to membrane lipids. This approach gives us some idea as to the order of the major glycopeptide segments of the molecule, but the overall ordering is only approximate since small peptides or those lacking sugar residues would not be detected by the methods used.

In addition to the four water-soluble glycopeptides described above, trypsin digestion of the isolated glycoprotein generates a water-insoluble peptide which does not contain carbohydrate but has instead a high content of non-polar amino acids[90]. A comparison of the amino acid sequence of this peptide with that of a cyanogen bromide fragment described below indicates that this tryptic peptide is derived from the middle third of the polypeptide chain and probably represents a part of the 'hydrophobic domain' of this glycoprotein.

Cyanogen bromide cleavage of the isolated glycoprotein[90] results in the production of at least five fragments, three of which overlap the tryptic peptides described above. Two of the CNBr fragments contain carbohydrate and represent fragments derived from the *N*-terminal or receptor region as depicted in Figure 5.11. The third CNBr fragment, labelled C-2, lacks homoserine and must therefore be derived from the *C*-terminal end of the original polypeptide chain. Partial amino acid sequence of this fragment and the insoluble tryptic peptide indicate that these two fragments are contiguous

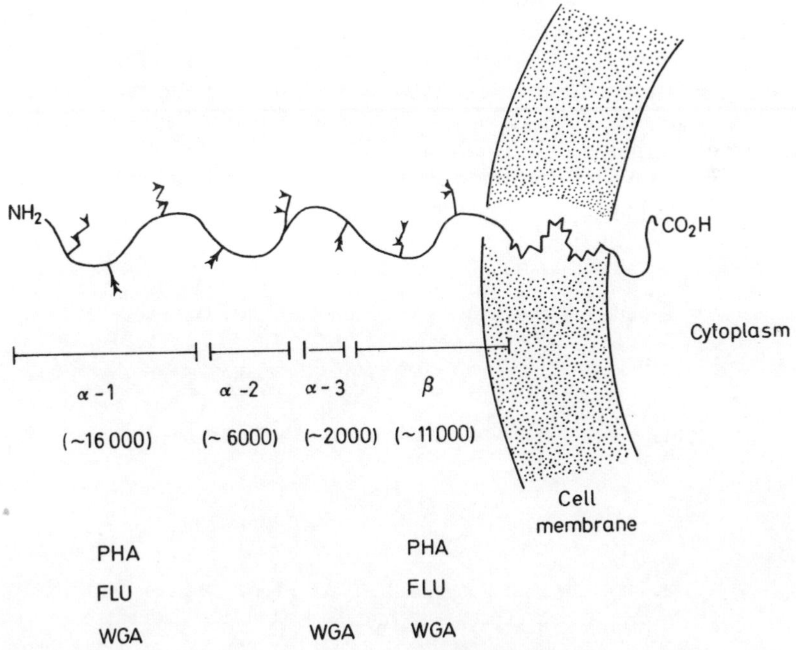

Figure 5.10 The probable order of tryptic glycopeptides in the glycophorin molecule. The two largest glycopeptides (α-1 and β) both carry receptors for a number of different lectins and for influenza viruses

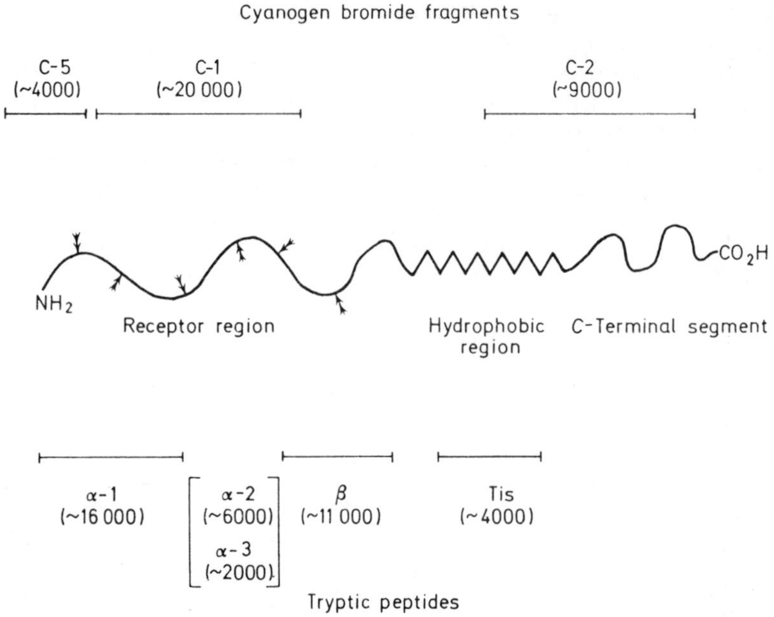

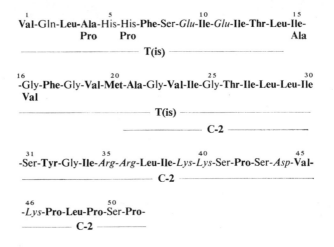

Figure 5.11 A provisional model for the segmental or regional differentiation of the glycophorin molecule based on the properties of peptide fragments derived by tryptic digestion and cyanogen bromide cleavage

along the polypeptide chain as shown in Figure 5.11[91] and represent a continuous sequence of 50 residues, which may represent approximately one quarter of the polypeptide chain of the glycophorin molecule. The sequence, which appears below, contains a stretch of 24 uncharged amino acids including a high proportion of the more hydrophobic residues.

1, 5, 10, 15
Val-Gln-**Leu-Ala**-His-His-**Phe**-Ser-*Glu*-**Ile**-*Glu*-**Ile-Thr-Leu-Ile**-
(below His 5: **Pro**; below His 6: **Pro**; below Ile 15: **Ala**)
———— **T(is)** ————

16, 20, 25, 30
-Gly-**Phe**-Gly-**Val-Met-Ala**-Gly-**Val-Ile**-Gly-**Thr-Ile-Leu-Leu-Ile**
(below Gly 16: **Val**)
———— **T(is)** ————
———— **C-2** ————

31, 35, 40, 45
-Ser-**Tyr**-Gly-**Ile**-*Arg*-*Arg*-**Leu-Ile**-*Lys*-*Lys*-Ser-**Pro**-Ser-*Asp*-**Val**-
———— **C-2** ————

46, 50
-*Lys*-**Pro-Leu-Pro**-Ser-**Pro**-
———— **C-2** ————

We speculate that this segment might be considered the 'hydrophobic domain' of the polypeptide chain since its amino acid sequence should permit it to interact via hydrophobic associations with either membrane lipids or hydrophobic regions of other proteins. This hydrophobic segment of the molecule seems to be considerably removed from the *C*-terminal end of the molecule, a point which may have some bearing on how the molecule might be orientated in the intact membrane as described below.

5.6.3 Orientation of glycophorin in the membrane

It is possible to conceive of at least four different ways in which a glycoprotein having the structure shown in Figure 5.11 might be orientated at the cell surface. These are illustrated in Figure 5.12. Each model depicts the sugar-containing segment of the molecule on the external surface of the cell, an orientation which is consistent with the known accessibility of the antigens and receptors to the external medium. This arrangement is also consistent with more recent studies showing the lack of binding of lectins to the inner surfaces of lysed red cells[92].

The mode of attachment of glycophorin to the membrane shown in Figure 5.12a is unlikely for several reasons. Although the polypeptide chain could easily fold so that the most hydrophobic residues are internalised, this would make them unavailable for associations with lipids of the membrane and, as a result, the binding of the glycoprotein to the membrane would most likely be via electrostatic interactions with the polar groups of the phospholipids. If this were so one would predict that the glycoprotein could be extracted from

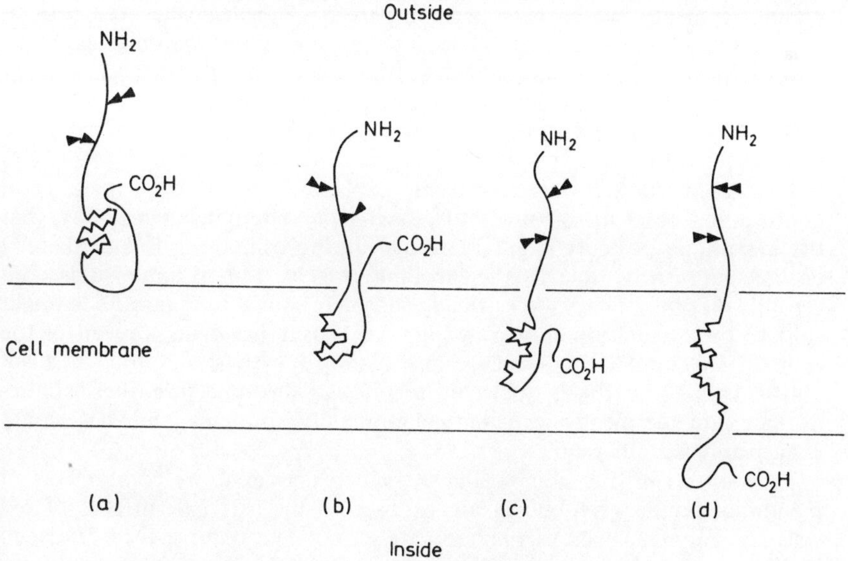

Figure 5.12 A glycoprotein with the tripartite structure shown in Figure 5.11 could be orientated in the cell membrane in at least these four different ways

the membrane with high ionic strength buffers or by manipulating the pH, and this is not the case. In addition, the results of labelling parts of the glycoprotein with radioactive reagents and antibodies are not consistent with this model.

The three remaining possibilities represent variations of the idea, originally proposed by Morawiecki[93] and Winzler[94], that this glycoprotein might be bound to the lipids of the membrane via hydrophobic associations between non-polar segments of the polypeptide chain and the membrane. This mode of attachment would probably be more stable under physiological conditions than that maintained by electrostatic interactions and this would also explain why this molecule is difficult to extract from membranes. The model proposed by Morawiecki and Winzler is essentially that shown in Figure 6.12c. Each has postulated that one end of the glycoprotein molecule was 'lipophilic' and was buried in the lipid regions of the membrane. However, the recently obtained sequence information described above indicates that a large hydrophobic segment of the glycoprotein is located not at the *C*-terminal end of the polypeptide chain but 40–60 amino acid residues away, the exact number is not known. In addition, the polypeptide chain between this hydrophobic segment and the *C*-terminal end contains a number of charged amino acids and, on thermodynamic grounds, one might predict that this segment would most likely reside in an aqueous environment. Thus, we feel that modes B and D are more consistent with the known structural features of the molecule.

Recently Bretscher has proposed that the major glycoprotein of the red cell membrane (the molecule we call glycophorin) extends completely across the membrane as illustrated in Figure 5.12d. This interpretation is based on experiments in which a water-soluble radioactive reagent was used to label exposed segments of membrane proteins[95]. Bretscher found that the 'hydrophobic segments' of the glycoprotein were not labelled when the reagent was added to intact red cells but these regions of the molecule could be labelled if the reagent was added to the membrane ghosts produced by osmotic lysis. He concluded that the altered permeability of the ghost membrane allowed the reagent to penetrate inside the cell and thereby label the internal segment.

This interpretation was a plausible explanation for these experimental findings, but other investigators suggested, as an alternative possibility, that the membrane proteins might rearrange during osmotic lysis and thereby become more accessible to the labelling reagent. For example, it is conceivable that the orientation of the glycoprotein shown in Figure 5.12c might shift to the orientation shown in Figure 5.12b during or as a result of the osmotic shock. Although the results obtained with studies of 'resealed' ghosts described earlier would tend to rule out this objection, these studies did not deal specifically with the membrane glycoproteins, and these could conceivably act differently.

To resolve this question, we have tried to determine which segments of membrane-bound glycophorin are exposed to the external surfaces of red cells by using the lactoperoxidase catalysed iodation procedure described by Phillips and Morrison[96] and Hubbard and Cohn[97]. Since this procedure is theoretically capable of iodinating all accessible tyrosine residues on proteins, we attempted to determine which portions of the glycoprotein are labelled by

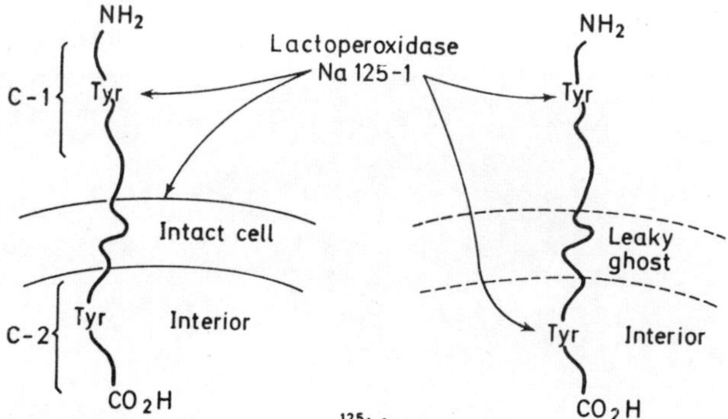

Labelling conditions	^{125}I Incorporated into C-1	C-2
Glycophorin in solution	+	+
Glycophorin on intact red cells	+	0
Glycophorin on 'leaky' ghosts	+	+
Glycophorin on 'sealed' ghosts	+	0

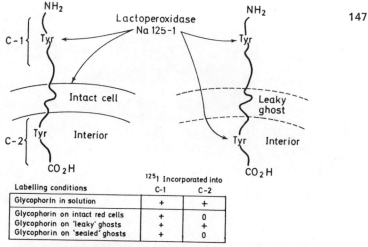

Figure 5.13 Lactoperoxidase can catalyse the iodination of tyrosine residues in the exposed (C-1) segment of glycophorin when intact cells are incubated with the enzyme and ^{125}I-NaI. The tyrosines in the *C*-terminal end of the molecule (the C-2 segment) are not iodinated unless the membrane is permeable enough to allow the the enzyme to gain access to the inside surface of the cell

extracting the molecules and analysing the radioactivity associated with individual cyanogen bromide peptides by autoradiography[90]. When intact red blood cells are exposed to the iodination reaction mixture only the *N*-terminal one-half of the glycophorin molecule becomes labelled, but when 'leaky' ghost membranes are incubated under the same conditions the label is distributed throughout the molecule including the *C*-terminal region. However, if the ghost membranes are 'resealed' after lysis by the addition of salt and divalent cations before iodination, then only the *N*-terminal segment becomes labelled. These experiments are illustrated diagrammatically in Figure 5.13.

Labelling studies were also carried out using antibodies directed against specific parts of the glycoprotein molecule with similar results. Antibodies prepared against the complete molecule readily agglutinated intact red cells while antibodies specific for a *C*-terminal fragment (C-2) did not react[98]. Both sets of observations indicate that part of the *C*-terminal end of the glycophorin molecule is normally located internal to the lipid barrier of the membrane. Once this barrier is modified by osmotic lysis, the *C*-terminal piece becomes accessible to the labelling reagents, probably due to their penetration across the membrane. Thus we tentatively conclude that the glycophorin molecule extends completely across the membrane.

5.6.4 Distribution of glycophorin molecules over the red cell surface

Freeze-etching and electron microscopy have been used to locate individual glycophorin molecules on the surface membranes of intact cells. Ferritin conjugates of both PHA or WGA can be used for this purpose since each of

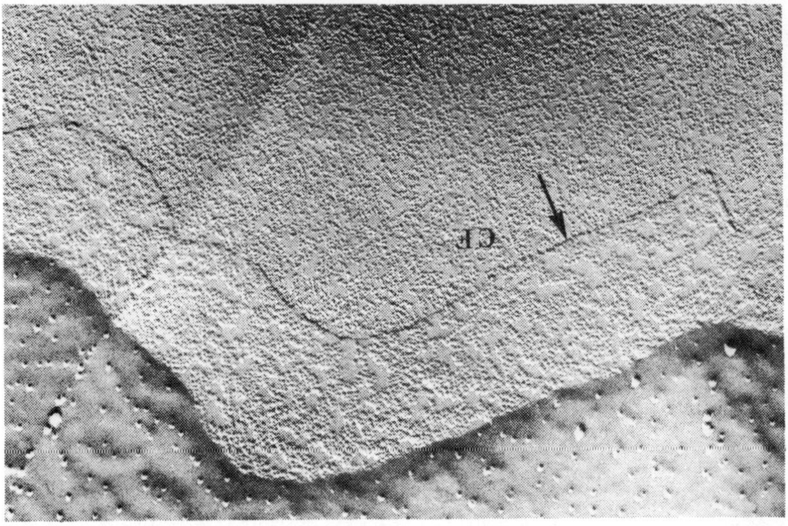

Figure 5.14 Electron micrograph of a red cell ghost labelled with wheat germ agglutinin (WGA) ferritin conjugates and then subjected to freeze-etching. The arrow marks the junction between the cleaved face (CF) and the external surface of the cell. The WGA-ferritin conjugates on the external surface have approximately the same distribution as the intramembraneous particle population, *ca.* 60 000 × (reduced $\frac{2}{3}$rds on reproduction)

these reagents binds to a specific receptor on the exposed segment of glycophorin. Figure 5.14 shows the pattern obtained when lectin–ferritin conjugates (either PHA or WGA) are incubated with either intact red cells or red cell ghosts. When such preparations are freeze-etched, it is possible to map the distribution of ferritin-conjugates over the external surfaces of the cells and compare their distribution with that of the intramembraneous particles present within the interior of the same membranes. The receptors for both PHA and WGA are distributed uniformly over the surface of the red cell membrane in a pattern which closely approximates that of the intramembraneous particle population. The relationship between glycoprotein molecules and the membrane particles has been studied in some detail in our laboratory[42] and by others[99], and the evidence suggests that the particles may represent an interaction between hydrophobic segments of the glycoprotein and membrane lipids or other membrane proteins. Tryptic digestion of ghost membranes was one of the means used to show this correlation, since it was observed that this treatment resulted in a marked rearrangement of the intramembraneous particles. By applying the ferritin–lectin conjugates to such trypsinised membranes it was possible to show clearly that receptor sites on the surfaces of such membranes also shifted to a new pattern which corresponded to that of the rearranged particles. This is illustrated in Figure 5.15.

Although the intramembraneous particles appear to be globular structures, with approximate diameters in the 70–80 Å range, these dimensions are obtained by measuring platinum–carbon replicas of freeze-etched preparations, and it is difficult to determine how much of the mass of these structures

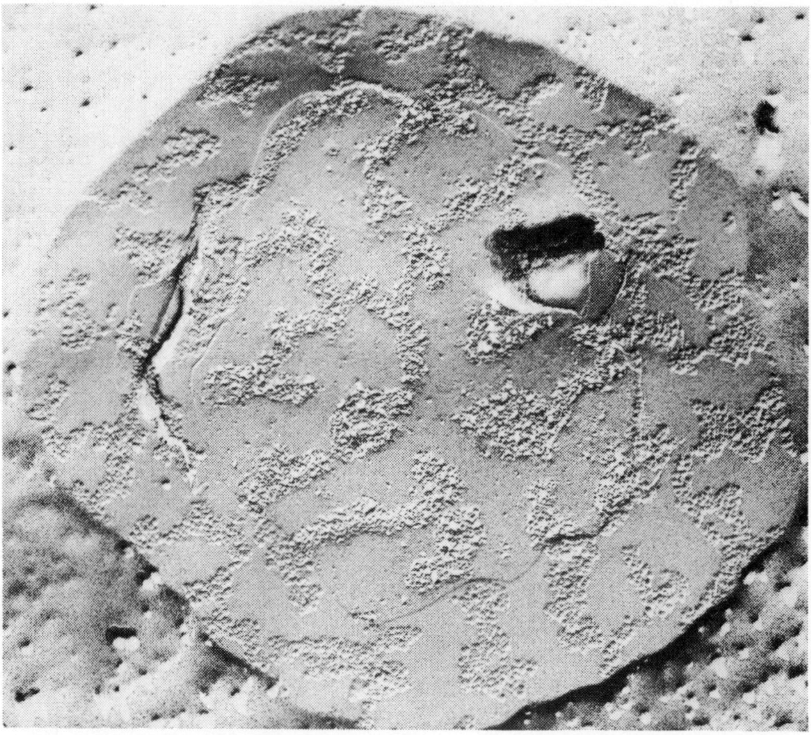

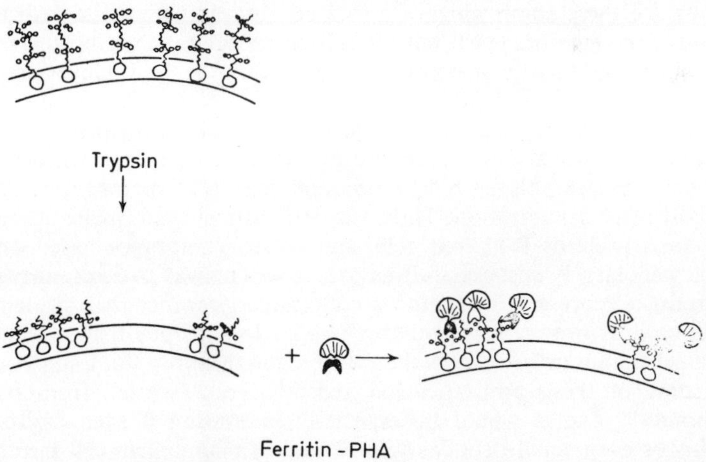

Figure 5.15 Electron micrograph of a red cell ghost subjected to tryptic digestion, followed by labelling with PHA-ferritin conjugates and then subjected to freeze-etching. Trypsin causes the intramembraneous particles to clump in a sinuous pattern, and the receptors for PHA on the external surface also shift to conform to this new pattern (a), *ca.* 70 000 × (reduced $\frac{8}{10}$ths on reproduction). An explanation for this result is shown diagrammatically in (b). (Taken from Tillack, Scott and Marchesi[42], *J. Exp. Med.*)

is due to the thickness of the replica. For this reason, it is not possible to estimate how much protein could be in the membrane at these sites, although it seems unlikely that the particle would be composed solely of the hydrophobic segment of glycophorin. Our current guess is that the particles may represent an aggregate of several different integral proteins including possibly the two other glycoproteins described by Tanner and Boxer[59].

5.7 ARE SOME PROTEINS LOCATED SOLELY WITHIN THE LIPID MATRIX

In the preceding section, evidence was reviewed which suggests that one or two membrane proteins may extend across the lipid region of the red cell membrane*. In the case of glycophorin, the major sialoglycoprotein of the membrane, it appears that a special segment of the polypeptide chain is composed predominantly of hydrophobic and non-polar amino acids which could interact with the hydrocarbon chains of the membrane lipids. Since the same protein also contains appreciable amounts of charged and polar amino acids and hydrophilic sugars, it is reasonable to suggest that this glycoprotein is orientated in the membrane so that the different segments are situated in milieu which promote thermodynamic stability. The models which best fit these criteria are illustrated in Figure 5.12. One variant was considered to be most likely on the basis of the labelling experiments described above, and it is considered the probable orientation of such amphipathic-type proteins in membranes. It is reasonable to ask whether all the membrane proteins which are associated with the lipids are solely of the amphipathic type, or are there some which are located entirely within the hydrophobic domain of the membrane? It has often been postulated that integral proteins exist which reside solely or predominantly within the lipid bilayer and which may play a key role in holding the membrane together. Attempts to isolate such proteins were carried out with the idea that these proteins would have a high affinity for lipids and also be insoluble in aqueous media. Unfortunately, most proteins are able to bind lipids and may also be insoluble in water especially after denaturation. Thus, the early attempts to isolate and characterise such proteins from red cells and other membranes based on these criteria were largely unsuccessful since there was no way to determine whether the products represented denatured contaminants rather than proteins of a special type. However, it is probably unwise to conclude that such proteins do not exist, and further efforts should be made to clarify this issue. A protein with some of these properties has recently been isolated from bacterial membranes[100] and it would be especially interesting if such hydrophobic proteins or even small peptides were found in mammalian cell membranes. These would be well suited to serve as membrane stabilisers of 'organisers' and possibly also as 'channel-formers' across the lipid along the lines suggested for the gramicidin like antibiotics[101].

* It is worth emphasising again that these results apply only to red blood cell membranes; comparable studies are now being carried out on other cell types, but at this time there are not enough clear-cut findings to make any generalisations.

5.8 THE ARRANGEMENT OF PROTEINS ON THE CYTO PLASMIC SURFACE OF THE PLASMA MEMBRANES: AN UNSOLVED PROBLEM

A great deal of attention and experimental work has been devoted to questions concerning how and where proteins and other receptors are arranged on the external surface membranes of cells. As described earlier, many microscopic techniques are now available for this purpose, and as a result of their application to this problem, the amount of information we now have along these lines is overwhelming. In contrast, we are completely ignorant as to how and where proteins are attached to the inside (or cytoplasmic) surfaces of the plasma membrane of any mammalian cell, red cells included. Most of the techniques now available for analysing the topography of proteins rely on the use of macromolecular tracers such as ferritin-conjugates and these cannot be used to label receptors on the inside surfaces of membrane unless some way is found to allow the markers to penetrate the membrane without drastically altering the arrangement of the attached proteins. There are many reasons for suspecting that the arrangement of proteins and specific enzymes on the cytoplasmic surfaces of cells may be more important in terms of providing clues about membrane functions than the more readily available data on the topography of the cell exterior. Many of the enzymes commonly associated with membrane functions such as the Na–K ATPase or adenyl cyclases are clearly located internal to the lipid barrier of the membrane. If we had some idea as to where and how they are attached to the membrane,

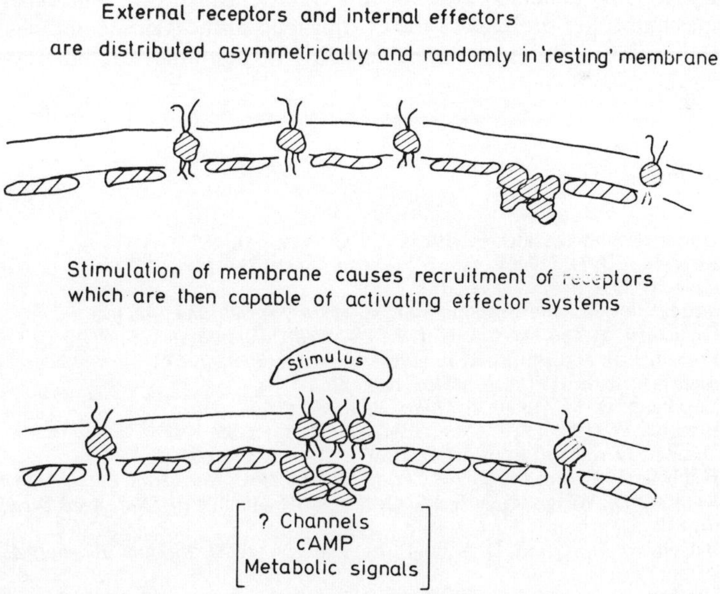
External receptors and internal effectors
are distributed asymmetrically and randomly in 'resting' membrane
Stimulation of membrane causes recruitment of
which are then capable of activating effector systems
Stimulus
? Channels
cAMP
Metabolic signals

Figure 5.16 One very hypothetical way in which mobile receptor units in the membrane could be 'organised' by a specific stimulus, such as a hormone, so as to initiate some metabolic response inside the cell

this could lead to a clearer understanding as to how their functions are regulated.

We also do not know how the stimulation of receptor molecules on the external surfaces of cells leads to a metabolic response inside the cell. Stimulation of cells by different types of hormones may be mediated by a common transducing mechanism which could ultimately connect to the same effector system such as an adenyl cyclase, and the specificity might reside primarily in the receptor function.

A hypothetical scheme which describes one way in which activation of surface receptors could lead to a functional response inside the cell is shown in Figure 5.16. Receptors for hormones or other stimuli are depicted as being distributed randomly over the surfaces of cells, possibly in a semi-mobile state, and connections between the internal or 'efferent' segments of their polypeptide chains and *effector enzyme systems* would be infrequent, transient, and non-productive. A stimulus in the form of a bound hormone might recruit and possibly immobilise a number of receptor molecules (as illustrated in Figure 5.16b) and once a critical mass was reached, subsequent contact between this aggregate of receptors and the effector enzymes might then be capable of triggering a response. This hypothetical idea postulates that clumps of such receptor molecules are present in the surface membranes of active cells (which have been observed) and these might be hooked up or in some way associated with effector systems inside the cell.

Since we do not as yet have any good way to study the distribution of effector enzymes on the inner surfaces of cells, it is hard to test this relatively simple idea.

Clearly, our understanding of the relationships of membrane-bound enzymes and other proteins to the cytoplasmic surfaces of membranes would be of enormous value to an understanding of the molecular biology of cell membranes.

References

1. Gorter, E. and Grendel, F. (1925). *J. Exp. Med.*, **41**, 439
2. Robertson, J. D. (1966). *Ann. N Y. Acad. Sci.*, **137**, 421
3. Korn, E. D. (1966). *Science*, **153**, 1491
4. Stoeckenius, W. and Engelman, D. M. (1969). *J. Cell Biol.*, **42**, 613
5. Danielli, J. F. and Davson, H. (1935). *J. Cell Comp. Physiol.*, **5**, 495
6. Finean, J. B. and Burge, R. E. (1963). *J. Molec. Biol.*, **7**, 672
7. Blaurock, A. E. (1971). *J. Molec. Biol.*, **56**, 35
8. Engelman, D. M. (1970). *J. Molec. Biol.*, **47**, 115
9. Branton, D. (1966). *Proc. Nat. Acad. Sci.* (*U.S.A.*), **55**, 1048
10. Deamer, D. W. and Branton, D. (1967). *Science*, **158**, 655
11. Hubbell, W. L. and McConnell, H. M. (1968). *Proc. Nat. Acad. Sci.* (*U.S.A.*), **61**, 12
12. Keith, A. D., Waggoner, A. S. and Griffith, O. H. (1968). *Proc. Nat. Acad. Sci.* (*U.S.A.*), **61**, 819
13. Barrett, M. D., Green, D. K. and Chapman, D. (1968). *Biochim. Biophys. Acta*, **152**, 20
14. Hubbell, W. and McConnell, H. M. (1971). *J. Amer. Chem. Soc.*, **93**, 2
15. Jost, P., Waggoner, A. S. and Griffith, O. H. (1971). *Structure and Function of Biological Membranes*, p. 83 (L. I. Rothfield, editor) (New York: Academic Press)
16. Lucy, J. A. (1964). *J. Theoret. Biol.*, **7**, 360

17. McConnell, H. M., Devaux, P. and Scandella, C. (1972). *Membrane Research*, p. 27 (C. F. Fox, editor) (New York: Academic Press)
18. Kornberg, R. D. and McConnell, H. M. (1971). *Proc. Nat. Acad. Sci.* (*U.S.A.*), **68,** 2564
19. Bretscher, M. S. (1972). *Nature New Biol.*, **236,** 11
20. Danielli, J. F. (1938). *Cold Spring Harbor. Symp. Quant. Biol.*, **6,** 190
21. Criddle, R. S., Bock, R. M., Green, D. E. and Tisdale, H. (1962). *Biochemistry*, **1,** 827
22. Green, D. E., Haard, N. F., Lenaz, G. and Sillman, H. I. (1968). *Proc. Nat. Acad. Sci.*, (*U.S.A.*), **60,** 277
23. Woodward, D. O. and Munkres, K. D. (1966). *Proc. Nat. Acad. Sci.* (*U.S.A.*), **55,** 872
24. Dreyer, W. J., Papermaster, D. S. and Kuhn, H. (1972). *Ann. N. Y. Acad. Sci.*, **195,** 61
25. Singer, S. J. and Nicolson, G. L. (1972). *Science*, **175,** 720
26. *Biological Membranes* (1968). (D. Chapman, editor) (New York: Academic Press)
27. *Structure and Function of Biological Membranes* (1971). (L. I. Rothfield, editor) (New York: Academic Press)
28. Fox, C. F. and Keith, A. D., editors (1972). *Membrane Molecular Biology* (Stamford, Conn.: Sinauer Assoc.)
29. Blasie, J. K. and Worthington, C. R. (1969). *J. Molec. Biol.*, **39,** 417
30. Cain, J., Santillan, G. and Blasie, J. K. (1972). *Membrane Research*, p. 3 (C. F. Fox, editor) (New York: Academic Press)
31. Lenard, J. and Singer, S. J. (1966). *Proc. Nat. Acad. Sci.*, **56,** 1828
32. Wallach, D. F. H. and Zahler, P. H. (1966). *Proc. Nat. Acad. Sci.* (*U.S.A.*), **56,** 1552
33. Urry, D. W., Medmiecks, M. and Bejnarowitz, E. (1967). *Proc. Nat. Acad. Sci.* (*U.S.A.*), **57,** 1043
34. Urry, D. W. and Krivacic, J. (1970). *Proc. Nat. Acad. Sci.* (*U.S.A.*), **65,** 845
35. Maddy, A. H. and Malcom, B. R. (1965). *Science*, **150,** 1616
36. Wu, C. W. and Stryer, L. (1972). *Proc. Nat. Acad. Sci.* (*U.S.A.*), **69,** 1104
37. Rambourg, A. and Leblond, C. P. (1967). *J. Cell Biol.*, **32,** 27
38. Martinez-Palomo, A. (1970). *Int. Rev. Cytol.*, **29,** 29
39. Nicolson, G. L. and Singer, S. J. (1971). *Proc. Nat. Acad. Sci.* (*U.S.A.*), **68,** 942
40. Pinto de Silva, P. and Branston, D. (1970). *J. Cell Biol.*, **45,** 598
41. Tillack, T. W. and Marchesi, V. T. (1970). *J. Cell Biol.*, **45,** 649
42. Tillack, T. W., Scott, R. E. and Marchesi, V. T. (1972). *J. Exp. Med.*, **135,** 1209
43. Shapiro, A. L., Vinuela, E. and Maizel, J. V. (1967). *Biochem. Biophys. Res. Commun.*, **28,** 815
44. Weber, K. and Osborn, M. (1969). *J. Biol. Chem.*, **244,** 4406
45. Maizel, J. V. (1969). *Fundamental Techniques in Virology*, p. 334 (K. Habel and N. P. Salzman, editors) (New York: Academic Press)
46. Fairbanks, G., Steck, T. L. and Wallach, D. F. H. (1971). *Biochemistry*, **10,** 2606
47. Carraway, K. L. and Shin, B. C. (1972). *J. Biol. Chem.*, **247,** 2102
48. Marchesi, V. T. and Steers, E., Jr. (1968). *Science*, **159,** 203
49. Clarke, M. (1971). *Biochem. Biophys. Res. Commun.*, **45,** 1063
50. Marchesi, V. T. and Palade, G. E. (1967). *Proc. Nat. Acad. Sci.* (*U.S.A.*), **58,** 991
51. Steers, E. and Marchesi, V. T. (1969). *Membrane Proteins*, p. 65 (Boston: Little, Brown and Co.)
52. Jacob, H. S., Ruby, A., Overland, E. S. and Mazia, O. (1971). *J. Clin. Invest.*, **50,** 1800
53. Singer, S. J. (1971). *Structure and Function of Biological Membranes*, p. 146 (L. I. Rothfield, editor) (New York: Academic Press)
54. Winzler, R. J., Harris, E. D., Pekas, D. J., Johnson, C. A. and Weber, P. (1967). *Biochemistry*, **6,** 2195
55. Cook, G. M. W. (1968). *Biol. Rev. Cambridge Phil. Soc.*, **43,** 363
56. Triplett, R. B. and Carraway, K. L. (1972). *Biochemistry*, **11,** 2897
57. Berg, H. C. (1969). *Biochem. Biophys. Acta*, **183,** 65
58. Bretscher, M. S. (1971). *J. Molec. Biol.*, **58,** 775
59. Tanner, M. J. A. and Boxer, D. H. (1972). *Biochem. J.*, **129,** 333
60. Bender, W. W., Garan, H. and Berg, H. C. (1971). *J. Molec. Biol.*, **58,** 783
61. Pinto da Silva, P. (1972). *J. Cell Biol.*, **53,** 777
62. Steck, T. L., Weinstein, R. S., Strauss, J. H. and Wallach, D. F. H. (1970). *Science*, **168,** 255
63. Nicolson, G. L., Marchesi, V. T. and Singer, S. J. (1971). *J. Cell Biol.*, **51,** 265
64. Marchesi, V. T. and Palade, G. E. (1967). *J. Cell Biol.*, **35,** 385

65. Tillack, T. W., Scott, R. E. and Marchesi, V. T. Unpublished observations
66. Steck, T. L. (1972). *J. Molec., Biol.*, **66,** 295
67. Winzler, R. J. (1970). *Int. Rev. Cytol.*, **29,** 77
68. Sweeley, C. C. and Dawson, G. (1969). *Red Cell Membrane, Structure and Function*, p. 172 (G. A. Jamieson and T. J. Greenwalt, editors) (Philadelphia, Toronto: J. B. Lippincott Co.)
69. Eylar, E. H., Madoff, M. A., Brody, O. V. and Oncley, J. L. (1962). *J. Biol. Chem.*, **237,** 1992
70. Rosenberg, S. A. and Guidotti, G. (1969). *J. Biol. Chem.*, **244,** 5118
71. Kathan, R. H., Winzler, R. J. and Johnson, C. A. (1961). *J. Exp. Med.*, **113,** 37
72. Blumenfeld, O. O., Gallop, P. M., Howe, C. and Lee, L. T. (1970). *Biochim. Biophys. Acta*, **211,** 109
73. Cleve, H., Hamaguchi, H. and Hutteroth, T. (1972). *J. Exp. Med.*, **136,** 1140
74. Marchesi, V. T. and Andrews, E. P. (1971). *Science*, **174,** 1247
75. Marchesi, V. T., Tillack, T. W., Jackson, R. L., Segrest, J. P. and Scott, R. E. (1972). *Proc. Nat. Acad. Sci.* (*U.S.A.*), **69,** 1445
76. Thomas, D. B. and Winzler, R. J. (1969). *Biochem. Biophys. Res. Commun.*, **35,** 811
77. Thomas, B. D. and Winzler, R. J. (1969). *J. Biol. Chem.*, **244,** 5943
78. Kornfeld, S. and Kornfeld, R. (1969). *Proc. Nat. Acad. Sci.* (*U.S.A.*), **63,** 1439
79. Kornfeld, R., Keller, J., Baenziger, J. and Kornfeld, S. (1971). *J. Biol. Chem.*, **246,** 3259
80. Kornfeld, S. and Kornfeld, R. (1971). *Glycoproteins of Blood Cells and Plasma*, p. 50 (G. A. Jamieson and T. J. Greenwalt, editors) (Philadelphia: J. B. Lippincott Co.)
81. Kraemer, P. M. (1971). *Biomembranes*, vol. 1, p. 67 (L. A. Manson, editor) (New York: Plenum Press)
82. Ginsburg, V., Hickey, C., Kobata, A. and Sawicka, T. (1971). *Glycoproteins of Blood Cells and Plasma*, p. 114 (G. A. Jamieson and T. J. Greenwalt, editors) (Philadelphia: J. B. Lippincott Co.)
83. Hughes, R. C. (1973). *Progr. Biophys. Molec. Biol.*, **26,** 189
84. Cunningham, L. W. (1971). *Glycoproteins of Blood Cells and Plasma*, p. 16 (G. A. Jamieson and T. J. Greenwalt, editors) (Philadelphia: J. B. Lippincott Co.)
85. Lenard, J. (1970). *Biochemistry*, **9,** 1129
86. Marchesi, V. T. (1970). *Fed. Proc.* (*Fed. Amer. Soc. Exp. Biol*), **29,** 600
87. Segrest, J. P., Jackson, R. L., Andrews, E. P. and Marchesi, V. T. (1971). *Biochem. Biophys. Res. Commun.*, **44,** 390
88. Jackson, L. J. and Seaman, G. V. F. (1972). *Biochemistry*, **11,** 44
89. Jackson, R. L., Segrest, J. P., Kahane, I. and Marchesi, V. T. (1973). *Biochemistry*, **12,** 3131
90. Segrest, J. P., Kahane, I., Jackson, R. L. and Marchesi, V. T. (1973). *Arch. Biochem. Biophys.*, **155,** 167
91. Segrest, J. P., Jackson, R. L., Marchesi, V. T., Guyer, R. B. and Terry, W. (1972). *Biochem. Biophys. Res. Commun.*, **49,** 964
92. Nicolson, G. L. (1972). *Membrane Research*, p. 53 (C. F. Fox, editor) (New York: Academic Press)
93. Morawiecki, A. (1964). *Biochim. Biophys. Acta*, **83,** 339
94. Winzler, R. J. (1969). *Red Cell Membrane, Structure and Function*, p. 157 (G. A. Jamieson and T. J. Greenwalt, editors) (Philadelphia: J. B. Lippincott Co.)
95. Bretscher, M. S. (1971). *Nature New Biol.*, **231,** 229
96. Phillips, D. R. and Morrison, M. (1970). *Biochem. and Biophys. Res. Commun.*, **40,** 284
97. Hubbard, A. L. and Cohn, Z. A. (1972). *J. Cell Biol.*, **55,** 390
98. Kahane, I. and Marchesi, V. T. Unpublished observations
99. Pinto da Silva, P., Branton, D. and Douglas, S. D. (1971). *Nature* (*London*), **232,** 194
100. Sandermann, H. and Strominger, J. L. (1971). *Proc. Nat. Acad. Sci.*, **68,** 2441
101. Urry, D. W., Goodall, M. C., Glickson, J. D. and Mayers, D. F. (1971). *Proc. Nat. Acad. Sci.* (*U.S.A.*), **68,** 1907

6
Role of the Surface in Production of New Cells

A. B. PARDEE and E. ROZENGURT
Princeton University

Aided by Grant CA-AI-11595 from the U.S. Public Health Service. This article was written at the Imperial Cancer Research Fund Laboratories, London, while the senior author was a Scholar of the American Cancer Society.

6.1 INTRODUCTION

Production of a new bacterial or eucaryotic cell is most evidently achieved when separation occurs. Separation is only the final process of a long series of events that were initiated specifically one or more cell generations earlier. Thus when we discuss cell production we must consider many events that occur throughout the replication cycle. These include the formation of the many molecules that will appear in the two cells—essentially this requires a doubling of all the original cell's components. We will be mostly concerned here with the role of the cell surface, and this role is a major one. Evidently, the surface undergoes dramatic changes during cell separation; these are essential to proper division of the cell's contents and the separation of the progeny cells. But more subtle roles of the cell surface have been discovered during the past decade. On the one hand, the replication of DNA and its equal partition between progeny cells appears to be membrane related, at least in bacteria. On the other hand, the membrane of cells of higher animals is thought to govern the decision of the cell either to make DNA and divide or to not divide, depending on signals that reach it from the outside environment and from other cells.

Most research on the role of the cell membrane in cell division has been done with a few bacterial species or with higher animals' cells in culture[1]. We will restrict this chapter to systems of these two sorts. We will first discuss problems of current interest with regard to bacterial division and then proceed to similar problems with animal cells. Our aim in this chapter is to present the main ideas regarding the above subjects, as we perceived them in early 1973. The literature to this date will not be referred to extensively, but will be limited to a few very recent references and reviews, from which earlier work can be located.

6.2 BACTERIAL DIVISION

6.2.1 Membrane–DNA reciprocal relations

Much of the current interest in the role of the membrane in DNA duplication started with the proposal that DNA is physically connected to the membrane in such a way that as the membrane grows the two daughter DNA molecules are separated, and are eventually segregated into the daughter cells[2]. A second concept which grew out of the original one is that the membrane is also important for both DNA initiation and its replication.

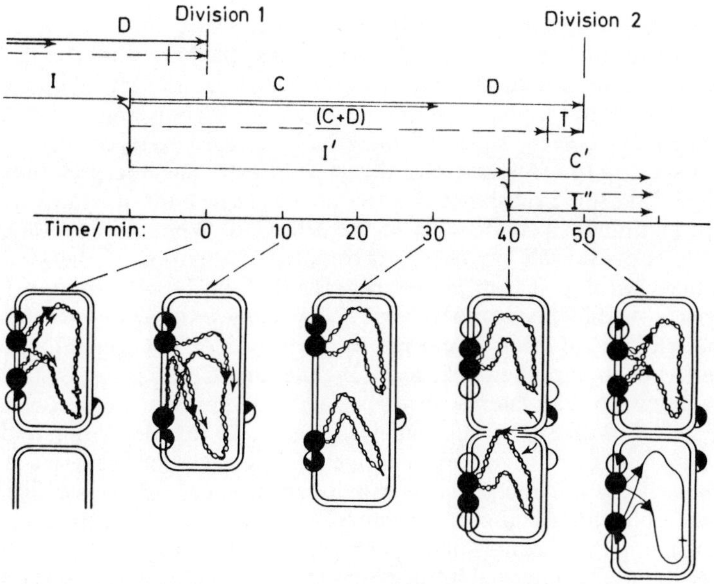

Figure 6.1 A schematic model of the *E. coli* division cycle. The upper part of the figure shows the timing of the periods of the cycle for *E. coli* growing with a 50 min division time. In the lower part are shown the state of the initiation process (circles on the left of the cells), of DNA replication (chains, with triangular replication points), and of the septation process (semi-circles on the right of the cells)

A reciprocal relation has also been proposed: DNA completion is usually first necessary to permit the membrane changes that divide bacteria. As a consequence of these two interrelationships, the duplication of DNA and its partition between daughter cells is coordinated with bacterial division; thus each daughter cell receives its full complement of genetic material. In the following sections, we will briefly discuss these processes in turn. A more extensively documented review has been prepared[3]. Many fundamental experiments on bacterial growth and macromolecular synthesis at different growth rates are summarised by Maaloe and Kjeldgaard[4].

6.2.2 Membrane events and DNA replication

6.2.2.1 DNA initiation and the I period

Each duplication of the bacterial cell is thought to proceed as a linear sequence of biochemical and morphological events, according to the model of Helmstetter and Cooper[5] (Figure 6.1). The first in this sequence is called the *I* period; during it some events take place that are required for initiation of DNA synthesis. The biochemistry of these events is not known and the discovery of their nature is one of the main problems of bacterial division. *I* is determined as the time interval between successive DNA initiations in a cell in exponential, balanced growth. It is equal in time to the interval between cell divisions, since according to the model each DNA initiation is followed after a constant time by a cell division. Thus, the length of *I* is highly dependent on growth rate; it can be as short as 20 min in very rich medium, or can require several hours in medium with a poor carbon source. DNA initiation takes place every time cell mass has increased by a constant amount (per DNA origin), irrespective of the growth rate[6]. The interpretation of this finding is that some substance or structure has to be built up gradually by the cell, in an amount proportional to the cell mass. When, after a time *I*, this quantity is made, DNA synthesis is initiated. Formation of the *I* substance continues, but now it is in preparation for the next DNA initiation.

What is the biochemical nature of *I*? Its synthesis is clearly not dependent on the synthesis of DNA or chromosome completion (as originally proposed in the Replicon hypothesis of Jacob, Brenner and Cuzin[2]). For instance, new DNA initiations can later be made possible if the mass is allowed to increase while DNA synthesis is prevented (for example by depriving a thymine requiring mutant of thymine). Also, new DNA initiations take place before old replications are completed in rapidly growing cells. This is called dichotomous replication, and can be pictured as several sets of replication points moving along the chromosome, spaced at time intervals equal to *I*. When *I* is shorter than the minimal time required for chromosome replication (*ca.* 41 min in *E. coli* B/r), dichotomous replication is observed. At these rapid growth rates, it is the frequency of DNA initiations and not the constant rate of replication at each replication point that determines the total rate of DNA synthesis per cell[4].

Although DNA synthesis does not seem to be involved in the next DNA initiation by bacteria, protein synthesis has been long known to be required. At least two proteins probably must be made—one by about 20 min and one by 5–10 min prior to DNA initiation—as reported by several workers[7]. A *dnaA* mutant is reported to make a thermosensitive (at 41°C) protein that is required for DNA initiation but not for further DNA replication[8]. The thermosensitive *dnaC* mutants also are defective in DNA initiation. RNA synthesis is also reported to be required at the time of DNA initiation[7]; this is consistent with studies *in vitro* showing that DNA synthesis starts with an RNA segment (for a review see Refs. 9 and 10). The roles of these proteins are not known, nor is their relation to the *dnaA* and *C* mutations. But the *I* period might be needed gradually to build up a structure that contains at least two proteins.

Does the membrane have a role in *I*? A likely process is the formation of some specialised structure which permits DNA to combine with the cell membrane for initiation. Some characteristics of DNA initiation make this possibility attractive. The DNA at the origin is preferentially isolated with the membrane of *Bacillus subtilis*[11], and attachment of DNA to membrane appears to be seen with the electron microscope[12]. Also the thermosensitive protein of a *dnaA* mutant is associated with the membrane[8]. This hypothesis is admittedly very speculative, but the alternatives proposed are even weaker. Thus, the *I* period has been proposed to be required to synthesise (gradually) a substance that titrates an inhibitor of DNA initiation, itself produced quickly at the time of the previous initiation, or to dilute this inhibitor[13]. Another scheme that could account for long-spaced events is based on periodic oscillations of metabolite concentrations in the cell, driven by the regular replication of certain genes. A key effector molecule could thereby reach a critical concentration that permits DNA initiation, at intervals that depend on nutrition[1,6]. Models of this sort that depend on free cytoplasmic molecules do not account for the unequal timing of DNA initiation in the two daughter cells produced upon division of *Caulobacter cresentis*, both of which inherit the same cytoplasm but different parts of the cell envelope[14], although their sizes differ slightly.

Replication of extra chromosomal segments of DNA, such as the *Flac* episome might be regulated similarly to chromosomal replication[15]. The episomal system can become integrated in a mutant whose chromosome initiation is defective and it then determines the replication of the chromosome[16]. Phage also provides model systems for investigating control of DNA replication[10].

6.2.2.2 DNA replication and the membrane

E. coli DNA replication starts at a specific position[2], at *ca.* 75 min on the genetic map. It proceeds bi-directionally from this origin and terminates at *ca.* 25 min[17]. Its replication proceeds at an apparently constant rate until termination. This rate is maximal at *ca.* 50 000 base pairs incorporated per minute per replication point, or a total of 41 min for complete chromosome replication. At growth rates slower than *ca.* 1 division per hour, DNA replication occupies about two-thirds of the generation time; it begins at about the time when the cell divides. This interval during which the DNA replicates is named the *C*-period[5]. Of course, other processes including cytoplasmic and membrane syntheses proceed during the *C*-period. The rate of DNA synthesis per site can be decreased up to threefold, independently of other changes, by limiting the concentration of thymidine supplied to a thymidine-requiring mutant[18]. As a result there are more DNA initiations during each cell cycle, dichotomous replication results, and the mass:DNA ratio is increased; greater mass:DNA ratio is seen in a greater cell diameter.

DNA replication appears to be associated with the membrane[10]. Newly synthesised DNA was very early shown to be preferentially associated with a membrane fraction[19]; such preparations can synthesise DNA[20]. The membrane that is associated with DNA can be separated from the rest, and seems to be different in having more phosphatidylethanolamine and a greater affinity for Mg ions[21]. Also the enzyme primarily responsible for DNA synthesis, polymerase III, is membrane bound[9,10]. This enzyme is coded for by the *dnaE* gene. It is only one of several proteins involved in DNA replication, because there are at least two other genes, *dnaB* and *dnaG* whose mutation stops DNA replication[16]. Synthesis of a membrane protein, *Y*, of molecular weight 44 000 dalton is diminished[22] when DNA synthesis by *E. coli* is blocked by any of a variety of techniques, the function of protein *Y* is unknown.

6.2.2.3 Separation of chromosomes

One of the two completed DNA copies normally goes to each of the two daughter cells. It is in this process that attachment of DNA to membrane would most evidently be advantageous. As suggested originally by Jacob *et al.*[2], each DNA copy could be attached to a site on the membrane; these sites could move away from one another as the membrane grows between them. Septation between the attachment points then would partition DNA between the daughter cells. However, separation and segregation presents a problem because the DNA strands of *E. coli* are each *ca.* 1200 μm long which is about one thousand times longer than the cell. But all this DNA appears to be condensed tightly, with loops held together by RNA links and with a connection to the membrane[23]. The *in vivo* diameter of DNA as observed with an electron microscope is about 0.5 μm[12]. In this condensed state, it is easy to imagine how the septum can separate two nuclei.

The DNA could be separated owing to attachment at its origin. The location of the actual site on the membrane to which the DNA is attached is unknown, as is the time at which the site is synthesised, and its chemical composition.

Mutants are known in which the only observed defect is that septa are not formed. Their DNA replication seems to be normal, and the DNA is found in clumps equally spaced along long snake-like cells. This suggests that the DNA-binding sites are made and located normally in the absence of septation[23,25], DNA-binding must be defective in some other mutants, in which DNA synthesis occurs but all of the DNA is deposited in the centre of the cell. In these cases, a mutation appears to uncouple DNA termination and septation.

6.2.3 Separation of bacterial cells

6.2.3.1 Timing of division events

After termination of a round of DNA synthesis there is normally a constant interval of at least 21 min before the separation of *E. coli B/r* cells. This interval is called the D period[5]. The total time required for a replication sequence is thus equal to $I + C + D$; the I period, whose length depends on nutrition, is followed by the C and D periods of constant minimal lengths. These sequences overlap in a way determined by the rule that a new I period starts as soon as the previous I period is completed (Figure 6.1). The model accounts well for many data at different growth rates, such as timings of DNA initiation and completion relative to division, quantities of DNA per cell, and larger mass per cell at higher growth rates[6].

The D period in each round is most readily observed with slowly growing cells, where it takes longer than 21 min and occupies the last part of the division cycle during which there is not DNA synthesis. With more rapidly growing cells DNA synthesis goes on throughout the entire cycle; but the DNA that is being made in the last 21 min before a division is not a prelude for that division but for the one to follow.

On the basis of these results on timing of events, and because when DNA synthesis is prevented only those cells divide that have already completed their DNA synthesis, completion of chromosome replication was proposed to initiate the cell division process. This model now seems too simple (see Refs. 6 and 26). As will be discussed below, a DNA completion is usually required for a division; but division also requires other biochemical events that occur throughout the *C* period (Figure 6.1). Several processes that are initiated early and go on in parallel through most of the cycle have been proposed to govern cell division. One is DNA synthesis itself, of duration *C*, and the other is of duration $C + D$ and is independent of DNA replication. Consistent with this idea are observations that the *D* period is shorter if DNA synthesis (*C* period) is lengthened by several methods. Also a thermosensitive *Salmonella typhimurium* mutant continues to divide after its DNA synthesis stops at 42°C, but its rate of division is highly dependent on nutrition; this shows that some process other than DNA synthesis now determines the division rate[26]. *B. subtilis* wild-type, in contrast to *E. coli*, continues to divide in the absence of DNA synthesis and forms DNA-less cells; its division time is apparently determined by a non-DNA mechanism[6].

A sufficient means to co-ordinate these major processes, DNA synthesis and cell division would be always to initiate the division process at some specific time of chromosome replication. This time would not necessarily be at chromosome termination nor would it even depend directly on replication of DNA; it could be that upon attachment of the DNA origin some change in membrane structure simultaneously initiates both events. In indirect support of this idea, elevated temperature both blocks DNA initiation and also uncouples division from DNA synthesis in *dnaA* and *dnaC* mutants[26]. An *E. coli dnaA* mutant (whose initiation is blocked) has a modified membrane and altered synthesis of a membrane protein[8]. The *dnaC* mutants also may have a modified membrane[26]. Also, cell mass increases more rapidly at the time of DNA initiation, as if new growth zones were formed at this time; the mass-increase effect can be still observed upon brief inhibition of DNA synthesis with nalidixic acid[27].

6.2.3.2 Sequential events of division

Since the processes required for septation and division are not all concentrated in the *D* period, we can ask what is known about those of the events that occur during *C*. Many mutants that are defective in division have been identified by numerous workers. Hirota *et al.*[28], have mapped some of the mutations; they have widely different structural consequences. Some do not form septa and grow into filaments, some form septa at unusual locations, etc. They might be classified into seven categories according to the process that is defective[24]. These mutants are in addition to ones in which a defect of DNA replication modifies division. The variety of chemical and physical agents that perturb division also point to a very easily disturbed process that has a number of steps.

Protein synthesis in the *C* period is the process so far found to be required for division. Synchronously growing cultures of *E. coli* exposed for short periods during *C* to protein synthesis inhibitors showed delayed division[29]. *E. coli* heated briefly to 45°C were delayed in division, and this delay was longest with the oldest cells, as if a heat-sensitive substance or structure essential for division (but not for DNA synthesis) had to be built up continually[30]. That this 'structure' probably contains protein was shown by the increased heat sensitivity of cells that earlier in their cycle had been grown with *p*-fluorophenylalanine, that is incorporated and increases the heat instability of proteins. A *S. typhimurium* mutant appears to be extra sensitive to heat-induced division delay[31]. It has an apparently normal membrane structure, but is defective in a lytic enzyme activity required for cell division.

The heat labile protein that must be made prior to *E. coli* division might very well be in the cell envelope. Heating causes *ca.* a 15% swelling of the cells[32]. When this swelling is osmotically prevented, then cell division is not delayed. Conversely, swelling caused solely by a negative osmotic shift produces the same delay as does heating. Possibly the effect of temperature on the normal cell is to derange a spatial relation of enzymes and substances in the cell envelope. Substances that decrease the stability of the membrane stimulate septation of *Agmenellum quadruplicatium* mutants[33]. The synthesis of a membrane protein *X* of molecular weight 50 000 dalton is increased under several conditions that block division[22]. Hirota *et al.*[16] also have observed membrane changes in division-defective mutants; some of these mutants are remediable by a high salt concentration in the medium. Division of *lon*⁻ mutants of *E. coli* is prevented after DNA synthesis is very slightly inhibited, for example with ultraviolet light in rich medium. Deranged DNA synthesis and cell division might be connected through a soluble material[34]. These mutants also showed a difference in their membrane protein pattern[35].

A new stage seems to be reached in the *E. coli* division process when DNA synthesis is completed (e.g. when the *D* period starts). Then the rate of wall synthesis increases up to twofold, and penicillin sensitivity rises sharply[36]. *B. subtilis* membrane synthesis rate doubles near the mid-point of chromosome replication[36a]. This is consistent with numerous models for division based on periodic production of excess cell envelope relative to mass; but this kind of explanation seems too simple to account for the special changes in cells observed prior to division.

The protein that is required for division needs to be synthesised for only a few minutes after DNA synthesis has terminated (as shown with inhibitors[6,36b]. Some mutants grow into filaments at 41°C; these contain normal quantities of DNA. They divide after the temperature is lowered; but not if protein synthesis is blocked with chloramphenicol[31]. Therefore, under these conditions protein synthesis is required up to the division event. But one such mutant does divide even in the presence of inhibitors of protein synthesis[25]. Chloramphenicol actually accelerates its division at 30°C and even causes it at 42°C, the non-permissive temperature. But other inhibitors that do not block peptide bond formation, such as puromycin, do not affect septation. Possibly some compound that is normally used up in peptide bond formation is needed to permit the final stages of septation, and accumulates in this mutant only when peptide bond formation is prevented. The results suggest a connection between protein synthesis and a very late division event.

Wall growth is required for cell division, since inhibitors of wall synthesis, such as penicillin (see later) prevent septation. Also mutants defective in wall[37] and phospholipid[38] synthesis do not divide normally. This is probably because the defective envelope structure cannot hold the shape of the inward growing wall, and there is swelling at the septum site. Similarly, wall-less propoplasts do not divide normally[39].

6.2.3.3 *Envelope and septum structure*

The synthetic and regulatory processes just discussed are directed toward a morphogenetic event, the formation of a septum. One problem is what submicroscopic or biochemical processes distinguish cell elongation from septation? Another is the nature of the ultrastructural basis for the position at which a septum forms. Bacterial envelopes are complex, as are their changes during division[40]. The *E. coli* envelope has at least three main layers: the innermost is the protein–phospholipid plasma membrane, next is the rigid peptidoglycan wall that gives shape to the cell, and outermost is a thin protein–lipopolysaccharide layer. *B. subtilis* has two layers in its envelope, plasma membrane and wall. Electron microscopy has not yet demonstrated structural differences between the longitudinally and inward growing parts of the envelope (see for example Ref. 41). However, selective inactivation of septation by chemicals or by heating certain temperature-sensitive mutants, suggests that special enzymes might be involved in making a biochemically different septum structure. The higher sensitivity of the septum site to low concentrations of penicillin weakens the cell selectively at this location and makes the membrane bulge. This penicillin concentration inhibits an endopeptidase[42]. The structure or availability for enzyme attacks of these peptide bonds appears to be different at the septum site. A mutant that forms septumless filaments (at 42°C) also does not produce these bulges with penicillin[31]. A structural difference is also indicated by more rapid degradation of sides than ends of *B. subtilis* wall preparations by an autolytic enzyme isolated from the bacteria[43].

The fine structural aspect of septation have been studied by Higgins and Shockman with the electronmicroscope, using *Streptococcus fecalis* whose septation process is simpler than that of the rod-shaped bacteria which elongate[44]. A new cell is approximately spherical; a wall band or ridge can be seen around its equator at which a mesosome[45] and DNA are attached. Inward membrane growth commences under this annular band, with wall being formed in the fold of its outer surface. The inward growing edge moves towards the centre of the cell, and simultaneously the following wall comes apart in two equal layers at an outer position. The result is that in a cross-section of the cell through the poles, the newly formed envelope material is seen as a Y–shaped structure, with the stem of the Y growing inward; and simultaneously being split apart at the junction into the two arms. The arms' tips are attached to the old envelope. Inward growth continues until the cell centre is reached, whereupon the cell is divided in two. By this mechanism septation and new-envelope synthesis are parts of the same process. In each round of replication the old envelope is retained as two hemispheres, one by each daughter cell, and each cell acquires a hemisphere of newly formed material. One anticipates finding only impermanent chemical differences between septum and envelope, since the former becomes the latter. Furthermore, the site of septation initiation is always at the predetermined location of the wall band, a structure that originated from the previous wall band; thus the location of the cell division site is determined, at this surface structure.

The picture is more complex with rod-shaped cells. The structural appearance of *E. coli* depends upon temperature and osmotic pressure[46]. Septation can be seen only late in the cycle. The electron microscope reveals inward growth of what appears to be a septum followed by formation of the entire multilayered envelope. Splitting of septum into two polar caps is delayed until just before division. Rod-shaped organisms do not have any visible site at which septation will occur, unlike *S. aureus* with its wall band. However, making use of the specific bulging of the *E. coli* envelope following penicillin treatment, Donachie and Begg[47] have shown that a sensitive site exists throughout the entire division process. It is a constant distance of 1.7 μm from the further (older) end in slowly growing cells. This result together with photographs of cells' growth on solid medium leads to the idea that a cell always has a preseptum site located at its newly formed end. At first its growth takes place unidirectionally beyond this point, until the cell's length doubles; then septation occurs at the original preseptum site, which is now at the cell's centre (Figure 6.1). The picture is slightly more complex for rapidly growing cells which behave like two minimal-medium cells that are joined at their newest ends. These cells are longer, and they undergo a 'dichotomous' envelope growth with two or more growing zones and several waves of septation. This form of growth would be expected, by analogy with DNA dichotomy, whenever the total time needed for septum formation is longer than the cell division time. Predetermined septum sites are seen even in a *S. typhimurium* mutant that at 38°C forms short filaments with most of its DNA at one position, which bud off cells lacking nuclear bodies. Penicillin-sensitive locations where septation later occurs are spaced along the filaments at orderly distances[26].

Growth of the envelope must occur in zones, as bands around the cell, according to the results of both Higgins and Shockman with *S. aureus*[44] and of Donachie and Begg with *E. coli*[47]. This zonal growth was found with some gram-positive coccal bacteria when surface antigens were labelled with fluorescent antibody and then distribution of fluorescence upon subsequent growth was followed in the microscope[48]. However, similar experiments with *E. coli* (in which growth of the lipopolysaccharide layer was probably being followed by this method) indicated diffuse intercalation of new material. These experiments and those in which radioactive envelope precursors were incorporated and then their distribution was followed by either radioautography or fractionation yielded contradictory results (see Ref. 49). Probably many of these components are redistributed after their incorporation. These problems of envelope assembly and reassortment are discussed by Fox in this book (see also Refs. 49 and 49a).

The mechanism by which septum sites are positioned on the cell is the problem of interest here. These sites are at least in part determined by the cell's genetic information, since some mutants' septa are displaced[16]. The structure and mechanism that determines the location of septa remains an unsolved problem. Evidence already mentioned suggests orderly, localised and timed septation from predetermined sites somewhere in the envelope[6,26,47]. Since septa appear after the walls of *B. subtilis* have been removed and then allowed to grow back[39], the information for specifying septum formation is not resident in the wall itself; however, the positioning of the septa might be determined by a wall structure.

The simplest mechanism for localising septum sites in rod-shaped cells is similar to the one proposed for cocci[44]. A special zone in the cell envelope could be duplicated and the copies passed on to the daughter cells. This process would require incorporation or assembly of special molecules into zones of the cell envelope. Although the incorporation of many envelope components appears to be random, these results do not rule out the localised incorporation of special molecules into one of the layers of the envelope. A strong claim for localised growth zones is based on unequal inheritance of β-galactoside permease by daughter cells of *E. coli*[49]. The large number of permease molecules originally present in induced cells were conserved into only four progeny cells upon further uninduced growth. It was concluded that the membrane grows in localised zones. Donachie and Begg[47] also proposed zonal growth.

6.2.3.4 Cell separation

Although the sites where septa will be formed are present throughout the cycle, and the preparation for septation occurs through most of the cycle, septa can be seen only late in *E. coli*. Thus, only a few per cent of the cells in an exponentially growing population have septa. In synchronised cultures a sharp peak in the percentage of septated cells is seen just before division. The entire time between the first appearance of a dark band at the septum site (seen in the microscope) and cell separation was only 2.5 min for *E. coli* in a very rich medium[50]. Little work has been done on effects of nutrition, etc. on the process. Its duration, calculated from the percentage of cells with a visible septum, varies from 5 to 10 min in various *E. coli* strains growing in glucose-salts medium[51]. Physiological independence of the daughter cells seems to happen only a few minutes before a septum is visible. The interval between this physiological separation and complete cell separation has been named the *T*-period[52]; for *E. coli* it was about 7 min in glucose–salts medium, leaving only *ca.* 14 min of the *D* period for the preparatory events. But, *B. subtilis* has a *T* period of *ca.* 138 min, relatively independent of growth rate. Hence there is dichotomous septation, so that chains are formed even in quite slowly growing cultures[53]. But the *T* period is quite short in a Mg^{2+}–containing medium[36a].

Cell separation requires autolytic enzyme attacks on the cell wall. These enzymes are found in the membrane, and can become firmly bound to the wall[54]. Some *B. subtilis* mutants with a temperature sensitive autolytic enzyme do not divide, and grow into long chains of cells[54a]; the purified autolysin dissociates the chains[55]. A *S. fecalis* mutant with an autolysin-resistant wall also forms long chains of cells[40]. Incorporation of ethanolamine in place of choline into the walls of *Diplococcus pneumoniae* resulted in chain formation; this compound apparently makes the walls resistant to autolysin[56].

6.2.4 Summary

Bacterial division is not one problem but many. As this article has indicated, division is at the cross-roads of several levels of structural complexity and of experimental approaches. The fascination of this problem perhaps lies in its potentiality for connecting these biological aspects. For example, many problems of cell regulation must be solved by providing observations on growth rates and nutrition with biochemical and molecular biological explanations. Similarly, what we know about gross cell size, shape and morphology must be connected with information on macromolecular and supramolecular structures. Hopefully, the bacterial cell provides an entity sufficiently complex for these events to occur, but not too complex for their investigation and understanding.

At present we can picture the events of the bacterial cycle as providing a set of sub-problems, classified according to the portion of the cycle in which they occur (*C*,*D*, etc.). The eventual aim is to describe everything in molecular terms. For example, the observable events of septation should be explained as building up of known macromolecular structures, with appropriate enzymes at strategic points for changing the growth pattern periodically. This molecular explanation of all parts of cell division will not be possible for some time yet. At present, we do well if we can link information on closely related levels. For example, we can hope to explain gross morphological observations in terms of structures seen with the electron microscope. We can list a heirarchy of experimental approaches, each underlying the previous ones. These include: (a) the cellular approach in which numbers of cells are counted under different experimental conditions, for example with synchronous cultures; (b) the gross morphological approach which considers cell size, shape, position of septa and various other major features; (c) the fine morphology approach in which the electron microscope is primarily used; (d) the study of macromolecules, particularly those that provide structure; (e) the replication of DNA; (f) enzymology of division and its regulation; (g) genetics of the division process.

As examples of this stepwise attack on the broad problem, if we consider the growth of the cell envelope (level b) how can we best fit the polar growth described by Donachie and Begg[47] with the several zones into which macromolecules (permeases)[49] are sequestered? Again, how do we account for the pattern of zonal growth in *S. fecalis* as seen by Higgins and Shockman[44] with information on the macromolecular structure of the cell wall[40]? Or, how do the enzymes and inhibitors of septation[42,54] relate to the mutations[16,37] that affect this process?

6.3 ANIMAL CELL DIVISION

Although many studies have been carried out with lower eucaryotes such as yeast, algae, Tetrahymena, Physarum, Paramecium and Stentor[57], very little of this work implicates the cell surface in relation to division. We therefore will pass on directly to studies made with vertebrate cells in culture, on which there is a voluminous literature. The general approach used with these cells is much more 'cellular' than is the 'biochemical' approach just discussed. This has been determined in part by the complexity of the cells, and until recently by the difficulty of obtaining sufficient material for analysis by conventional biochemical means.

We are confronted with a quite different set of questions when we turn to the study of eucaryotic cell division. These cells, unlike bacteria, have two alternative states: growth leading to the production of more cells or the resting condition in which specialised products of differentiation are often formed. The switch from one of these states to the other is influenced by numerous factors including nutrition, virus infection and density of the cell population (cell–cell interaction); each of these factors poses separate problems in the regulation of cell division. The surface membrane is implicated in the control of the switch between the dividing and non-dividing states. This is the subject of this part of the chapter.

6.3.1 The mammalian cell cycle

6.3.1.1 Events during the cycle

Mammalian cells in culture proceed from one mitosis to the next by a poorly understood sequence of events. The interval between two successive mitoses (the cell cycle) has usefully been divided into four stages mainly following the early observation that DNA synthesis takes place only during a discrete part of the cell cycle. Between the period of mitosis (M phase) and DNA synthesis (S phase) are the 'gap' intervals G_1, the period between mitosis and initiation of DNA synthesis and G_2, the interval between completion of chromosome replication and mitosis. The cell cycle is generally assumed to be a sequence of ordered events. However, which events are crucial for the progress through the cell cycle and particularly through the 'gap' intervals, G_1 and G_2, remain major problems in cell biochemistry.

The M period has been described in terms of a morphological sequence that culminates in segregation of the chromosomes into the daughter cells[58]. In contrast to the pronounced morphological changes, the M period appears to be relatively quiescent with regard to metabolism. The rate of protein synthesis declines markedly[59] and very low levels of RNA synthesis have been detected during M[60], possibly due to a reduction in template activity for RNA synthesis of mitotic chromatin[61]. The energy requirements for mitosis have been studied; it was suggested that small quantities of energy are necessary for the progress through mitosis except for metaphase[62].

Little is known regarding the sequence of events during the G_1 period. Continuous protein synthesis is required to complete this phase of the cell cycle[63,64]. The duration of G_1 differs by up to fourfold from one cell to another in a growing culture, indicating a loose coupling in the interval between division and DNA synthesis. After division a decision is made in the cell either to proceed through G_1 toward DNA synthesis or to stay in a quiescent state, called G_0. Cells arrested in G_0 can be induced to divide again when an appropriate stimulus is applied. Smith and Martin[64a] provide evidence that all cells enter a Go-like state, and escape from this state with constant probability (P) per unit time. Conditions that permit rapid cell growth are ones with a high P value, whereas resting (G_0) cells have a low P value. This model accounts quantitatively for the variability of G_1 and other cell cycle properties. Comparisons of newly synthesised proteins by cells in

G_0 or other phases of the cell cycle support the concept that the cells arrested in G_0 enter into a state distinct from any phase of the normal cell cycle[65]. Thus, at least two different kinds of events may operate depending upon the physiological state of the cell; one in continuously dividing cells and other in cells arrested for some time in the G_0 state. Accordingly, we will discuss separately the surface changes that take place in cells traversing the cell cycle and those observed when the cells move into the resting state or reinitiate growth from it.

The S period during which DNA replication takes place can also be subdivided since some parts of particular chromosomes are made prior to others[66]. The unravelling of the recognition signals involved at specific times constitutes an important task to define the mechanism that determines sequence of events during S. Contradictory results have been published on the topology of DNA replication. Recent evidence shows that DNA replicates at sites distributed throughout the cell nucleus and not associated preferentially with the nuclear envelope[67,68]. A marked increase in the activity of several enzymes related to the synthesis of the precursors for DNA replication, particularly the ribonucleotide reductase activity[69], has been found, and structural proteins like histones are also made during the S phase[70]. Cell-cycle dependent chromatin changes were detected by several techniques; it has been suggested that they reflect the degree of complexing between DNA and chromosomal proteins[71].

RNA and protein synthesis seem to be required during the G_2 phase in order that the cells go into the following mitosis[63,64]. Changes in the pattern of chromatin proteins have been reported to occur during this phase[72].

The normal progression of the cell cycle can be interrupted at different stages by a variety of compounds. Reversal of the block usually is followed by a partially synchronised wave of growth. Cells arrested in G_1 (or G_0) can be produced by omitting isoleucine from the culture medium[73] or by serum starvation[74]. Cells arrested near the G_1/S boundary are produced by adding compounds, i.e. hydroxyurea or 5-fluorodeoxyuridine, that block the synthesis of DNA precursors. High levels of thymidine are frequently used for this purpose, but were shown to inhibit DNA synthesis incompletely[75] and to modify the synthesis of some enzymes but not others[76]. Limitations of these sorts are important to keep in mind when an inhibitor is used for producing synchronized populations of cells. Cells can also be arrested in mitosis by compounds like colchicine, colcemid, vinblastine, etc. that dissociate the mitotic spindle. Unfortunately no procedure is yet available to obtain cell populations specifically arrested in G_2; this hampers the biochemical characterisation of this phase of the cell cycle.

6.3.1.2 Surface changes during the cell cycle

Some striking membrane changes have been detected during the cell cycle using a variety of experimental approaches. Fox *et al.*[77] used wheat germ agglutinin conjugated with fluorescein isothiocyanate to examine this lectin binding to the cell surface during the cell cycle. These experiments revealed that normal cells are only labelled during M and early G_1. Recently this cyclic surface phenomenon was confirmed with a different lectin, fluorescein conjugated concanavalin A[78]. Furthermore, in Hela cells the expression of 'H' blood group activity was maximal in M and early G_1[79], and antigenic sites of mouse lymphoma cells were also expressed in early G_1[80] or M[80a]. All these results suggest that the cell membrane undergoes a change in its configuration during M and early G_1.

Biochemical studies have shown that the incorporation of several membrane precursors is sharply increased shortly after division of KB cells[81]. The incorporation into cell surface of glycosamine was also reported to be periodic, and the rate increased immediately after division[82]. The freeze–cleavage technique has been used to show a marked change in the number of intramembraneous globular particles during the transition from metaphase to early G_1[83]. This observation provides direct evidence for a structural change of the membrane during this stage of the cell cycle.

A functional change in the membrane has also been seen during this same period. A twofold increase in sodium ion flux approximately 2 h after division was demonstrated in L 51787 mouse cells[84]. Transport rates of several compounds including an amino acid and pyrimidines increased twofold within an hour after division[85]. Furthermore, the specific activity of adenyl cyclase, a membrane bound enzyme, is markedly increased during M and early G_1[86].

Until recently very few changes of any sort had been described during the G_1 period of the cell cycle. On the basis of morphological, biochemical, structural and functional experiments there is sufficient information to consider now an early G_1 stage during which several specific changes in membrane properties are already evident. We would like to propose, as a working hypothesis that, in contrast to the S phase in which nuclear DNA replication is the predominant event, the early G_1 period is mainly concerned with membrane changes which are important for the progression of the cell cycle.

6.3.2 The resting state

6.3.2.1 Cessation of growth at confluence

Soon after cells become completely confluent the rate of DNA synthesis and cell division is markedly reduced, in most cultures of diploid cells and some other cell lines like mouse 3T3 cells. Growth tends to stop with the formation of a confluent monolayer under the usual culture conditions. This phenomenon has been called density dependent inhibition of growth[86a]. In contrast, transformed cells usually continue to multiply long after they reach the confluent state and they form a multilayer sheet. The partial escape of malignant cells from growth controls *in vivo* is often postulated to involve a failure of density-dependent inhibition of growth[87,88,88a]. Accordingly, the significance of density-dependent inhibition resides in the possibility that it is a manifestation of an inherent growth regulatory mechanism, the loss of which may be a factor contributing to malignancy. In spite of its importance, the nature of the density-dependent inhibition of growth remains incompletely understood. The evidence available indicates that the cessation of growth is a complex phenomenon in which the inhibitory influence of contact between cells (called topoinhibition)[88] and also the availability of general nutrients and growth factors supplied by the serum play an important role. Which of these factors is primarily involved in growth restriction at confluence still remains unresolved. Other factors, particularly pH[89], may influence the maximal degree of growth.

Pardee suggested that the cell membrane might be responsible for control of growth[90], and recently Holley[91] has proposed that the availability of critical nutrients inside the cell is a crucial factor controlling growth. Nutrients are defined here in a broad sense, including macromolecular serum factors as well as low molecular weight factors like essential amino acids or ions.

According to this hypothesis a high cell density somehow reduces the local availability, uptake, or utilisation of these factors. Support for this possibility comes from the demonstration that high levels of serum can overcome the growth block in density-inhibited 3T3 cultures[92], whose final saturation density is largely dependent upon the concentration of serum present in the medium[93]. Furthermore, the requirement for serum to initiate DNA synthesis is increased as the cell population density increased[94].

The 'wound-healing' experiment was classically considered as a demonstration of the operation of cell–cell contacts in growth restriction at confluence. When a scratch is made in a confluent, non-growing monolayer of cells, some of the remaining cells move into the wound and then divide. Since the dividing cells in the wound and the highly inhibited cells in the layer share the same medium, a uniform depletion of essential nutrients cannot be responsible, and freedom from contact was considered the important factor. However, the cells in the wound also are exposed to a different local environment, and this could be responsible for growth. In any event, migration alone is not sufficient. Recently, a serum fraction that promotes movement in 3T3 cells has been separated from another fraction that induces DNA synthesis[95]. In the presence of this partly purified movement factor, the density inhibited cells migrated into a wound but they were not committed to DNA synthesis unless a complete serum was added, presumably providing a positive signal for the initiation of the cell cycle. This result suggests an alternative interpretation of the wound-healing experiment; it rather indicates that the cells that have moved into the wound might have a better efficiency to utilise factors present at limiting concentrations in the depleted medium.

All these experiments are consistent with the possibility that contact or proximity among cells may limit the utilisation of growth factors, perhaps by reducing their uptake from the medium. A direct test of this hypothesis must await the availability of sufficiently purified serum growth factors. On the other hand, reductions in transport of several low molecular nutrients into contact-inhibited cells have been demonstrated and will be discussed in the next section.

Transformed cells are known to have a very low serum requirement for the initiation of DNA synthesis as contrasted with normal cells[88,94]. This difference is not simply of a quantitative character but rather of qualitative nature[96]. For instance, medium exposed to 3T3 cells for several days has very little capacity to support growth of these cells but can support growth of transformed 3T3 cells[93]. Furthermore, the removal of certain globulins from serum severely reduced its capacity to support growth of 3T3 cells, but this serum was fully competent for growth of their transformed derivatives[97]. Using this selective factor free medium, Scher and Nelson-Rees[98] isolated a new type of SV40 transformed 3T3 cell which is capable of growing in factor-free medium but retains the characteristic low saturation density of the parental cell line. This interesting result indicates that this serum factor does not restrict growth at confluence. Since the addition of serum induces DNA synthesis in confluent cultures of this transformed cell line[98], cell contact may reduce the uptake of another factor(s) that become limiting at confluence. Clearly more experiments are necessary to establish the relationship between proximity or cell–cell contact and nutrient uptake.

All these results clearly show that independently of the detailed mechanism(s) the cell surface plays a central role in the phenomenon of density-dependent inhibition of growth. In what follows we will discuss surface changes that occur when cells move into the resting state (G_0), when they are stimulated to reinitiate growth, and when the density dependent mechanism of growth control is broken by transformation with oncogenic viruses.

6.3.2.2 Surface changes associated with the cessation of growth

Chemical and functional membrane changes were found when the cells moved from the growing to the quiescent state. Membrane glycoproteins and glycolipids change markedly under this condition. A fucose containing glycopeptide of apparently higher molecular weight is reported to be present in growing but not in non-growing cells[99]. Also, the concentrations of certain 'higher glycolipids' (more than two saccharides per ceramide) increased when cell growth had stopped at confluence[100-102]. The changes include the incorporation of the terminal α-galactosyl residue of ceramide trihexoside in a density-inhibited BHK line and the terminal sialosyl residue of disialosyl hematoside of human diploid cells. Some glycolipid variations were found in non confluent cells made quiescent by serum starvation, suggesting that, at least partly, the modification of this pattern does not require cell–cell contact[103]. In addition to glycolipid changes, the turnover of certain phospholipids showed large changes after growth of 3T3 cells at confluency[104]. Thus, it appears that extensive changes take place in the structure of the cell membrane when density-inhibited cell lines grow to confluency. The significance of the surface changes described is still not clear. Cell fractionation techniques should be used to characterise the cell membranes in which these glycolipid changes occur, since the great majority of the studies were performed on whole cells. A most important point is to establish whether such changes are the cause or the effect of inhibition of growth.

As mentioned before, there are also transport changes when the cells go into the resting state. A small decrease in transport of amino acids[105] and a fivefold decrease in phosphate uptake[106] have been observed when 3T3 cells became density inhibited. The inhibition of amino acid transport depends upon cessation of growth rather than cell contact. In chick fibroblasts uridine[107] and glucose[108] transport were shown to be sharply reduced when the cells reached the confluent state. Another functional change in the cell membrane with density-inhibited plasmocytoma cells was observed as a large decrease in the activities of two plasma membrane enzymes, Na^+ and K^+-activated ATPase and 5′-nucleotidase, after the cells became confluent[109].

Although at the present time there is not information sufficient for attributing a specific regulatory role to the permeability changes described, they certainly show in a clear-cut way that the nutrient availability at confluency is reduced[109a]. Depletion of certain low molecular weight nutrients such as glutamine and isoleucine[73], tryptophan[110], zinc[111], or phosphate[91] arrest cell growth in G_1 suggesting that limitation of some nutrients can, indeed, restrict growth under certain experimental conditions. On the other hand, the addition of purine nucleosides and nucleotides enhance the DNA synthesis in quiescent fibroblasts induced by serum[111a,111b]. More studies on different transport systems should help to understand the relative roles of nutrient availability on cell regulation, and means of the mechanisms whereby some transport systems are tightly coupled to the growth state of the cell.

6.3.3 The reinitiation of cell growth

Resting cells can be stimulated by a variety of agents to recommence their DNA synthesis and division. Many of the agents act on the cell surface, and among the earliest events associated with the reinitiation of growth are changes in the cell membrane.

Trypsin and other proteolytic enzymes at low concentrations reinitiate growth of density-dependent inhibited 3T3 cells[112]. Burger has demonstrated that this trypsin treatment also makes the cells agglutinable by wheat germ agglutinin. Trypsin attached to beads is also effective, which shows that the proteolytic enzyme triggers growth by acting on the cell surface[112]. Neuraminidase was also shown to induce growth in density-inhibited chick fibroblasts, presumably removing sialic acid residues from the cell surface[113]. As discussed earlier, addition of serum to density-inhibited cells initiates DNA synthesis. Whether or not serum factors act primarily on the cell membrane is not known mainly because there is not available a sufficient pure serum growth factor to test this possibility. The efficiency of serum to induce DNA synthesis in quiescent fibroblasts is greatly increased in the presence of insulin[94]. Surface receptor sites for this hormone have been detected in human fibroblasts growing in tissue culture[114]. Phorbol ester, a potent tumour promoter which binds to the cell membrane, releases 3T3 cells from density-dependent inhibition[115]. A variety of other surface active agents can stimulate division of fibroblasts[116]. In general, all the stimuli described induce DNA synthesis after a lag of several hours (depending on the particular cell system) that is usually longer than the G_1 period, suggesting that the cells are not simply released from a stage of the G_1 process.

Early membrane changes, particularly increases in several uptake systems, have been found after growth is reinitiated by different stimuli. Increases in transport of pyrimidine nucleosides and phosphate were observed within 15 min of addition of fresh serum to density-inhibited 3T3 cells[117]. These transport changes account for the increased incorporation into RNA, previously considered to be the earliest changes in serum-stimulated cells. A large increase was observed in glucose uptake by confluent embryo fibroblasts exposed to trypsin and serum[108]. Neuraminidase and insulin-stimulated chick fibroblasts also show an early increase in glucose transport[113]. An increased rate of phospholipid turnover is also related to the initiation of division of 3T3 cells by serum[104].

Stimulation of homogeneous populations of resting lymphocytes by plant lectins has provided results that parallel the above. The lectins bind to specific receptor sites of the lymphocyte cell membrane and induce rapid changes in membrane properties. A twofold increase in amino acid transport by lymphocytes was observed only 5 min after exposure to lectins[118]. The uptake of uridine and glucose[119] and the K^+ influx[120] are also increased shortly after lectin stimulation. Furthermore, rapid phospholipid changes have been observed in this system[121], new membrane synthesis, measured by incorporation of glycosamine into the glycoproteins, also occurs within a few hours after stimulation[122].

Other important early changes, presumably related to the cell membrane, are alterations in the intracellular concentrations of cyclic nucleotides, cyclic AMP and cyclic GMP. We will refer to them in Section 6.3.5. All these experiments show that surface agents modify the cell growth pattern and induce early striking changes in membrane properties.

6.3.4 Surface changes and viral transformation

Transformation dramatically alters the cell surface in several ways, including changes of structural, compositional and functional nature. One of the most definitive changes is that tissue culture cells transformed by tumour viruses and chemical carcinogens are agglutinated by several plant lectins more readily than are the parental untransformed cells. Wheat germ agglutinin and concanavalin A have been mostly used for these studies[123]. These lectins do not compete in binding experiments[124], and separate receptors for each one have been extracted from cells[125]. These results clearly indicate that these lectin receptor sites are chemically and topographically distinct from each other.

When agglutination by wheat germ agglutinin and the final cell density achieved were compared in the 3T3 cell system there was an excellent correlation[112]. Particularly interesting, the flat revertant isolated by Pollack which exhibits low saturation density is correlatively weakly agglutinated by WGA[112]. Furthermore, BHK cells transformed by a thermosensitive mutant of polyoma virus[126] or SV40 transformed 3T3 cells that, because of a cellular mutation express the transformed phenotype at 32°C but not at 39°C[127], were agglutinated by lectins much more readily when the cells were maintained for at least one cell cycle at the permissive temperature for transformation.

A brief proteolytic treatment, capable to induce one cycle of growth in density-inhibited cells, as mentioned before, also increases the susceptibility to lectins. Most remarkably, if the surface of the transformed cell is covered with a concanavalin A preparation that has been previously partly hydrolysed with trypsin, density-dependent inhibition is re-established[112]. This experiment is essentially the opposite result of 'uncoating' and re-establishment of growth by proteolytic enzymes such as trypsin. These results further emphasise the involvement of the cell surface in the phenomenon of density-dependent inhibition of growth.

The importance of proteolytic activity for the expression of the transformed phenotype has been recently stressed in experiments reported by Schnebli and Burger[128]. Several inhibitors of protease activity were found to markedly depress the growth rate and the final saturation density of transformed cells but not of normal cells. It was suggested that the inhibitors block a protease-like activity that is required for the escape of density-dependent inhibition of growth characteristic of transformed cells[128]. Furthermore, cultures of embryo fibroblasts from different species transformed by DNA or RNA viruses have fibrinolytic activity under appropriate conditions[128a].

The nature of the differential agglutinability is still under discussion. It has been proposed that transformation produces a configurational change of the membrane that exposes the agglutinin sites which otherwise are cryptic in untransformed cells[112].

An important prediction of such a model is that the number of binding sites for the agglutinins should increase upon transformation, during M and after proteolytic treatment. Several groups have performed saturation analysis using radioactive lectins and found no significant differences in the total number of sites when transformed cells were compared with the parental cell lines[124,239]. Although a significant difference in [^{3}H]Con A binding was seen when the assay was conducted at 0°C this variation may not be sufficient to account for the differential agglutinability[130]. On the other hand, visualisation by electron microscopy of the ferritin conjugated concanavalin A molecules bound to the membrane suggest that transformation modifies the topgraphical distribution of the lectin receptor sites on the membrane[131]. Transformed cells seem to have their receptor sites mainly clustered while of those untransformed cells appear dispersed. This change in distribution could explain why transformed cells are readily agglutinated by lectins even if they do not have more exposed lectin binding sites that the normal, parental cells. The clarification of this point will be of great interest not only for understanding the surface changes produced by transformation but also will give some general information about the molecular structure of the cell surface.

Various antigens appear on the surface of transformed cells, particularly the tumour specific transplantation antigens (TSTA) and the surface antigens (S). At least some of the S antigens are present in normal cells, as has been shown after proteolytic treatment[132].

Changes in the chemical composition of the cell surface have also been described. There was a marked reduction of complex neutral glycolipids in transformed hamster cell lines[100,101] and a decrease in the activity of one of the synthetic enzymes required to elongate the carbohydrate moiety of glycolipids[133].

In addition, SV40 transformed mouse cell lines showed a decrease in the levels of several gangliosides, glycolipids containing sialic acid[134]. More recently, however, no correlation was found between the presence of particular glycolipids and the physiological properties of the cells[135].

Changes in the glycoprotein profile of the cell surface have been observed

Table 6.1 Changes in membrane-linked properties of normal and transformed cells under different growth conditions

	*Sparse**	*Confluent*	*Confluent + serum*	*Transformed*	*Ref.*
Transport					
Nucleoside	+++	+	+++	+++	117
Phosphate	+++	+	+++	+++	117, 106
Amino acids	++	±+		+++	105, 139
Glucose	+++	+	+++	++++	108, 139, 140, 141
High mol. wt. fucose glycopeptide	++	+		++++	136, 138
Chain length of sugar glycolipids	+	+++		+	100, 101, 102
Lectin agglutinability	+	−		++++	112
cAMP concentration	+	+++	+	+	152, 152a 153, 155,
Growth inhibition by protease inhibitors	−	−		+++[2]	128

* These properties of sparse and confluent transformed cells are not different.

in a variety of cell systems. The membranes of transformed cells are enriched in fucose-containing glycopeptides of higher molecular weight as judged by gel chromatography[99]. Rapidly dividing cells[99] or cells arrested in mitosis[136] displayed a similar change in glycoprotein pattern. In chick embryo fibroblasts transformed by a temperature sensitive mutant of Rous sarcoma virus, the early-eluting glycopeptide was only observed at 35°C, the temperature permissive for transformation[139]. It appears that the chromatographic behaviour of this glycopeptide material is altered drastically by enzymatic removal of sialic acid residues and that a specific sialic transferase is present in greater amounts in transformed or dividing cells[138].

An important point is that many of these membrane changes are also observed when growing cells are compared to resting cells. In general, cells in mitosis and early G_1 display the membrane changes that were initially thought to be only observed in transformed cells. This agrees with the suggestion made by Burger[112] that transformation by oncogenic viruses causes a fixation of the cell surface in the normal mitotic state, and may be associated with the presence of proteolytic activity apparently located on the cell surface of transformed cells[128].

Transport also changes in transformed cells. Foster and Pardee[105], showed an approximately threefold increase in uptake of certain amino acids by polyoma-transformed 3T3 cells as contrasted to the parental line. These differences in amino acid transport were recently confirmed[139]. Hatanaka *et al.*[130] have reported a striking increase in the rate of glucose uptake by mouse or rat cells transformed by murine sacroma virus and in chicken fibroblas transformed by Rous sarcoma virus. Glucose transport by chicken cells transformed by a temperature sensitive mutant of Rous sarcoma virus is significantly increased to the permissive temperature[141]. Glucose transport was also found to be increased in cell lines transformed by DNA viruses[139] in contrast to an early report stating that glucose uptake was changed only after transformation by RNA viruses[142]. In addition to these changes, transformed cells do not exhibit the reductions in several transport systems observed when normal cells approach the confluent state (Section 6.3.2.2).

The numerous and striking changes in the membrane (summarised in Table 6.1) following transformation certainly strongly suggest that these changes are central to the appearance of unhibited growth and malignancy. A basic question at present is which are the causal relationships of these surface changes observed in transformed cells to the characteristic unrestrained growth displayed by the malignant cells.

6.3.5 Cyclic nucleotides and growth regulation

We have discussed the likelihood that the membrane has a central role in growth regulation: (a) there is a tight coupling between the uptake of several metabolites and ions and the growth state of the cell; (b) there are chemical and structural changes under different conditions that could be important for cellular adhesion and recognition; and (c) the membrane receives signals from the environment. The question that concerns us now is how the environmental information received by the plasma membrane is transferred to the intracellular milieu in order to produce a proper growth response and how the membrane function itself (transport phenomena, for example) is controlled. We will see that both processes are possibly mechanistically interrelated.

Evidence is accumulating very rapidly to support the notion that cyclic AMP is involved in the control of growth, morphology and differentiation of tissue culture cells. Indeed, some aspects of the transformation process seem to be accounted for by an alteration in cAMP metabolism. The role of this nucleotide as a second messenger in hormone action and as a regulator of a variety of cellular processes is now well known[143]. Particularly interesting, the metabolism of this nucleotide is intimately related to the plasma membrane. cAMP is synthesised by the membrane bound enzyme adenylate cyclase and hydrolysed to 5′-AMP by specific phosphodiesterases, an isoenzyme of which is also associated with the cell membrane. Furthermore, plasma protein kinases sensitive to cAMP activation are being identified in the membrane. These enzymes provide a cAMP receptor system potentially capable of modifying surface properties by phosphorylation of certain membrane proteins[144,145].

Bürk[146] was the first to show that the addition of caffeine and theophylline, well known inhibitors of phosphodiesterase activity, depress growth of BHK cells. Johnson *et al.*[147] and Hsie and Puck[148] have shown that the addition of dibutyryl cyclic AMP (dbcAMP) produces a dramatic morphological change of a variety of tissue culture cells. Particularly interesting, the morphology of some transformed fibroblasts appeared more normal in the presence of dbcAMP. This effect is reversed by compounds that bind to microtubular proteins; this suggests their involvement in the morphology change, possibly through phosphorylation of the microtubular subunits. In addition to dbcAMP and inhibitors of phosphodiesterase activity, several of the prostaglandins (which are a family of compounds that activate the adenylate cyclase activity of cultured fibroblasts) modify morphology and decrease cell growth[149]. Addition of dbcAMP lowered the saturation density of untransformed 3T3 cells but it did not decrease the final saturation density of virus-transformed 3T3 cells[150] or CHO cells[151]. This result suggests that transformed cells have a defect in response to cAMP, rather than only in cAMP formation or degradation. A comparison of the kinetic properties of the different cAMP-modified protein kinase activities present in normal and transformed cells is an important problem on which there is not yet information available.

The intracellular levels of cAMP were studied under a variety of conditions. According to one report[152] the cAMP concentration increases when 3T3 cells susceptible to density-dependent inhibition of growth have reached the confluent state. In addition, significant changes in cAMP levels were found when the saturation density of WI38 fibroblasts was modified by changing the pH of the medium[152a]. This result provided further evidence to support the early suggestion made by Bürk[146], that cAMP is the intracellular mediator of the density dependent inhibition of growth. On the other hand, no change was observed in cAMP levels when 3T3 cells reached confluency[153]. This discrepancy should be clarified since these contradictory results lead to different models of the role of cAMP in the mechanism of growth restriction at confluence.

The levels of cAMP are closely inversely correlated with the growth rates of a variety of cell lines, and were found to be significantly lower in transformed cells[152,153]. Chick embryo fibroblasts transformed by a temperature sensitive mutant of Rous sarcoma virus showed a dramatic drop in cAMP concentration only a few minutes after the cultures were switched to the permissive temperature for transformation[154]. This interesting result suggests that cAMP concentration is somehow controlled by a viral gene product.

Serum, trypsin and insulin, the addition of which reinitiate DNA synthesis and cell division in G_0-arrested fibroblasts (Section 6.3.3), each induce a rapid fall in cAMP level concentration[153,155]. Burger *et al.*[156] have proposed that these early changes in cAMP concentration may act as a 'trigger' for the initiation of the pathway to DNA synthesis. Opposite effects of serum and cAMP on the initiation of the cell cycle of CHO cells have been found; thus many of the effects of serum may be related to the cAMP system[151]. Accordingly, 3T3 cells arrested before confluency by serum starvation display a doubling in cAMP level[157] and the initiation of growth produced by serum is inhibited by the addition of dbcAMP. In addition, the effect of insulin on growth of BHK cells is also counteracted by dbcAMP[157a].

If cAMP is involved in the regulation of the progression of the cell cycle its intracellular concentration should change during growth of synchronised cultures. Several groups have recently-reported oscillations in the cAMP level during the cell cycle in normal and transformed cells synchronised by different procedures[156,158]. The most remarkable change was a drop in cAMP levels during mitosis and early G_1, much more evident in normal than in transformed cells. In conjunction with the results obtained with serum and proteolytic enzymes it seems that a transient decrease in cAMP levels may act as a trigger signal for the initiation of the cell cycle. It should be pointed out that a drop in cAMP concentration, although necessary, may not be sufficient to commit the cell to DNA synthesis, since when cells were exposed to fresh serum for only a few minutes, the serum did not induce much DNA synthesis although it produced a fall in cAMP (unpublished observations). Thus, other factors are possibly involved for the progression through the later steps of the G_1 phase of the cell cycle.

The enzymological basis of the cAMP concentration changes, and in particular the basis of cAMP-dependent defective metabolism in transformed cells, is still not clear. Bürk[146] showed that polyoma-transformed BHK cells have a reduction in adenylate cyclase, a result that was confirmed for this cell line but not for the 3T3 mouse cell system[159]. The comparisons have been based on enzyme activity. But a thorough kinetic study of the adenylate cyclase system from normal and transformed cells is highly desirable in order to clarify the role of this enzyme in transformation.

In addition to cAMP, cGMP seems to be involved in growth regulation. A rapid and striking increase in cGMP concentration was observed in lymphocyte cultures exposed to purified lectins[160]. cGMP was proposed to act as a trigger in this system, but the addition of this cyclic nucleotide has not yet been shown to bring about changes similar to those produced by lectins. If the guanylate cyclase is found to be in the cytoplasm, as in other tissues, it will be of greater interest to determine what sort of connection exists between the cell surface, to which the lectin binds, and the enzyme that synthesizes cGMP in the cell. Whether or not cGMP has a role in growth regulation of fibroblasts is at present not known; but certainly is a possibility that should be explored.

The addition of dbcAMP or compounds that modify the intracellular concentration of cAMP produces changes in several surface properties. The agglutinability by plant lectins was decreased[161] and the adhesivity to the substratum was increased by the addition of dbcAMP[162]. Furthermore, the transport of certain amino acids is reduced by dbcAMP in CHO cells[151]. Recently, Rozengurt and Jimenz de Asua have shown that the early increase in uridine transport induced by serum and insulin in quiescent mouse fibroblasts (Section 6.3.3.) is blocked by prostaglandins and by inhibitors of phosphodiesterase activity[163]. These results suggest that even the early transport changes induced by serum are possibly mediated by cAMP, and provides a likely explanation for the tight coupling between uridine transport and the growth state of the cell. In fact, normal growing cells have a high uridine transport level and low intracellular concentration of cAMP while the reverse is true at confluence.

At present it is not clear how cyclic nucleotides control so many different cellular functions. The hypothesis of protein modification by phosphorylations catalysed by cAMP-dependent protein kinases provides a framework to interpret some of the results. Briefly, different cell proteins were shown to be in equilibrium between phosphorylated and dephosphorylated forms. Nuclear proteins[164], ribosomal proteins[165], microtubular proteins[166], membrane proteins[144,145] and enzymes related to synthesis and degradation of glycogen[167] have been found to be modified by phosphorylation. Although the physiological significance of some of these phosphorylation reactions is not yet clear, the cAMP oscillations detected during the cell cycle may be triggering an ordered series of phosphorylations and dephosphorylations of structural proteins and enzymatic activities. Investigation of this possibility could give information for understanding the subtle mechanisms that control the progression through the cell cycle.

6.4 SUMMARY

For both bacteria and animal cells the normal growth cycle is most simply conceived of as a sequence of events, each dependent on the preceding one. Analysis of bacterial growth suggests that these events are structured in overlapping sequences, and not as cycles. That is, the final event of one cycle does not initiate the initial event of the next cycle. Conditions for initiation of each cycle are built up periodically, becoming active approximately when the cell mass reaches a critical value; then the rest of the cycle follows in a fixed order and with a well determined time sequence. Thus, several processes of the division cycle can take place at the same time in one cell, for instance preparation for the next DNA initiation, DNA synthesis in progress, and septation which separates DNA already completed. Furthermore, several different processes can be necessary for progress into the next phase of a cycle. For example, both DNA completion and protein synthesis are necessary for initiation of *E. coli* septation.

Cells of higher animals proceed through their cycle of division by the familiar stages G_1, S, G_2 and M (see page 167). The requirements for initiation of each stage are known in only a general way; considerable work remains to be done before even the level of understanding reached with bacteria is achieved. The key event for initiation of a cycle appears to be in the G_1 period. Cells stop at this point under many non-optimal conditions; a special process seems to be required to permit them to continue through G_1 and then on into the S phase.

One of the puzzles of the animal cell cycle is the extreme variability of cycle length with different cells. This variability is found to mainly be in the G_1 phase. Smith and Martin[64a] have provided a most attractive explanation, proposing that all cells go into a quiescent state after division (the *A* state, which is very similar to G_0), and that the probability of escape from this state for each cell is constant per unit time (just as radioactive decay is probabilistic). Thus, their model provides a quantitative basis for variable G_1. This proposal awaits a biochemical basis for its functioning.

Throughout studies of the cell cycle with both bacteria and animals one perceives a close interplay between internal and surface events. As documented in the second section, bacterial membrane events seem important for DNA initiation and replication; conversely DNA completion is important for septation to occur. With animal cells the membrane is evidently involved in the terminal events of cytokinesis, but also in the critical event that determines whether (or how soon, according to the Smith and Martin model) the cells will go on into another cycle. The membrane has a central role in the decision to grow or not to grow. The many data on this matter can be organised by use of a hypothetical scheme, shown in Figure 6.2. We can best discuss Figure 6.2 in stages, indicated by the numbers in circles.

As the cell proceeds through its replication sequence (1), it arrives at a point in G_1 which we call the restriction point *R* (2). Passage beyond this point has an unusually high requirement for nutrients (3). In particular, it is influenced by the intracellular concentration of cAMP (4), which exerts a negative effect on transport. The cAMP concentration is determined by its rates of synthesis and removal (5); these processes can be modified by various drugs, hormones, etc. Furthermore, dibutyryl-cAMP added to the medium (6) can mimic to some extent the effects of cAMP in the cell. Although it is possible to define several states of the membrane we propose, for simplicity, two major states *P* (for productive) and *Q* (for quiescent). (7) Dulbecco[168] has also proposed that the membrane may exist in two configurations: the growing and resting states. The former has higher transport activity for several substrates, and a higher lectin agglutinability than does the *Q* state. The *P* state is proposed to be less active in cAMP synthesis than is the *Q* state, because of low activity

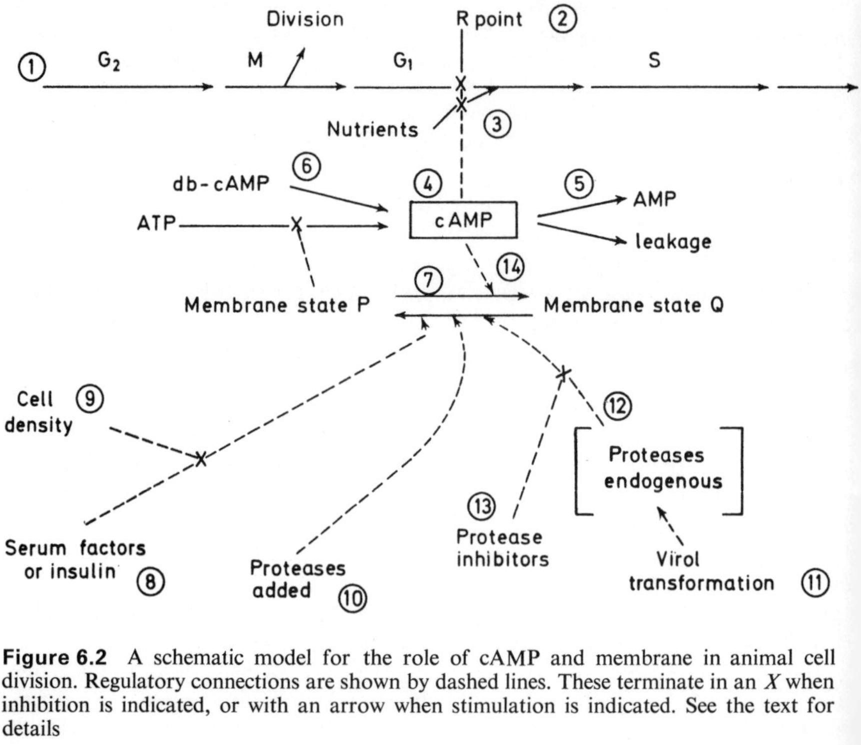

Figure 6.2 A schematic model for the role of cAMP and membrane in animal cell division. Regulatory connections are shown by dashed lines. These terminate in an *X* when inhibition is indicated, or with an arrow when stimulation is indicated. See the text for details

of membrane-associated cyclase (5), or to permit the leakage of the cyclic nucleotide to the medium. Thus, cells in the *P* state have a lower cAMP concentration and so proceed past the *R* point with a higher probability. Factors in serum (8) shift the cells toward the *P* state; densely crowded cells (9) do not respond to serum factors as readily. Added proteases (10) quickly activate the cells; viral transformation (11) appears to create protease activity (12), as indicated by the inhibitory effects of protease inhibitors (13) on growth of these transformed cells. Finally, cAMP itself shifts the cells in the direction of the *Q* state (14). This creates an autocatalytic regulatory cycle. An agent such as serum causes cAMP to drop, and this in turn shifts more of the membrane to the *P* form with a consequent further drop in cAMP, etc. This is a plausible mechanism for activation of the entire membrane surface, somewhat akin to the propagation of activation along a nerve membrane. The activated state of the membrane could persist until some later metabolic events elevate the cAMP level, with a resulting autocatalytic shift of the entire cell's membrane to the *Q* state. This sort of mechanism could account for periodic variations in cAMP level and membrane properties at stages of the cell cycle.

This scheme is an over-simplification even at the present level of knowledge. Two membrane states are possibly insufficient to explain the different characteristics of the cell surface under various growth conditions, and the proposal that transformation affects membrane properties via endogenous proteases is of tentative nature considering the little information available. In spite of these limitations and that many of the results and ideas in this brief review are sure to change in the near future, we believe it should be useful in helping to bring about more speedily just these changes which are the essentials of progress in science.

References

1. Mitchison, J. M. (1971). *The Biology of the Cell Cycle* (London: Cambridge University Press)
2. Jacob, F., Brenner, S. and Cuzin, F. (1963). *Cold Spring Harbor Symp. Quant. Biol.*, **28**, 329
3. Pardee, A. B., Wu, P. C. and Zusman, D. R. (1973). *Bacterial Surfaces* in the press (L. Leive, editor) (New York: Plenum Press)
4. Maaloe, O. and Kjeldgaard, N. O. (1966). *The Control of Macromolecular Synthesis* (New York: W. A. Benjamin)
5. Helmstetter, C. E., Cooper, S., Pierucci, O. and Revelas, E. (1968). *Cold Spring Harbor Symp. Quant. Biol.*, **33**, 809
6. Donachie, W. D., Jones, N. C. and Teather, R. (1973). In the press (23rd S.G.M. 1973: Cambridge)
7. Messer, W. (1972). *J. Bacteriol.*, **112**, 7
8. Blau, S. and Mordoh, J. (1972). *Proc. Nat. Acad. Sci.* (*U.S.A.*), **69**, 2895
9. Gross, J. D. (1972). *Current Topics in Microbiol. Immunol.*, **57**, 39
10. Klein, A. and Bonhoeffer, F. (1972). *Annu. Rev. Biochem.*, **41**, 301
11. Quinn, W. G. and Sueoka, N. (1968). *Cold Spring Harbor Symp. Quant. Biol.*, **33**, 695
12. Ryter, A. (1968). *Bacteriol. Rev.*, **32**, 39
13. Pritchard, R. H., Barth, P. T. and Collins, J. (1969). *Microbiol Growth* 263 (19th S.G.M. 1969: Cambridge)
14. Degnan, S. T. and Newton, W. A. (1972). *J. Molec. Biol.*, **64**, 671
15. Helinski, D. R. (1973). *Ann. Rev. Microbiol.*, **27**, 437

16. Hirota, Y., Mordoh, H., Scheffler, I. and Jacob, F. (1972). *Fed. Proc.* (*Fed. Amer. Soc. Exp. Biol.*), **31,** 1422
17. McKenna, W. G. and Masters, M. (1972). *Nature* (*London*), **240,** 536
18. Zaritsky, A. and Pritchard, R. H. (1971). *J. Molec. Biol.*, **60,** 65
19. Goldstein, A. and Brown, B. J. (1961). *Biochem. Biophys. Acta*, **53,** 19
20. Firshein, W. (1972). *J. Molec. Biol.*, **70,** 383
21. Ballesta, J., Cundliffe, E., Daniels, M. J., Silverstein, J. L., Susskind, M. M. and Schaechter, M. (1972). *J. Bacteriol.*, **112,** 195
22. Inouye, M. (1972). *J. Molec. Biol.*, **63,** 597
23. Worcel, A. and Burgi, E. (1972). *J. Molec. Biol.*, **71,** 127
24. Mendelson, N. H. and Cole, R. M. (1972). *J. Bacteriol.*, **112,** 994
25. Zusman, D. R., Inouye, M. and Pardee, A. B. (1972). *J. Molec. Biol.*, **69,** 119
26. Shannon, K. P., Spratt, B. G. and Rowbury, R. J. (1972). *Molec. Gen. Genet.*, **118,** 185
27. Ward, C. B. and Glaser, D. A. (1971). *Proc. Nat. Acad. Sci.* (*U.S.A.*), **68,** 1061
28. Hirota, Y., Ricard, M. and Shapiro, B. (1971). *Biomembranes*, Vol. 2, 13, (L. A. Manson, editor) (New York: Plenum Press)
29. Pierucci, O. and Helmstetter, C. E. (1969). *Fed. Proc.* (*Fed. Amer. Soc. Exp. Biol.*), **28,** 1755
30. Smith, H. S. and Pardee, A. B. (1970). *J. Bacteriol.*, **101,** 901
31. Ahmed, N. and Rowbury, R. J. (1971). *Microbios.*, **4,** 181
32. Wu, P. C. and Pardee, A. B. (1973). *J. Bacteriol.*, **114,** 603
33. Ingram, L. O. and Fisher, W. D. (1973). *Biochem. Biophys. Res. Commun.*, **50,** 200
34. James, R. and Gillies, N. E. (1973). *J. Gen. Microbiol.*, **76,** 429
35. Leighton, P. M. (1972). *Can. J. Microbiol.*, **18,** 1611
36. Hoffman, B., Messer, W. and Schwarz, U. (1972). *J. Supramol., Structure*, **1,** 29
36a. Sargent, M. (1973). *J. Bacteriol.*, in the press
36b. Jones, N. C. and Donachie, W. D. (1973). *Nature New Biol.* (*London*), **243,** 100
37. Reeve, J. N., Groves, D. J. and Clark, D. J. (1970). *J. Bacteriol.*, **104,** 1052
38. Henning, U., Dennert, G., Rehn, K. and Deppe, G. (1969). *J. Bacteriol.*, **98,** 784
39. Landman, O. E., Ryter, A. and Fréhel, C. (1968). *J. Bacteriol.*, **96,** 2154
40. Rogers, H. J. (1970). *Bacteriol. Rev.*, **34,** 194
41. Green, E. W. and Schaecter, M. (1972). *Proc. Nat. Acad. Sci.* (*U.S.A.*), **69,** 2312
42. Hartmann, R., Höltje, J. and Schwarz, U. (1972). *Nature* (*London*), **235,** 426
43. Fan, D. P., Pelvit, M. C. and Cunningham, W. P. (1972). *J. Bacteriol.*, **109,** 1266
44. Higgins, M. L. and Shockman, G. D. (1971). *CRC Critical Reviews in Microbiology*, **1,** 29
45. Reusch, Jr., V. M. and Burger, M. M. (1973). *Biochem. Biophys. Acta*, **300,** 79
46. Steed, P. and Murray, R. G. E. (1966). *Can. J. Microbiol.*, **12,** 263
47. Donachie, W. D. and Begg, K. J. (1970). *Nature* (*London*), **227,** 1220
48. Cole, R. M. (1965). *Bacteriol. Rev.*, **29,** 326
49. Kepes, A. and Autissier, F. (1972). *Biochem. Biophys. Acta*, **265,** 443
49a. Ryter, A., Hirota, Y. and Schwarz, U. (1973). *J. Molec. Biol.*, **78,** 185
50. Hoffman, H. and Frank, M. F. (1965). *J. Bacteriol.*, **89,** 212
51. Wu, P. C. and Pardee, A. B. (1973). in preparation
52. Clark, D. J. (1968). *Cold Spring Harbor Symp. Quant. Biol.*, **33,** 823
53. Paulton, R. J. L. (1971). *Nature New Biol.*, **231,** 271
54. Forsberg, C. W. and Ward, J. B. (1972). *J. Bacteriol.*, **110,** 878
54a. Forsberg, C. and Rogers, H. J. (1971). *Nature* (*London*), **229,** 272
55. Fan, D. P. (1970). *J. Bacteriol.*, **103,** 494
56. Briles, E. B. and Tomasz, A. (1970). *J. Cell. Biol.*, **47,** 786
57. Padilla, G. M., Whitson, G. L. and Cameron, J. L. (editors) (1969). *The Cell Cycle* (New York: Academic Press)
58. Nicklas, R. B. (1971). *Advan. Cell Biology*, **2,** 225
59. Hodge, L. D., Robbins, E. and Scharff, M.D. (1969). *J. Cell Biol.*, **40,** 497
60. Zylber, E. A. and Penman, S. (1971). *Science*, **172,** 947
61. Stein, G. and Farber, J. (1972). *Proc. Nat. Acad. Sci.* (*U.S.A.*), **69,** 2918
62. Tobey, R. A., Petersen, D. F. and Anderson, E. C. (1969). *Biochemistry of Cell Division* p.39, (R. Baserga, editor) (Springfield: C. C. Thomas)
63. Schneiderman, M. H., Dewey, W. C. and Highfield, A. P. (1971). *Exp. Cell Res.*, **67,** 147

64. Doida, Y. and Okada, S. (1972). *Cell Tissue Kinetics*, **5,** 15
64a. Smith, J. A. and Martin, L. (1973). *Proc. Nat. Acad. Sci.* (*U.S.A.*), **70,** 1263
65. Becker, H. and Stanners, C. P. (1972). *J. Cell Physiol.*, **80,** 51
66. Watson, J. D. (1971). *Advan. Cell Biology*, **2,** 1
67. Fakan, S., Turner, G. N., Pagano, J. S. and Hancock, R. (1972). *Proc. Nat. Acad. Sci.* (*U.S.A.*), **69,** 2300
68. Huberman, J. A., Tsai, A. and Deich, R. A. (1973). *Nature* (*London*), **241,** 32
69. Turner, M. K., Abrams, R. and Liebcrman, I. (1968). *J. Biol. Chem.*, **243,** 3725
70. Mueller, G. C. (1969). *Fed. Proc.* (*Fed. Amer. Soc. Exp. Biol.*), **28,** 1780
71. Pederson, T. (1972). *Proc. Nat. Acad. Sci.* (*U.S.A.*), **69,** 2224
72. Bhorjee, J. S. and Pederson, T. (1972). *Proc. Nat. Acad. Sci.* (*U.S.A.*), **69,** 3345
73. Enger, M. D. and Tobey, R. A. (1972). *Biochemistry*, **11,** 269
74. Bürk, R. R. (1970). *Exp. Cell. Res.*, **63,** 309
75. Bostock, C. J., Prescott, D. M. and Kirkpatrick, B. J. (1971). *Exp. Cell Res.*, **68,** 163
76. Churchill, J. R. and Studzinski, G. P. (1970). *J. Cell. Physiol.*, **75,** 297
77. Fox, T. O., Sheppard, J. R. and Burger, M. M. (1971). *Proc. Nat. Acad. Sci.* (*U.S.A.*), **68,** 244
78. Shoham, J. and Sachs, L. (1972). *Proc. Nat. Acad. Sci.* (*U.S.A.*), **69,** 2479
79. Kuhns, W. J. and Bramson, S. (1968). *Nature* (*London*), **219,** 938
80. Cikes, M. and Friberg, Jr. S. (1971). *Proc. Nat. Acad. Sci.* (*U.S.A.*), **68,** 566
80a. Thomas, D. B. (1971). *Nature* (*London*), **233,** 317
81. Gerner, E. W., Glick, M. C. and Warren, L. (1970). *J. Cell Physiol.*, **75,** 275
82. Onodera, K. and Sheinin, R. (1970). *J. Cell Sci.*, **7,** 337
83. Scott, R. E., Carter, R. L. and Kidwell, W. R. (1971). *Nature New Biol.*, **233,** 219
84. Jung, C. and Rothstein, A. (1967). *J. Gen. Physiol.*, **50,** 917
85. Sander, G. and Pardee, A. B. (1972). *J. Cell Physiol.*, **80,** 267
86. Makman, M. H. and Klein, M. I. (1972). *Proc. Nat. Acad. Sci.* (*U.S.A.*), **69,** 456
86a. Stoker, M. G. P. and Rubin, H. (1967). *Nature* (*London*), **215,** 171
87. Martz, E. and Steinberg, M. S. (1972). *J. Cell Physiol.*, **79,** 189
88. Dulbecco, R. (1970). *Nature* (*London*), **227,** 802
88a. Dulbecco, R. and Stoker, M. G. P. (1970). *Proc. Nat. Acad. Sci.* (*U.S.A.*), **66,** 204
89. Ceccarini, C. and Eagle, H. (1971). *Proc. Nat. Acad. Sci.* (*U.S.A.*), **68,** 229
90. Pardee, A. B. (1964). *Cancer Inst. Monogr.*, **14,** 7
91. Holley, R. W. (1972). *Proc. Nat. Acad. Sci.* (*U.S.A.*), **69,** 2840
92. Todaro, G. J., Matsuiya, Y., Bloom, S., Robbins, A. and Green, H. (1967). *Growth Regulating Substances for Animal Cells in Culture* (V. Defendi and M. Stoker, editors) p.87, (Philadelphia, Pennsylvania: Wistar Inst. Press)
93. Holley, R. W. and Kiernan, J. A. (1968). *Proc. Nat. Acad. Sci.* (*U.S.A.*), **60,** 300
94. Clarke, G. D., Stoker, M. P. G., Ludlow, A. and Thornton, M. (1970). *Nature* (*London*), **227,** 798
95. Lipton, A., Klinger, I., Paul, D. and Holley, R. W. (1971). *Proc. Nat. Acad. Sci.* (*U.S.A.*), **68,** 2799
96. Paul, D., Lipton, A. and Klinger, I. (1971). *Proc. Nat. Acad. Sci.* (*U.S.A.*), **68,** 645
97. Jainchill, J. L. and Todaro, G. J. (1970). *Exp. Cell Res.*, **59,** 137
98. Scher, C. D. and Nelson-Rees, W. A. (1971). *Nature New Biol.*, **233,** 263
99. Buck, C. A., Glick, M. C. and Warren, L. (1971). *Biochemistry*, **10,** 2176
100. Hakomori, S. (1970). *Proc. Nat. Acad. Sci.* (*U.S.A.*), **67,** 1741
101. Robbins, P. W. and Macpherson, I. (1971). *Proc. Royal Soc.* (*London*), **B177,** 49
102. Sakiyama, H., Gross, S. K. and Robbins, P. W. (1972). *Proc. Nat. Acad. Sci.* (*U.S.A.*), **69,** 872
103. Critchley, D. R. and Macpherson, I. (1973). *Biochem. Biophys. Acta*, in the press
104. Cunningham, D. D. (1972). *J. Biol. Chem.*, **247,** 2464
105. Foster, D. O. and Pardee, A. B. (1969). *J. Biol. Chem.*, **244,** 2675
106. Weber, M. J. and Edlin, G. (1971). *J. Biol. Chem.*, **246,** 1828
107. Weber, M. J. and Rubin, H. (1971). *J. Cell Physiol.*, **77,** 157
108. Sefton, B. M. and Rubin, H. (1971). *Proc. Nat. Acad. Sci.* (*U.S.A.*), **68,** 3154
109. Leliévre, L., Prigent, B. and Paraf, A. (1971). *Biochem. Biophys. Res. Commun.*, **45,** 637
109a. Griffiths, J. B. (1972). *J. Cell Sci.*, **10,** 515
110. Brunner, M. (1973). *Cancer Res.*, **33,** 29
111. Rubin, H. (1972). *Proc. Nat. Acad. Sci.* (*U.S.A.*), **69,** 712

111a. Clarke, G. D. and Smith, C. (1973). *J. Cell Physiol.*, **81,** 125
111b. Schor, S. and Rozengurt, E. (1973). *J. Cell Physiol.*, **81,** 339
112. Burger, M. M. (1971). *Current Topics in Cellular Regulation*, **3,** 135
113. Vaheri, A., Ruoslahti, E. and Nordling, S. (1972). *Nature New Biol.*, **238,** 211
114. Gavin III, J. R., Roth, J., Jen, P. and Freychet, P. (1972). *Proc. Nat. Acad. Sci.* (*U.S.A.*), **69,** 747
115. Sivack, A. (1972). *J. Cell. Physiol.*, **80,** 167
116. Vasiliev, J. M., Gelfand, M., Guelstein, V. I. and Fetisova, E. K. (1970). *J. Cell Physiol.*, **75,** 305
117. Cunningham, D. D. and Pardee, A. B. (1969). *Proc. Nat. Acad. Sci.* (*U.S.A.*), **64,** 1049
118. Van der Berg, K. J. and Bestel, I. (1973). *Exp. Cell Res.*, **76,** 63
119. Peters, J. H. and Hausen, P. (1971). *Europ. J. Biochem.*, **19,** 509
120. Quastel, M. R. and Kaplan, J. G. (1970). *Exp. Cell Res.*, **63,** 230
121. Fisher, D. B. and Mueller, G. C. (1968). *Proc. Nat. Acad. Sci.* (*U.S.A.*), **60,** 1396
122. Hayden, G. A., Crowley, G. M. and Jamieson, G. A. (1970). *J. Biol. Chem.*, **245,** 5827
123. Sharon, N. and Lis, H. (1972). *Science*, **177,** 949
124. Ozanne, B. and Sambrook, J. (1971). *Nature New Biol.*, **232,** 156
125. Janson, V. K., Sakamoto, C. K. and Burger, M. M. (1973). *Biochem. Biophys. Acta*, **291,** 136
126. Eckhart, W., Dulbecco, R. and Burger, M. M. (1971). *Proc. Nat. Acad. Sci.* (*U.S.A.*), **68,** 283
127. Noonan, K. D., Renger, H. C., Basilico, C. and Burger, M. M. (1973). *Proc. Nat. Acad. Sci.* (*U.S.A.*), **70,** 347
128. Schnebli, H. P. and Burger, M. M. (1972). *Proc. Nat. Acad. Sci.* (*U.S.A.*), **69,** 3825
128a. Ossowski, L., Unkeless, J. C., Tobia, A., Quigley, J. P., Rifkin, D. B. and Reich, E. (1973). *J. Exp. Med.*, **137,** 112
129. Cline, H. J. and Livingston, D. C. (1971). *Nature New Biol.*, **232,** 155
130. Noonan, K. D. and Burger, M. M. (1973). *J. Biol .Chem.*, **248,** 4286
131. Nicholson, G. L. (1971). *Nature New Biol.*, **233,** 244
132. Häyri, P. and Defendi, V. (1970). *Virology*, **41,** 22
133. Cumar, F. A., Brady, R. O., Kolodny, E. H., McFarland, V. W. and Mora, P. T. (1970). *Proc. Nat. Acad. Sci.* (*U.S.A.*), **67,** 757
134. Brady, R. O. and Mora, P. T. (1970). *Biochem. Biophys. Acta*, **218,** 308
135. Sakiyama, H. and Robbins, P. W. (1973). *Fed. Proc.* (*Fed. Amer. Soc. Exp. Biol.*), **32,** 86
136. Glick, M. C. and Buck, C. A. (1973). *Biochemistry*, **12,** 85
137. Warren, L., Critchley, D. and Macpherson, I. (1972). *Nature* (*London*), **235,** 275
138. Warren, L., Fuhrer, J. P. and Buck, C. A. (1972). *Proc. Nat. Acad. Sci.* (*U.S.A.*), **69,** 1838
139. Isselbacher, K. J. (1972). *Proc. Nat. Acad. Sci.* (*U.S.A.*), **69,** 585
140. Hatanaka, M., Augl, C. and Gilden, R. V. (1970). *J. Biol. Chem.*, **245,** 714
141. Martin, G. S., Venuta, S., Weber, M. and Rubin, H. (1971). *Proc. Nat. Acad. Sci.* (*U.S.A.*), **68,** 2739
142. Hatanaka, M., Todaro, G. J. and Gilden, R. V. (1970). *Int. J. Cancer*, **5,** 224
143. Sutherland, E. W. (1972). *Science*, **177,** 401
144. Rubin, C. S. and Rosen, O. (1973). *Biochem. Biophys. Res. Commun.*, **50,** 421
145. Johnson, E. M., Tetsufumi, U., Maeno, H. and Greengard, P. (1972). *J. Biol. Chem.*, **247,** 5650
146. Bürk, R. R. (1968). *Nature* (*London*), **219,** 1272
147. Johnson, G. S., Friedman, R. M. and Pastan, I. (1971). *Proc. Nat. Acad. Sci.* (*U.S.A.*), **68,** 425
148. Hsie, A. and Puck, T. T. (1971). *Proc. Nat. Acad. Sci.* (*U.S.A.*), **68,** 358
149. Johnson, G. S. and Pastan, I. (1971). *J. Nat. Cancer Inst.*, **47,** 1357
150. Johnson, G. S. and Pastan, I. (1972). *J. Nat. Cancer Inst.*, **48,** 1377
151. Rozengurt, E. and Pardee, A. B. (1972). *J. Cell Physiol.*, **80,** 273
152. Otten, J., Johnson, G. S. and Pastan, I. (1971). *Biochem. Biophys. Res. Commun.*, **44,** 1192
152a. D'Armiento, M., Johnson, G. S. and Pastan, I. (1973). *Nature New Biol.*, **242,** 78
153. Sheppard, J. R. (1972). *Nature New Biol.*, **236,** 14
154. Otten, J., Bader, J., Johnson, G. S. and Pastan, I. (1972). *J. Biol. Chem.*, **247,** 1632

155. Otten, J., Johnson, G. S. and Pastan, I. (1972). *J. Biol. Chem.*, **247,** 7082
156. Burger, M. M., Bombik, B. M., Breckenridge, B. and Sheppard, J. R. (1972). *Nature New Biol.*, **239,** 161
157. Seifert, W. and Paul, D. (1972). *Nature New Biol.*, **240,** 281
157a. Jimenez de Asua L., Surian, E., Flavia, M. and Torres, H. (1973). *Proc. Nat. Acad. Sci.* (*U.S.A.*), **70,** 1388
158. Sheppard, J. R. and Prescott, D. M. (1972). *Exp. Cell Res.*, **75,** 293
159. Perry, C. V., Johnson, G. S. and Pastan, I. (1971). *J. Biol. Chem.*, **246,** 5785
160. Hadden, J. W., Hadden, E. M., Haddox, M. K. and Goldberg, N. D. (1972). *Proc. Nat. Acad. Sci.* (*U.S.A.*), **69,** 3024
161. Sheppard, J. R. (1971). *Proc. Nat. Acad. Sci.* (*U.S.A.*), **68,** 1316
162. Johnson, G. S. and Pastan, I. (1972). *Nature New Biol.*, **236,** 247
163. Rozengurt, E. and Jimenez de Asua, L. (1973). (Manuscript in preparation)
164. Langan, T. A. (1969). *J. Biol. Chem.*, **244,** 5763
165. Eil, C. and Wool, I. G. (1971). *Biochem. Biophys. Res. Commun.*, **43,** 1001
166. Murray, A. W. and Froscio, M. (1971). *Biochem. Biophys. Res. Commun.*, **44,** 1089
167. Walsh, D. A., Brostrom, C. O., Brostrom, M. A., Chen, L., Corbin, J. D., Reimann, E., Soderling, T. R. and Krebs, E. G. (1972). *Advances in Cyclic Nucleotide Research*, **1,** 33
168. Dulbecco, R. (1971). *Growth Control in Tissue Culture*. (G. E. W. Wolstenholme and J. Knight, editors) p. 71 (London: Churchill Livingstone)

7
Membranes in Nerve Impulse Conduction

C. W. COTMAN and W. B. LEVY
University of California

7.1 INTRODUCTION

Nerves, whose processes extend over great distances, use an electrical impulse called an action potential to relay both incoming and outgoing signals throughout the cell. Synaptic potentials impinging on dendrites some distance from the cell body are relayed in part to the cell body via an action potential, and an action potential initiated at the cell body relays information to the synapses.

7.2 NATURE OF THE EXCITATION PROCESS

The action potential is a transient change in the membrane potential that is conducted longitudinally without decrement along the excitable membranes of the cell. It arises from a combination of several stimulus-generated changes in ionic permeability of the membrane in conjunction with the cable (core-conductor) properties of the nerve cell. At rest, the unexcited nerve cell maintains a potential of approximately 65 mV across the cell membrane (inside more negative than the outside). This resting, polarising potential arises mainly as the result of a concentration gradient of potassium across the membrane; potassium is concentrated inside the cell. The cell membrane at rest is more permeable to K than any other cation or anion so that a concentration diffusion potential is established, following the laws established by Nernst and others.[1] The membrane potential (V), between solution 1 and 2, is controlled by the total ion gradient in combination with their permeability ratios.

$$V \text{ (zero current)} = V_R = \frac{RT}{F} \ln \frac{\varepsilon P_i (M_i)' + \varepsilon P_i (X_j)''}{\varepsilon P_i (M_i)'' + \varepsilon P_i (X_j)'}$$

where $'$ and $''$ indicate solution 1 and 2

M_i = cation activities in solution 1

X_j = anion activities in solution 2

P_{ij} = respective ion permeabilities (e. g. U_i K_i, the ion mobility $\times$ its partition coefficient)

Thus, V may depend upon any single ion in a mixture of ions if that ion has a very high permeability in comparison with those of other ions at similar concentrations or has a high concentration relative to those of other ions with similar permeabilities. At rest, the axon K permeability, as indicated by conductance* measurements, is about twelve times greater than that of sodium. Consequently, despite having a concentration gradient for Na opposite to that of K, the resting potential is determined by the K gradient. In the case of the giant squid axon, anion permeability is so low that it can be neglected.

The action potential is initiated by a stimulus that induces a critical depolarisation of about 20 mV (i. e. acts to make the polarising resting potential about 20 mV less negative). When this critical, or threshold depolarisation is reached, as the result of injection of a current or local application of a potential, a series of reactions is initiated that creates the action potential. First, there is a rapid increase in Na permeability, causing an abrupt inward, Na current. This increase in Na permeability, during the early phase puts Na ions in control of the membrane potential and the potential is driven toward the Na equilibrium, or resting value, of about +40 mV. Much more slowly, the K permeability also increases. With the rise in K permeability the membrane potential drops once more toward zero and then, because the Na permeability also decreases (turns off), the potential returns to the original

* Conductance, expressed in reciprocal ohms per centimetre (mho cm^{-1}) 'measures' the product of permeability and concentration. In this sense the membrane is modelled as an electric circuit and the conductance is defined at $I/(V - V_X)$ where I is the non-capacitive current.

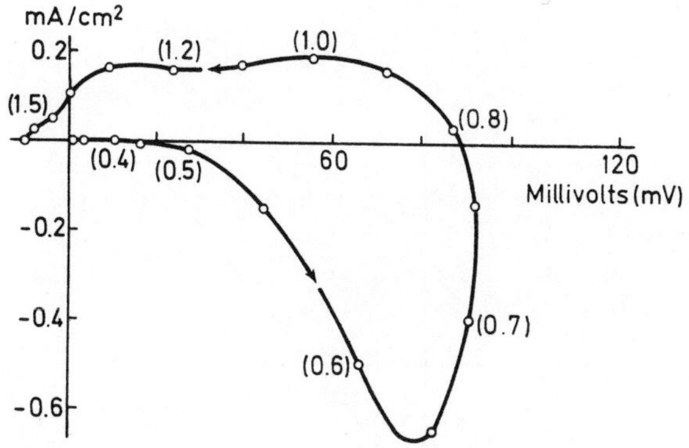

Figure 7.1 Relation between membrane potential and current during a propagating impulse. Times are given in parentheses. Inward currents are negative, outward positive. (From Cole[66], by courtesy of University of California Press)

resting value, determined by the K equilibrium. In the absence of the K conductance change, the excited membrane would reside at the Na equilibrium potential, but the coordinated K conductance change closes the cycle by returning the membrane to the resting potential. Thus the action potential is created by a precisely coordinated sequence of changes in Na and K permeability. The flow and time course of current changes during an action potential is illustrated in Figure 7.1. A cycle is completed in about 1.5 ms, with the initial Na conductance change requiring approximately 0.6 ms. The action potential is dependent only upon the properties of the axon membrane together with the ionic gradients, and independent of other cellular components. Action potentials can be fired repeatedly over long periods of time in axons where the axoplasm has been removed.

An action potential is an 'all-or-none' phenomenon. A stimulus short of the threshold will produce no trace of an action potential. Any stimulus greater than the threshold initiates an action potential, but there is no difference between the action potential produced by a minimal stimulus and one much larger. Why is an action potential 'all-or-none'? The answer to this question lies at the very basis of the distinctive properties of excitable membranes since inexcitable membranes only produce a graded potential (e.g. proportional to the applied stimulus). The basis for the 'all-or-none' nature is that the Na permeability and membrane potential are linked regeneratively: Na permeability is voltage dependent and increases when the potential becomes more positive. We can illustrate the role of the dependence of conductance on membrane potential by considering the behaviour of the membrane under different levels of depolarisation. Small depolarisations below threshold result in a slight increase in Na conductance, but once the depolarising pulse is removed, the membrane returns to its resting potential because the outward K currents over-ride the inward Na currents. At threshold, the inward Na currents just balance the outward K currents. Once the stimulating pulse drives the membrane potential beyond threshold, a rapid increase in Na conductance in comparison with K conductance allows Na transport to take

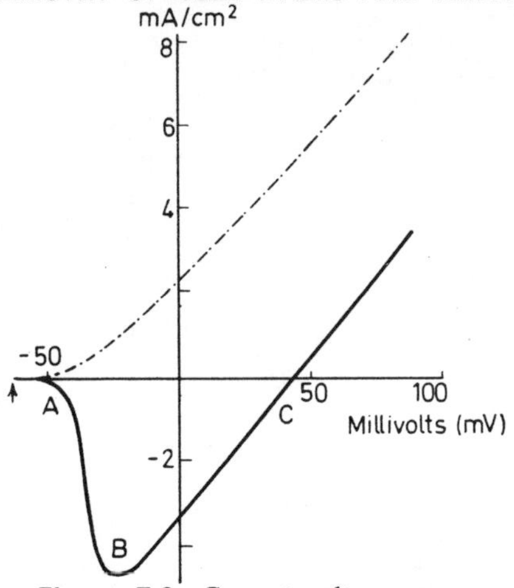

Figure 7.2 Current voltage curves obtained by the voltage clamp technique for a squid axon. Solid line illustrates the peak early current (primarily Na current). In the excitation region (A–B) the current shows a negative resistance Dashed line illustrates the steady state (primarily K current). (From Cole[67], by courtesy of University of California Press)

control of the membrane potential. The control is regenerative and allows Na to enter at an accelerating rate. The steep link between membrane potential and conductance moves the membrane to the Na equilibrium potential in an 'all-or-none' fashion.

The conductance properties of an axon membrane at different values of the membrane potential are measured by the voltage-clamp technique. The voltage clamp effectively holds the action potential at different values of membrane potential and prevents the spatial spread of the potential so that conductance can be measured. The voltage clamp maintains a potential constant across the membrane by a feed-back mechanism that varies the amount of current as the conductance changes. The amount of current required to maintain the potential constant is measured and plotted as a function of the voltage. The current–voltage (*I–V*) curve shows directly the variation of conductance changes with potential.

Excitable systems have steeply non-linear *I–V* characteristics which stem from the property that the conductance changes as a function of the membrane potential. By contrast, inexcitable membranes have 'ohmic' *I–V* characteristics (e.g. their conductance is constant and voltage independent). The *I–V* curve obtained plotting the currents against the clamping voltage for a squid axon is illustrated in Figure 7.2. Small positive changes in membrane voltage result in a small outward current essentially proportional to the clamping voltage. The *I–V* characteristics are ohmic until threshold is reached at A. At threshold an outward current is observed, and the voltage shifts the conductance (dI/dE) negative. Between −50 and −25 mV the conductance changes rapidly, and it is this steep negative conductance change which in the squid giant axon makes the axon excitable. At B the inward Na current decreases and the I–V relationship represent the ohmic behaviour of the membrane in the uniformly excited state. The Na conductance is relatively constant and independent of voltage in this range. The K conductance like Na conductance is also voltage dependent. Under the conditions illustrated, its I–V curve shows a 'rectifying' characteristic so conductance increases at about −25 mV in a nearly linear manner (Figure 7.2).

In summary, an action potential develops from a change in permeability of the membrane to Na and K ions. The permeability sequence is controlled by the membrane potential and the kinetics of the individual cationic processes. Critical depolarisation of the membrane past threshold initiates an increase in the permeability of the membrane to Na and K. Both permeability changes begin at about the same value of membrane potential, but the Na permeability changes more rapidly and moves the membrane potential positive toward the Na equilibrium potential. K permeability is slower to occur and its increase, together with a spontaneous decrease in Na permeability, restores the membrane potential to its resting level.

The action potential is conducted along an axon because the potential in a small increment spreads passively along the nerve and initiates the same processes responsible for the action potential further along the fibre. This passive (or electrotonic) spread of potential occurs because the low resistance of the axoplasm allows the spread of currents through the axon core. The initial large and rapid voltage change triggered by the increase in Na conductance provides sufficient current for the potential to spread to adjacent membrane sectors and trigger an action potential. In fact, one of the advantages of Na as the initial permeant ion is that its equilibrium potential is sufficiently positive to generate the currents required to elicit action potentials in adjacent membrane sectors.

The axon can be represented as a core-conductor which permits a detailed analysis of properties affecting conduction. The parameter describing the decay of potential is the space constant which is given by $\sqrt{aR_m/2R_i}$ where R_m is the membrane resistance per unit area, R_i the axoplasm resistivity, and a is the axon radius. The rate of passive spread of potential is determined by the values of membrane resistance (R_m) and membrane capacitance (C_m); the time constant specifying the rate of passive spread is given by the product, R_mC_m. The generation of an action potential is independent of membrane capacitance, but both the membrane resistance and capacitance are involved in propagative conduction.

We can summarise by outlining the central questions which require a biochemical explanation. In terms of the mechanism of generation of an action potential three major questions can be singled out: (1) What is the molecular basis of ion selectivity in the axonal membrane? (2) In what way do the molecular-structural properties which control permeability come under direction of the membrane potential? (3) How do the various individual components coordinate at the molecular level to develop the action potential? We can expect that the answers to these key questions will ultimately emerge from a hierarchy of chemical-structural changes which cannot be described in full without detailed knowledge of membrane structure. In this context it will be clear that our discussion of membrane excitability is more a state of the art than of detailed description of a well understood biochemical-structural phenomenon.

7.3 MOLECULAR BASIS OF ION SELECTIVITY

Na channels are in terms of a model at present 'short narrow pores that open and close in an all-or-nothing fashion'[2]. The Na channel has a low permeability to K, and the K channel is relatively impermeable to Na. The Na:K permeability ratio at rest is approximately 1:12 [3].

Experimental analysis clearly indicates that the Na and K channels are separate pathways. The evidence supporting this conclusion, recently summarised by Hille[2], is derived from the selective action of drugs. K conductance can be blocked by certain quaternary ammonium compounds without affecting the Na conductance. Tetrodotoxin (TTX), a toxin isolated from gonads and liver of the puffer fish, blocks Na conductance without a measurable effect on K channels. Currents carried by other ions through the Na channel are also blocked by TTX.

TTX is one of the deadliest poisons known and it is not surprising that Na conductance is blocked at very low concentration (10^{-7}–10^{-8} M)[4,5]. TTX (shown below) contains a guanidinium group and a unique hemilactal group. Studies of structure–activity relationships suggest that the guanidinium moiety enters the Na channel while other groups, probably hydroxyl, become associated with reactive membrane molecules, perhaps via hydrogen

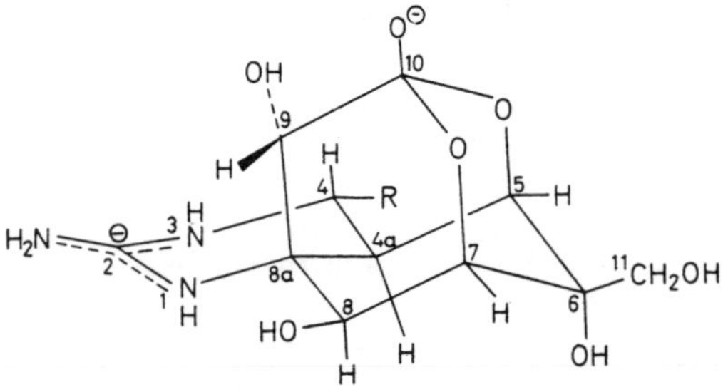

R = OH

(From Narahashi *et al.*[68], by courtesy of American Association for the Advancement of Science)

bonds[5]. Any modifications in structure generally reduce the activity. Guanidine itself can penetrate the Na channel to produce action potentials under certain conditions[6], and the role of the guanidinium group in TTX may reflect this property[5].

Because TTX is an effective inhibitor at such low concentrations and because the equilibrium dissociation constant is low (10^{-8}–10^{-9} M), TTX binding has been used as a tool to set an upper limit to the density of Na channels. The number of TTX molecules bound to the unit area of nerve membrane has been measured by bioassay and with tritiated TTX. Both measurements on a variety of nerves show that the number of sites per μ^2 is amazingly small. Lobster axons contain 13 [7] to 30 sites per μ^2 [8]. Rabbit vagus nerve contains $75/\mu^2$, crab nerve $49/\mu^2$ [9], and garfish axons $4/\mu^2$ [10]. Red blood cells lack high affinity TTX sites. Various workers have noted that the density is exceedingly small, but perhaps most descriptive is the comparison made by Keynes[11] who notes that 1 μ^2 contains 2×10^6 phospholipid molecules but only 13–75 Na channels per μ^2! Estimates of Na channels from TTX binding are upper estimates and, although probably not in major errors, may yet have to be reduced. Independent estimates from Na conductance give a similar value for the density of Na channels. So few channels are required because the net flux of ions is very rapid, of the order of 400 ions ms^{-1}. Hille points out that nearly every Na ion which strikes the mouth of the channel must pass through[2].

A comparison of TTX binding to the outside and inside surface of axons provides definitive data on the asymmetry of the Na channel. The axoplasm of the giant squid axon can be squeezed out so that the effect of TTX can be studied selectively by internal perfusion. TTX applied internally is ineffective and it has been estimated that the difference in sensitivity of the external and internal surfaces of the squid axon membrane to TTX is 300–3000-fold or more[5].

Benzer and Raftery[10] have studied the effect of certain enzymes on TTX binding to gain additional insight into the nature of the Na channel and TTX sites. Nerve membrane fragments from garfish olfactory nerve were incubated with various enzymes. Most enzymes do not reduce binding (ribonuclease, dexoyribonuclease, hyaluronidase, phospholipase C, phospholipase D, trypsin). Binding was inhibited by phospholipase A and two proteases, chymotrypsin and pronase. Binding becomes susceptible to trypsin if the membrane is pretreated with phospholipase A. Pretreatment with phospholipase A also increases the extent of inhibition obtained with pronase and chymotrypsin. Previously it has been demonstrated that the ability of phospholipase A to block action potentials in axons derives not from its destruction of phospholipids *per se* so much as from inhibition by products generated in the phospholipase A reaction[12, 13]. These substances, which probably possess detergent-like action, might expose TTX sites to increased proteolysis. Binding is also considerably reduced by heat treatment or exposure of the membrane to acid pH. Benzer and Raftery conclude the membrane binding component is a protein or a phosphoprotein. The results with phospholipase A suggest the active site for TTX is buried in the membrane. Further studies on the solubilized component are required to establish with certainty the nature of the TTX binding component.

The data obtained by Benzer and Raftery agrees in part with data obtained on blockage of the action potential by enzymes. The action potential is blocked by treatment with phospholipase A[13], but not by treatment with phospholipases B or D. Phospholipase C effects depend on the axon[13]. Externally applied proteases have only limited effects on excitability. Trypsin and chymotrypsin block the action potential in squid axons when applied internally, but mixtures of non-specific endopeptidase do not, despite their greater action[14]. Perfusion of a giant squid axon with a solution of proteolytic enzyme selectively prevented the inactivation of Na conductance[15].

Studies of TTX binding and alterations of the Na channel in the same system could provide additional insight into the properties of the Na channel. A careful analysis of the effect of internally and externally applied enzymes on the TTX binding polypeptide could provide data on the organisation of the molecule in the membrane. The finding that the Na channel is attacked by internally applied enzymes and externally applied TTX, and that TTX binds to this protein in the inactive state, implies that the protein or protein complex spans the membrane, even in the resting state. In this context it is interesting that freeze fracture studies have revealed intramembraneous particles in myelinated axon plasma membrane at nodes of Ranvier[16]. Such particles are not preset in adjacent membrane regions which are insulated by myelin and not involved in propagation. It may be these intramembraneous particles reflect ion channels or they may be large molecular complexes such as the Na–K–ATPase.

Ion selectivity is accounted for through precise structural arrangements which have been described in part by comparing permeabilities of various ions which will substitute for Na in the Na channel. From these studies it is evident that no one single property of a pore completely accounts for its discriminating properties. The channel behaves as if it were about 3×5 Å in cross-section[2, 17]. Both small metal and organic cations permeate the membrane. The permeability of many organic ions can be accounted for on size differences and by hypothesising the presence of oxygen groups in the channel wall, which coordinate with permeant cations. A steric theory alone does not account for why the large hydroxylamine is as permeable as small Li nor the difference in Na and K permeabilities[18]. Recent considerations suggest that the Eisenman electrostatic model[19, 20] for ion exchange may provide the theoretical base for the required ion selectivity. In this view the relative permeabilities of ions are proportional to the relative affinities of the ions for the binding sites. The free energy associated with moving the cation from water to the site is determined by differences in the coulombic force of attraction betweencationand binding site(s) and betweencation and water. Chandler and Meves[3] first suggested the Na channel behaves like a high field strength system which accepts partially hydrated ions. Hille examined the permeability of various metal ions and found that the permeability sequence agreed quite well with that predicted by the electrostatic theory of ion binding[18]. Additional support for the theory comes from the suggestion that the Na channel contains a high field strength anion. The existence of an anion, possibly a carboxyl group from a protein acidic amino acid, is based upon the observation that titration curves follow those of a single acid group and show that the Na conductance is reduced to one-half of normal when the pH is lowered from physiological to pH 5.2[21]. The titration data must as yet be viewed with caution because neither alterations in surface charge nor pH dependent shifts in channel conformations which might affect permeability have been rigorously excluded.

We can summarise our current understanding of the Na channel. Na channels are very sparse. The channel characterises as a small pore which restricts permeability to large cations. Ions otherwise not restricted by size are selected on the basis of their ion exchange properties. It appears as if groups are placed in suitable positions which sequentially dehydrate cations by substitutive coordination (via oxygen atoms) and bind them electrostatically (via carboxyl groups) cations. Structural data on channels, in addition to providing clues on the nature of the channel, will ultimately be of use as aids in synthesising model compounds which can be tested in bilayers. The molecular basis of ion selectivity used by the axon will be advanced considerably once the physical characteristics of the TTX binding molecules (Na channels) are obtained. Simple considerations, such as the size of the TTX binding component, would provide clues on whether the component resembles one of the small peptides known to confer a voltage dependent ion permeability to lipid bilayers in addition to whether the molecule or molecular complex might be able to span the membrane.

K channels in the squid axon appear to be about 50 times more abundant than Na channels[2]. Armstrong[22] studied the interaction of quaternary ammonium derivatives with K channels in the squid axon and suggested that K channels are gramicidin-like tunnels with a large opening at their inner end. A hydrated K fits into the opening of the channel, but only a dehydrated K ion permeates the pore itself. Goodall and Sachs[23] have reported substances they extracted from rat brain or eel electroplax which selectively alter K conductance when added to a lipid bilayer. The identity of the active component in the extract is not yet known.

7.4 GATING MECHANISM

The mechanism by which the channels open and close under direction of the membrane potential is unknown. Ion selectivity experiments have not yet provided clues on the voltage dependence of permeability, and whether groups involved in ion selectivity also share a role in ion gating is not known. With the steepness of the potential transition it is often assumed that several charges or dipoles are involved simultaneously in the process[11, 24]. Shifts in several charges could occur in a single component or through an interaction of several singly-charged components. If, when the gate opens, charges collectively undergo a net movement across or in the membrane, a pulse of current would be generated (the so-called gating current). Recent measurements by Armstrong and Bezanilla[25] may have identified a gating current so that it appears probable that an early component of conductance change involves the movement of dipoles or charges. Despite many observations of changes (birefringence, light scattering, fluorescence emission after staining the membrane with fluorescent dyes) that are indicators of membrane structural changes associated with the excitation process, Keynes and coworkers conclude that none of these changes can yet be directly related to membrane conductance[9, 11, 26]. General membrane structure, however, appears to be altered by applied voltages. The membrane probably alters thickness, owing to compressive forces exerted by attraction between opposite charges on either side of the membrane. In addition the entropy of the membrane appears to decrease when the potential is reduced[27]. These presumed entropy changes inferred from measurements of ΔH, however, could originate from proton transfer processes[28].

The gating mechanism may ultimately be found to reside in a number of physical chemical phenomena. Bass and Moore[28, 29] have raised the interesting possibility that the nerve action potential is gated by proton transfer processes through a Wien effect. According to the Wien theory, the dissociation constant of a weak electrolyte depends upon the electric field strength. In this model depolarisation would elicit an increase in the pK value of acidic residues which become an instantaneous sink for protons. The proton shift alters protein conformation, or alternatively, the lining of the channel itself. The Wien effect model can account for a number of observations which are not readily accommodated by other models[28].

7.5 COMPOSITIONAL DATA ON NEURONAL MEMBRANES

Detailed compositional data on highly purified axonal plasma membranes, is not available. Gross compositional information on nerve membranes may in any event, cast little light upon the specific nature of the permeability mechanisms, in view of the apparently minuscule contribution that the transport sites would make toward the total composition. This information can, however, provide some help in assessing limiting conditions and potential ancillary control mechanisms. Camejo and co-workers[30] fractionated the first stellar nerves of squid and isolated a fraction they concluded was enriched in axonal plasma membranes. The composition of this fraction is shown in Table 7.1. Lipids comprise 70% of the membrane constituents, with cholesterol and polar lipids, phosphatidylcholine (PC) and phosphatidylethanolamine (PE) being predominant. Phosphatidylserine (PS) is the major charged lipid, but is not itself very abundant. Other charged molecules might arise from sialic acid containing molecules or proteins, but at present no data are available. The fatty acid side chains C_{16} and C_{23} are prevalent. The molar ratio of cholesterol to phospholipids is approximately 1.0, which is

Table 7.1 Composition of axon plasma membrane fraction isolated from squid axon (Compiled from Camejo *et al.*[30], by courtesy of Elsevier)

Component	*% Total*
protein	30
cholesterol	19
polar, lipids, total	41
sphingomyelin	4
phosphatidycholine	19
phosphatidylethanolamine	14
phosphotidylserine	4
fatty acids	4
hydrocarbons	6

Table 7.2 Lipid classes present in synaptic plasma membrane fractions (From Cotman *et al.*[31] and Breckenridge *et al.*[32], by courtesy of American Chemical Society and Elsevier)

Class	*% of Total*
Phosphatidylethanolamine	28
Phosphatidylcholine	34
Phosphatidylserine	10
Phosphatidylinositol	2
Sphingomyelin	3
Lysophosphatidyl-choline	1
Cholesterol	19
Ceramide	2

Table 7.3 Composition of acyl chains (weight %) in phospholipids of synaptic plasma membranes (From Cotman *et al.*[31], and Breckenridge[32], by courtesy of American Chemical Society and Elsevier)

	*Lipid**				
Side chain	PE	PC	PS	PI	*Sphingomyelin*
16:0	7.5	50.7	0.5	10.0	5.2
16:0	0.3	1.0	—	—	—
17:0	0.4	0.2	—	—	—
18:0	23.7	12.6	48.6	38.4	87.5
18:1	6.6	24.2	7.6	8.5	0.1
18:2	0.2	0.6	—	0.3	—
20:0	—	—	—	—	1.9
18:3	0.4	1.0	—	—	—
20:3	0.3	0.1	—	—	—
20:4	18.0	5.6	2.0	36.9	—
22:4	8.4	0.8	3.9	—	—
22:5	1.5	—	3.2	—	—
22:6	32.9	3.4	34.1	4.7	—
24:0	—	—	—	—	1.3
24:1	—	—	—	—	0.9

* PE, phosphatidylethanolamine; PC, phosphatidylcholine; PS, phosphatidylserine; PI, phosphatidylinositol. Dash indicates chains were not detectable

similar to values obtained from erythrocytes and myelin. At present the purity of the squid axon membrane preparation is not established so the compositional data must be regarded as tentative.

Analysis of synaptic plasma membranes (SPM) is much more complete. A portion of this membrane can be classed as excitable since action potentials pass to the synaptic area and since Ca influx preceding transmitter release is regenerative. The lipid-to-protein ratio is 1.0, and the molar ratio of cholesterol to polar lipids is 0.44[31,32]. Polar lipids are less abundant as is cholesterol by comparison to plasma membrane preparations from squid. Gangliosides make up about 5% of the total constituents[32]. PE and PC are the most abundant polar lipids and the charged lipids are minor constituents (Table 7.2) The fatty acid composition is highly distinctive. PE and PS contain an unusually high quantity of long chain highly unsaturated fatty acids (Table 7.3). It is significant that of all the various membranes analysed only retinal rod outer segments have an equally high proportion of unsaturated long chain fatty acids[33]. In both SPM and rod segments, cholesterol is low. It may be that the high proportion of unsaturated fatty acids and the correspondingly low proportion of cholesterol permits a high degree of fluidity which is certainly compatible with the demonstration of fluid properties of retinal rod outer segments[34].

The polypeptide constituents of SPM have been analysed and found to consist of a large number of discrete chains with molecular weights ranging from 40 000–200 000 daltons[35]. Polypeptides of 50 000 and 100 000 are dominant. With the possible exception of the 100 000 chain, which may be Na–K–ATPase, the composite polypeptides have not been identified. The total amino acid composition of the membrane protein is rich in charged amino acids[36]. Acidic amino acids prevail suggesting that the charge contribution from the proteins may be negative.

Clearly, solid analytical data on a well-defined and highly purified axon plasma membrane is badly needed. The ideal system of choice should be suitable for both detailed neurophysiological and chemical analysis. Squid axon is not particularly well suited for chemical study since the connective tissue surrounding the axon is very abundant and squid not readily available. The olfactory nerve of the garfish may offer large quantities of axons without interfering connecting tissue[37]. Unfortunately plasma membrane fragments of these axons have not yet been purified. Complete analytical data on axonal membranes would be a valuable addition toward studies aimed at reconstructing the excitability properties of axons. Studies of model membranes will ultimately involve testing the various natural constituents for a possible role in excitability. Thiamine, for example, is associated with nerve membranes and appears to play a role in an unknown aspect of axon physiology[38]. Similarly PS destruction via enzymatic conversion to other phospholipids alters the action potential[39], but as will be illustrated below has only a minor role in excitability studies in lipid bilayers.

Some membrane constituents will prove to serve the action potential directly or its recovery process while others will provide for ancillary functions of the axon. As noted above the Na–K channels are few and the majority of constituents are probably relegated to roles other than the excitation process itself. It is interesting to ask what is the minimum number of natural constituents required to support excitation? There is no answer at present but model membrane studies suggest the requirements are few.

7.6 STUDIES ON BILAYERS

A lecithin bilayer in the presence of the well characterised peptide alamethicin exhibits a steeply voltage-dependent conductance[40-42]. A negative resistance region in the current–voltage characteristics, similar to that seen in axons, can be developed at high alamethicin concentrations or by introducing an ion concentration gradient[42,43]. Similar observations have been made with monoazomycin, a polyene-like antibiotic[44,45].

The negative conductance behaviour in these bilayer systems is well understood and its analysis provides some insight into the bases of negative conductance behaviour in bilayers as well as axons. Negative conductance behaviour exists for situations provided the voltage dependent conductance (g) is strong (e.g. dg/dV is large) and provided $V < V_X$, where V is the potential difference across the membrane and V_X is the value of the diffusion potential existing in the presence of a salt gradient[44]. That is, because the membrane current is given by:

$$i = (V - V_X)\, g$$

the change in conductance is:

$$di/dV = g + (V - V_X)\, dg/dV$$

For $di/dV < 0$ it follows that:

$$g + (V - V_X)\, dg/dV < 0$$

If dg/dV is large (e.g. strong voltage-dependent conductance), then for $V < V_X$ we can satisfy the above relationships and obtain a negative conductance. Thus, for example, Muller and Finkelstein[44] observed the appearance of a negative conductance in the I–V characteristic of a monoazomycin doped bilayer by introducing a positive V_X into the system. No significant alteration of g-V characteristic is required or seen. Eisenberg and co-workers report similar observations with alamethicin[42].

The negative conductance characteristic of Na channel behaviour in axons can be viewed in the manner described above. Other unique properties are not required. The strong voltage dependent conductance of the Na channel and the relationship between the axon's resting potential and the Na equilibrium potential provides the negative conductance behaviour which serves to establish a zone of 'instability' at threshold. Excitation drives the membrane into this zone and sets off an action potential.

The negative resistance characteristic of alamethicin-lipid bilayers can be further modified with protamine which can absorb in or near the channel in consequence of a stimulus driven current to change the channel selectivity from weakly cationic to anionic and produce potential changes analogous to action potentials[40]. Excitability inducing material (EIM) discovered by Mueller and Rudin[46], which is a rather poorly characterised protein complex of bacterial origin[47], also confers a voltage-dependent conductance on lipid bilayers. The steady-state conductance voltage curves from EIM and for a squid axon are qualitatively similar. Basic proteins can also modify the EIM conductance to develop action potential analogues in which cation–anion permeability transitions are apparently implicated rather than Na–K transitions[46]. In bilayers doped with either alamethicin or EIM the induced conductance develops by discrete steps which appear to originate from the voltage dependent opening and closing of individual pores or channels[42,43,49,50]. The conductance values are similar to the values obtained for the conductance of a single Na channel in axons[51]. The nature of the channel opening and closing process is still under active investigation.

In general excitation effects induced in bilayers are dependent primarily upon a factor (protein, polypeptide) introduced into the lipid matrix[44,48,50]. Lipids of the membrane do, however, exert some control over the kinetic and steady-state properties of the protein channels[40,45,52,53]. Prominent and distinctive effects of different lipids can be expected in view of their diverse binding, screening, organisational and fluid properties.

No processes totally like the action potential have been observed yet in bilayers. In an action potential, the increase in Na conductance is coupled to an inactivation process and an increase in K conductance. Detailed reconstitution experiments will ultimately be required to determine the minimum components required for an action potential. Components not directly involved in the voltage dependent specific ionic conductance are probably required to organise and stabilise the system and perhaps coordinate the cycle.

7.7 SURFACE CHARGES OF AXONAL MEMBRANES

Surface charges of the axonal membrane determine in part the electric field profile and the localised ion environment, two factors which can modify the excitation process. The electric potential profile across a nerve membrane depends on the field generated by the fixed surface charges of the membrane as well as the potential created by selective ion permeabilities. The density and intrinsic properties (sign, binding ability and distance from the membrane) of the fixed surface charges determine their contribution to the field. The internal and external surface of the squid axon membrane appear to be negative and based on certain assumptions, approximations on the density of charges have been made. The average separation of negative charges appears to be 11–15 Å on the external surface of the axon membrane[54,55]. Chandler, Hodgkin and Meves[56] estimated the internal membrane charge separation to be 27 Å. These fixed charges provide for the field across the membrane illustrated in Figure 7.3. The resting potential measured between the internal and external bulk phases is −65 mV and as described above originates from the selective permeability to K. The fixed surface charges create at the outside a potential of −46 mV and at inside −13 mV. The actual field across the membrane is −32 mV, which for a 100 Å thick membrane, represents an enormous potential, 32 000 Vcm^{-1}. The field probably exists only across the thinner lipid core so is still higher.

Surface charges may have a significant effect on excitability properties such as current–voltage relationships. Gilbert[55] notes that ions might regulate the controlling potential for ionic conductance without necessarily influencing the resting potential. It is evident from Figure 7.3 that the inner and outer surface charges can be altered without affecting the potential from a to d. Divalent anions, for example, can reduce the negative surface potential by a screening process[54,57]. Studies on bilayers constructed from PS, one of the major charged lipids of nerve, established that increases of Ca tenfold (between 10^{-3} and 10^{-1} M) decrease the surface potential 27 mV as predicted by a theoretical analysis of screening effects on the surface potential[57]. It has

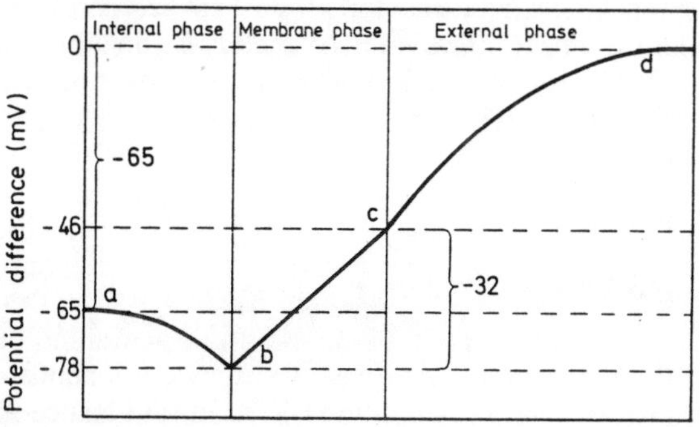

Figure 7.3 Potential difference across squid axon membrane. The distances are not to scale. The potential across the membrane as drawn does not indicate complex intramembrane potentials which exist. (From Ehrenstein[51], by courtesy of van Nostrand, Reinhold Company)

been noted that increases in Ca (2–20 mM) in fact shift the axon conductance-voltage curve along the voltage axis by 21 mV[55]. Almost identical behaviour has been observed in the conductance–voltage curves for the non-linear lipid bilayer systems using EIM[52,53] or monazomycin[45]. By analogy to studies on bilayers, McLaughlin *et al.*[44]. and others suggest this shift may be due to a screening effect of divalent ions. Since changes in field on one side of the membrane probably have little affect on the other, the membrane field is further increased by increases in external divalents. The role of divalent screening effects depends on the presence of high density of negative surface charges[57], an assumption not yet directly established.

Although it is clear that surface charges directly contribute to the field at the membrane surface the precise contribution of fixed charges to the field at the axon membrane is still uncertain. Little is known about their organisation, nor has the identity of the charges been established. It is unlikely that PS, the major charged lipid, is present in quantities large enough to account for the expected surface charge. Pure PS bilayers have a charge density of approximately 1 charge/38 $Å^2$ [57], which is less than estimations of external or internal surface charge. Evidently protein carboxyl groups and sialic acids account for the major surface charges on the external surface while probably only protein groups contribute to the internal surface. In erythrocytes, sialic acid residues provide most of the total surface anions[58,59], but in axons the charge around the channel is probably not due to sialic acid. Gilbert and Ehrenstein[60] calculated that the pK for the average axonal surface charge is about 4.6, which is similar to the pK_a of free carboxyl groups of glutamic and aspartic acid residues in proteins. These data suggest that proteins contribute a major portion of the surface charge, but since estimates of charge density at present come from conductance measurements, the charge density measurements apply only to the membrane in the vicinity of the channels, not over the entire surface. The field is in fact probably not uniform over the surface[2], and we can imagine the presence of a complex surface potential, particularly around the channels. Sialic acids might be sufficiently distant from the membrane surface to enable the channels' protein carboxyl groups to set the surface charge. Direct biochemical studies on the type and organisation of surface charges would clearly provide a valuable adjunct to the electrical analysis of the excitation process.

7.8 CONCLUSION

Research over the past years has provided a precise description of the phenomena underlying the action potential. The main tool has been the voltage clamp. Voltage clamp studies have resulted in a quantitative statement of ionic conductances during an action potential. The present challenge is to define the fundamental mechanisms underlying the selectivity and voltage dependence of ionic conductances. Certainly some progress has been made. It appears reasonably certain that channels are specific sites rather than any generalised lattice process. We can define the probable number and permeability properties of channels and assemble a hypothetical model, but the nature of the molecules still cannot be assessed. Specific channels open and close under orders from the membrane potential, and the activation properties differ for the Na and K channels. The mechanistic inter-relationship between potential and permeability are unknown, but it is known that membrane structural properties are altered at approximately the same time as the potential changes. The entropy of the membrane increases and the membrane alters thickness under the influence of the membrane potential. The membrane capacitance is altered only very slightly. Solid data are required on the nature of molecules or groups causing gating. Advances in the chemical modification of specific membrane groups[61] (both internal and external) should provide new insights into the selectivity and gating processes.

The properties of ion selectivity and the voltage dependence of permeability may reside in a single molecule or a molecular complex. Adjacent membrane structures are passive or actively involved. At present there is no way to decide. Components need to be isolated and their function tested in defined bilayers. Natural components of axon membranes have not been tested, perhaps in part because no pure axonal plasma membrane fraction is available.

Several polypeptides or proteins, which produce voltage dependence conductance characteristics in bilayers, illustrate the relative simplicity of systems required to produce axon-like characteristics and provide insight into the possible nature of voltage dependent conductance changes. They also demonstrate that several different molecular mechanisms can produce related effects. To determine which of these, if any, is related to the physiological conductance mechanisms used in axons, physiological type properties must be distinguished from others by applying rigorous kinetic and structural criteria known for the axon.

The Hodgkin–Huxley model describes kinetic characteristics of conductance changes and the appropriate molecules as Na channel candidates need to at least approximate the known steep conductance changes. In addition, characteristics of selectivity and pharmacological properties need evaluation. Hodgkin and Huxley used first-order kinetics to describe the conductance changes, and although their formulation predicted the gating current, the rate constants do not provide information about a chemical mechanism. It is important to note that the Hodgkin–Huxley equations are not unique. Other solutions fit the voltage-clamp data[62, 63]. The voltage-clamp data can be quantitatively fitted to a Wien effect model for the pH dependence of an enzyme, acetylcholinesterase[63, 64], despite the fact that acetylcholinesterase itself is probably not involved directly in the excitation process[65]. Quantitative solutions of mechanism based on physical properties of components known to be centrally involved need to be developed as the relevant components are identified.

It is unlikely that the physiological mechanism of the action potential will be elucidated in the near future. Nonetheless we can look into the future and consider the significance of elucidating the molecular mechanism of the excitation process. The action potential is ideally suited for inter-relating membrane function with membrane properties, and it is likely new concepts about the fundamental properties of membrane will arise. We will gain a detailed understanding of the action of various anesthetics and poisons. The 'all-or-none' nature of the action potential leaves little room for the study of control properties. In brain the action potential subserves the problem of information transfer, and integration and control occurs via synapses and some membrane properties. Yet little is known about potential plastic properties of action potentials, particularly in excitable dendrites, and it is possible that at least thresholds are regulated. Change is only one critical aspect of nervous function. Stability is essential. The mechanism of the action potential will provide an incredibly detailed biophysical–chemical conceptualisation of this reliable and precise means of information transfer.

Acknowledgements

We are grateful to Dr Ross Bean and Dr Walter J. Moore for their critical comments on the manuscript and much helpful discussion. We also appreciate very much the helpful discussion of Dr Moishe Eisenberg. We thank Mrs Pat Lemestre for secretarial aid.

References

1. Cole, K. S. (1968). *Membranes, Ions and Impulses*, 196 (Berkeley and Los Angeles: University of California Press)
2. Hille, B. (1970). *Progr. Biophys. Mol. Biol.*, **21,** 1
3. Chandler, W. K. and Meves, H. (1965). *J. Physiol.* (*London*), **180,** 788
4. Narahashi, T., Moore, J. W. and Scott, W. R. (1964). *J. Gen. Physiol.*, **47,** 965
5. Narahashi, T. (1972). *Fed. Proc.*, **31,** 1124
6. Tasaki, I., Singer, I. and Watanabe, A. (1965). *Proc. Nat. Acad. Sci.*, **54,** 763
7. Moore, J. W., Narahashi, T. and Shaw, T. I. (1967). *J. Physiol.* (*London*), **188,** 99
8. Hafemann, D. R. and Houston, A. A. (1971). *Fed. Proc.*, **30,** 255
9. Keynes, R. D., Ritchie, J. M. and Rojas, E. (1971). *J. Physiol.* (*London*), **213,** 235
10. Benzer, T. I. and Raftery, M. A. (1972). *Proc. Nat. Acad. Sci.*, **69,** 3634
11. Keynes, R. D. (1972). *Nature* (*London*), **239,** 29
12. Condrea, E. and Rosenberg, P. (1968). *Biochim. Biophys. Acta*, **150,** 271
13. Rosenberg, P. and Condrea, E. (1968). *Biochem. Pharmacol*, **17,** 2033
14. Takenaka, T. and Yamagishi, S. (1969). *J. Gen. Physiol.*, **53,** 81
15. Rojas, E. and Armstrong, C. (1971). *Nature, New Biol.*, **229,** 177
16. Livingston, R. B., Pfenninger, K., Moor, H. and Akert, K. (1973). *Brain Res.*, **58,** 1
17. Hille, B. (1971). *J. Gen. Physiol.*, **58,** 599
18. Hille, B. (1972). *J. Gen. Physiol.*, **59,** 637
19. Eisenman, G. (1962). *Biophys. J.*, **2,** 259
20. Eisenman, G. (1969). *Nat. Bur. Stand.* (*US*), Spec. Publ. 314
21. Hille, B. (1968). *J. Gen. Physiol.*, **51,** 221
22. Armstrong, C. M. (1971). *J. Gen. Physiol.*, **58,** 413
23. Goodall, M. C. and Sachs, G. (1972). *Nature, New Biol.*, **237,** 252
24. Hodgkin, A. L. and Huxley, A. F. (1952). *J. Physiol.*, **117,** 500
25. Armstrong, C. M. and Bezanilla, F. (1973). *Nature* (*London*), **242,** 459
26. Cohen, L. B., Hille, B., Keynes, R. D., Landowne, D. and Rojas, E. (1971). *J. Physiol.*, **218,** 205
27. Howarth, J. V., Keynes, R. D. and Ritchie, J. M. (1968). *J. Physiol.*, **194,** 745
28. Bass, L. and Moore, W. J. (1973). *Progr. Biophys. Molec. Biol.*, **27,** 143
29. Bass, L. and Moore, W. J. (1968). *Struct. Chem. Molec. Biol.*, 356 (A. Rich and N. Davidson, editors) (San Francisco: W. H. Freeman and Co.)
30. Camejo, G., Villegas, G. M., Barnola, F. V. and Villegas, R. (1969). *Biochim. Biophys. Acta*, **193,** 247
31. Cotman, C. W., Blank, M. L., Moehl, A. and Synder, F. (1969). *Biochem.*, **8,** 4606
32. Breckenridge, W. C., Gombos, G. and Morgan, I. G. (1972). *Biochim. Biophys. Acta*, **266,** 695
33. Neilsen, N. C., Fleischer, S. and McConnell, D. G. (1970). *Biochim. Biophys. Acta*, **211,** 10
34. Cone, R. (1973). *Photoreceptor Membranes*, paper presented at Squaw Valley Membrane Conference
35. Banker, G., Crain, B. and Cotman, C. W. (1972). *Brain Res.*, **42,** 508
36. Cotman, C. W., Mahler, H. R. and Hugli, T. (1968). *Arch. Biochem. Biophys.*, **126,** 821
37. Easton, D. (1971). *Science*, **172,** 952
38. Itokawa, Y., Schulz, R. A. and Cooper, J. R. (1972). *Biochim. Biophys. Acta*, **266,** 293
39. Cook, A. M., Low, E. and Ishijmi, M. (1972). *Nature, New Biol.*, **239,** 150
40. Mueller, P. and Rudin, D. O. (1968). *Nature* (*London*), **217,** 713
41. Cherry, R. J., Chapman, D. and Graham, D. E. (1972). *J. Mem. Biol.*, **7,** 325
42. Eisenberg, M. (1973). *J. Mem. Biol.*, (in press)
43. Eisenberg, M. (1972). Ph.D. Thesis, California Institute of Technology
44. Muller, R. U. and Finkelstein, A. (1972). *J. Gen. Physiol.*, **60,** 263
45. Muller, R. U. and Finkelstein, A. (1972). *J. Gen. Physiol.*, **60,** 285
46. Mueller, P. and Rudin, D. O. (1963). *J. Theor. Biol.*, **4,** 268
47. Kushnir, L. D. (1968). *Biochim. Biophys. Acta*, **150,** 285
48. Bean, R. C., Shepherd, W. C., Chan, H. and Eichner, J. T. (1969). *J. Gen. Physiol.*, **53,** 741
49. Ehrenstein G., Lecar, H. and Nossal, R., (1970) *J. Gen. Physiol.*, **55,** 119

50. Gordon, L. G. M and Haydon, D A. (1972). *Biochim. Biophys. Acta*, **255,** 1014
51. Ehrenstein, G. (1971). *Biophysics and Physiology of Excitability Membranes*, 463 (W. J. Adelman, editor) (New York: Van Nostrand Reinhold Co.)
52. Bean, R. C., Shepherd, W. C. and Eichner, J. T. (1971). *Biogenic Amines and Physiological Membranes in Drug Therapy*, Vol. v (J. H. Biel and L. G. Abood, editors) Medicinal Research Series (New York: Marcel Dekker, Inc.)
53. Bean, R. C. (1972). *J. Mem. Biol.*, **7,** 15
54. Gilbert, D. L. and Ehrenstein, G. (1969). *Biophys. J.*, **9,** 447
55. Gilbert, D. L. (1971). *Biophysics and Physiology of Excitability Membranes*, 359 (W. J. Adelman, editor) (New York: Van Nostrand Reinhold Co.)
56. Chandler, W. K., Hogkin, A. L., Weves, H. (1965). *J. Physiol.*, **180,** 821
57. McLaughlin, S. G. A., Szabo, G. and Eisenman, G. (1971). *J. Gen. Physiol.*, **58,** 667
58. Weiss, L. and Mayhew, E. (1967). *J. Cell Physiol.*, **69,** 281
59. Seaman, G. V. F., Vassar, P. S. and Kendall, M. J. (1969). *Experientia*, **25,** 1259
60. Gilbert, D. L. and Ehrenstein, G. (1970). *J. Gen. Physiol.*, **55,** 822
61. Wallach, D. F. H. (1972). *Biochim. Biophys. Acta*, **265,** 61
62. Jain, M. K., Marks, R. H. L. and Cordes, E. H. (1970). *Proc. Nat. Acad. Sci.*, **67,** 799
63. McIlroy, D. K. (1970). *Math. Biosci.*, **7,** 313
64. Bass, L. and McIlroy, D. K. (1968). *Biophys. J.*, **8,** 99
65. Katz, B. (1966). *Nerve Muscle and Synapse*, 90 (New York: McGraw Hill Book Co.)
66. Cole, K. S. (1968). *Membranes, Ions and Impulses*, 338 (Berkeley and Los Angeles: University of California Press)
67. Cole, K. S. *ibid*, 314
68. Narahashi, T., Moore, J. W. and Poston, R. N. (1967). *Science*, **156,** 976

8
The Actions of Penicillin and other Antibiotics on Bacterial Cell Wall Synthesis

J. L. STROMINGER
Harvard University

8.1 INTRODUCTION

One of the primary observations on the effects of penicillin on the bacterial cell was the observation by Lederberg[1] of the conversion of these cells in suitably protective medium to spherical forms termed protoplasts or spheroplasts. This effect is similar to that produced by the action of lysozyme. By coincidence, lysozyme was also discovered by Alexander Fleming, just 50 years ago, several years before his discovery of penicillin. Although these two agents produce a similar end result, they act by different mechanisms. Penicillin inhibits the biosynthesis of bacterial cell walls while lysozyme brings about their solubilisation as the consequence of the hydrolysis of glycosidic linkages within the structure. In both cases the end result is that bacteria lose their protective coating and rigid shape, and become roughly spherical.

A second important early observation was that enormous amounts of uridine nucleotides accumulate in some penicillin-treated bacteria[2]. These compounds were later shown by Park and Strominger[3] to contain constituents of bacterial cell walls and to be intermediates in their synthesis.

8.2 STRUCTURE OF BACTERIAL CELL WALLS

The cell walls of bacteria show considerable diversity among species. In particular, their well-known division into Gram-positive and Gram-negative species is correlated with important differences in the structures of cell walls of these two general types of bacteria. In general, Gram-negative organisms

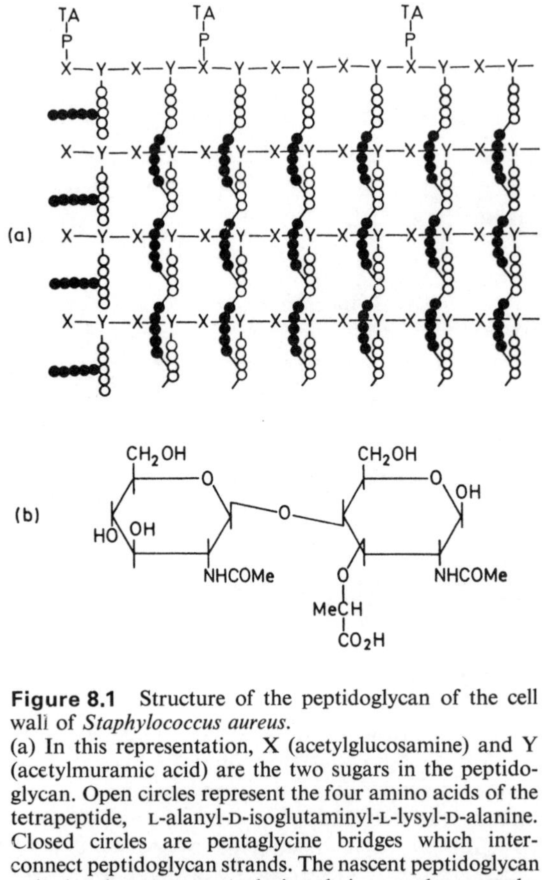

Figure 8.1 Structure of the peptidoglycan of the cell wall of *Staphylococcus aureus.*
(a) In this representation, X (acetylglucosamine) and Y (acetylmuramic acid) are the two sugars in the peptidoglycan. Open circles represent the four amino acids of the tetrapeptide, L-alanyl-D-isoglutaminyl-L-lysyl-D-alanine. Closed circles are pentaglycine bridges which interconnect peptidoglycan strands. The nascent peptidoglycan units bearing open pentaglycine chains are shown at the left of each strand. TA—P is the teichoic acid antigen of the organism which is attached to the polysaccharide through a phosphodiester linkage.
(b) The structures of X, *N*-acetylglucosamine and Y, *N*-acetylmuramic acid which are linked by β 1–4 linkages and alternate in the glycan strand

have walls of far greater complexity than the Gram-positive organisms. However, there is a basic similarity in the structure of the peptidoglycan, the complex and gigantic macromolecule that is found in one layer of both of these types of cells.

The peptidoglycan is a polymeric structure of a glycan composed of amino sugars in one dimension and cross linked through branched polypeptides in the others (Figure 8.1)[4-8]. The amino sugars are *N*-acetylglucosamine and *N*-acetylmuramic acid (which is the 3-*O*-D-lactic acid ether of *N*-acetylglucosamine). These amino sugars strictly alternate in the polymer forming the glycan strands. The carboxyl group of the lactic acid moiety of the acetylmuramic acid is substituted by a tetrapeptide which in *Staphylococcus aureus* has the sequence: L-alanyl–D-isoglutaminyl–L-lysyl–D-alanine. All of the muramic acids are substituted in this way to form peptidoglycan strands. Some close variant of this structure is characteristic of all bacterial species. The peptidoglycan strands are further linked to each other by means of an interpeptide bridge. In the genus *S. aureus*, this bridge is a pentaglycine chain which extends from the terminal carboxyl group of the D-alanine residue of one tetrapeptide to the ε-NH_2 group of the third amino acid, L-lysine, in another tetrapeptide. The third dimension is probably built up by bridges extending in different planes. This gigantic macromolecule has the mechanical stability required for the cell wall.

8.2.1 Biosynthesis of the peptidoglycan

The stages of synthesis of the peptidoglycan have been worked out over a period of almost 20 years, and although much of this information is not new, the essential features of it will be reviewed. The biosynthesis can conveniently be broken down into three stages. The first stage (Figure 8.2) involves the synthesis of uridine diphosphate *N*-acetylmuramyl-pentapeptide, the major compound which accumulates in penicillin-treated cells. First, the condensation of *N*-acetylglucosamine 1-phosphate with UTP leads to the formation of UDP-*N*-acetylglucosamine. A specific transferase catalyses a reaction with phosphoenolpyruvate to give the 3-enolpyruvyl ether of UDP-*N*-acetylglucosamine. The pyruvyl group is then reduced to lactyl by an NADPH-linked reductase, thus forming the 3-*O*-D-lactyl ether of *N*-acetylglucosamine; this compound is known as UDP-*N*-acetylmuramic acid. Conversion of this compound to its pentapeptide form occurs by the sequential addition of the requisite amino acids. Each step requires ATP and a specific enzyme which assures the addition of the amino acids in the proper sequence. The addition of L-alanine occurs first, followed by D-glutamic acid (later amidated to D-isoglutamine), L-lysine (attached by its α-amino group to the γ-carboxyl group of the glutamic acid) and finally the addition of the dipeptide, D-alanyl–D-alanine, as a unit. The latter dipeptide is formed by two enzymatic reactions: conversion of L-alanine to D-alanine by a racemase, followed by the linking of the two alanine residues in an ATP-requiring reaction to form D-alanyl–D-alanine. These reactions, all of which occur in the soluble compartment of the bacterial cells, have been fairly well characterised[8,9].

The second stage occurs in the membrane. It is the polymerisation of *N*-acetylglucosamine and *N*-acetylmuramyl-pentapeptide to form the linear peptidoglycan strands. This reaction turned out to be far more complicated than was at first apparent. Initially it was believed that the polysaccharide polymers were formed by simple and direct transglycosylations involving UDP-acetylmuramyl-pentapeptide and UDP-*N*-acetylglucosamine[10, 11]. However, when the stoichiometry of this reaction was examined in detail using a

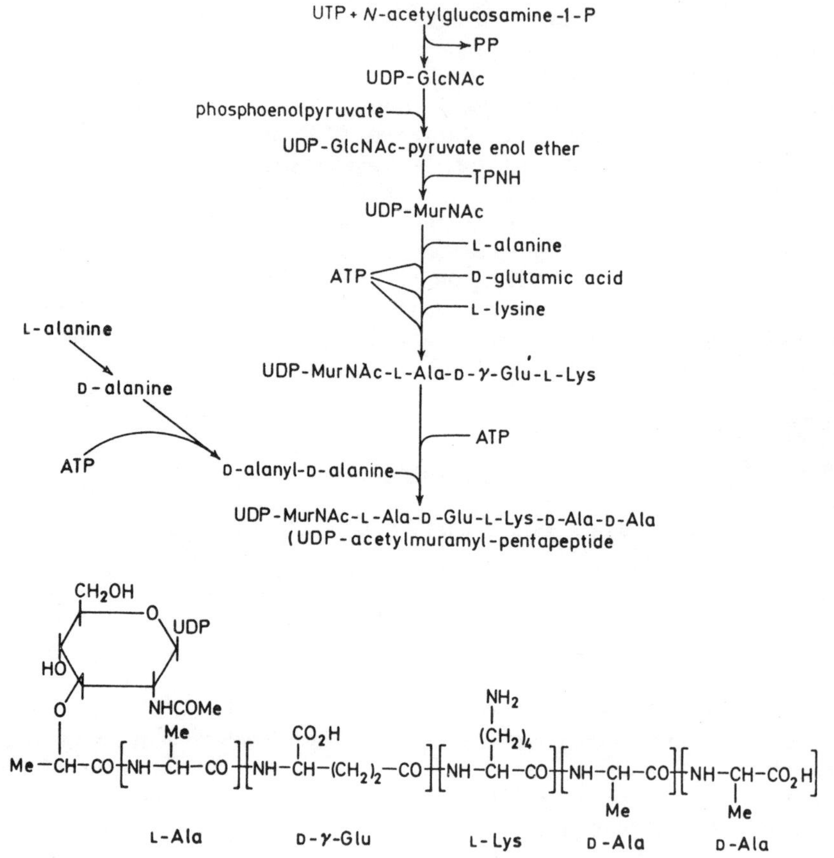

Figure 8.2 The first stage of cell wall synthesis: formation of UDP-*N*-acetylmuramyl-pentapeptide (structure shown)

variety of radioactive precursors the products turned out to be somewhat unexpected. UDP was derived from UDP-*N*-acetylglucosamine but the products formed from UDP-acetylmuramyl-peptapeptide were UMP and P_i[12, 13]. This result was inconsistent with a direct transglycosylation reaction. When the substrates and products were separated from one another by paper chromatography, there was always a trace of radioactivity present at the solvent front. For some time this radioactive 'contaminant' was ignored, but it proved to be the key to the mechanism of the reaction. The 'contaminant' was a lipid intermediate in the reaction sequence (Figure 8.3)[12, 14]. First, UDP-acetylmuramyl-pentapeptide reacts with a phospholipid in the membrane transferring *phospho*-acetylmuramyl-pentapeptide, a reaction which also leads to the formation of UMP. *N*-Acetylglucosamine is then added to the lipid intermediate via a typical glycosylation from UDP-*N*-acetylglucosamine; UDP is released. The disaccharide–pentapeptide unit is then transferred from the lipid intermediate to the peptidoglycan and lipid pyrophosphate is generated. Elimination of one P_i regenerates the lipid phosphate

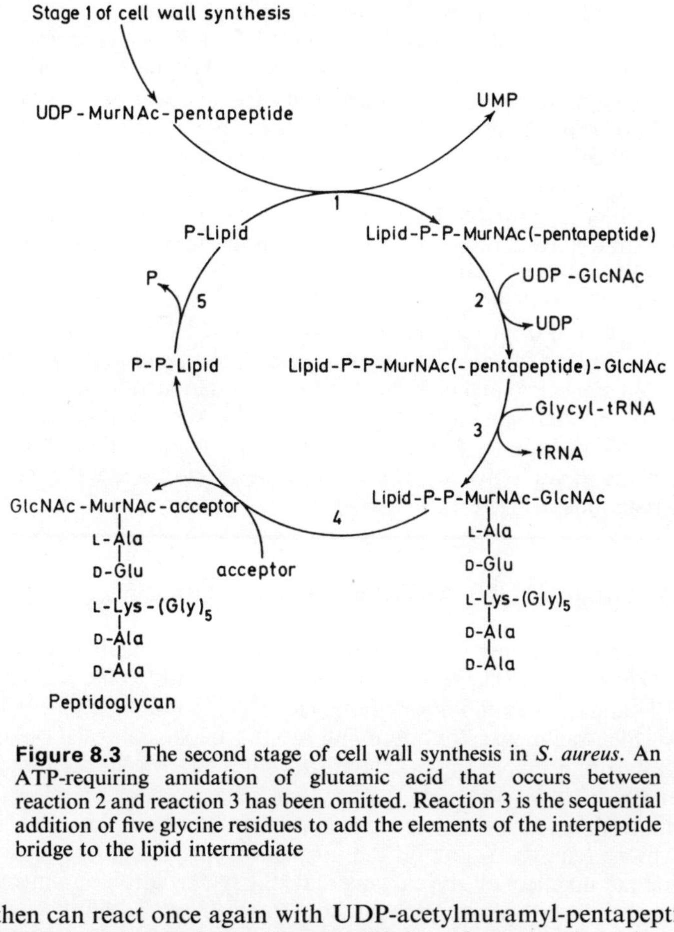

Figure 8.3 The second stage of cell wall synthesis in *S. aureus*. An ATP-requiring amidation of glutamic acid that occurs between reaction 2 and reaction 3 has been omitted. Reaction 3 is the sequential addition of five glycine residues to add the elements of the interpeptide bridge to the lipid intermediate

which then can react once again with UDP-acetylmuramyl-pentapeptide and participate in another cycle resulting in addition of a new unit to the growing peptido glycan strand. Bacitracin (see below) is a specific inhibitor of the dephosphorylation of the pyrophosphate form of the lipid.

The next problem was the identification of the specific phospholipid of the membrane involved in the peptidoglycan synthesising reaction. It took several years to separate the phospholipid in a pure form, since it represented only about 0.1% of the total phospholipid of the membrane. It took a much shorter time to identify it from its mass spectrum, an illustration of the power of the tools of modern biochemistry. The compound was a C_{55}-isoprenyl alcohol, consisting of eleven isoprene units ending in an alcoholic function, to which the pyrophosphate and disaccharide–pentapeptide are attached[15]. The same alcohol was also identified as the lipid component of an intermediate in bacterial lipopolysaccharide synthesis[16].

Since the identification of this isoprene alcohol, several other syntheses of polysaccharides have been shown to occur through the intermediate action of isoprene alcohols. For example, the synthesis of mannan in *Micrococcus lysodeikticus* occurs by direct transfer of the sugar to C_{55}-isoprenyl phosphate although this reaction sequence is of a different type from the one involved in cell wall synthesis[17]. Lipid intermediates with 18 to 20 isoprene units known as dolichols have been known for some time to be present in mammalian liver. The phosphate esters of these alcohols are involved in the synthesis of some types of glycans in liver[18, 19]. Recent studies in our laboratory of synthesis of surface glycoproteins in human lymphocytes show that these cells contain an extremely active system for the synthesis of dolichylphosphomannose, which is apparently used in synthetic reactions[20]. The isoprene alcohols seem to have a rather general function in the synthesis of a variety of substances in bacteria, plants, and animals. They probably serve a membrane transport function[12]. The syntheses involved are concerned with the formation outside the cell of products that are synthesised from *intermediates inside* the cell. Possibly the function of the isoprene alcohol is to act as a carrier for the activated sugar fragments from the inside to the outside of the cell.

The third and final stage of cell wall synthesis is the reaction cross linking the peptidoglycan strands. These reactions occur outside of the cell at the site of pre-existing peptidoglycan. These terminal reactions are the penicillin-sensitive reactions of bacterial cell wall synthesis (see below).

8.3 MECHANISM OF ACTION OF BACITRACIN

The bacitracins are a mixture of peptide antibiotics originally isolated from *Bacillus licheniformis*. These compounds have slightly differing structures (Table 8.1) and growth-inhibitory potencies. The major component is bacitracin A. Bacitracin was found to inhibit the dephosphorylation of C_{55}-isoprenyl pyrophosphate, thus causing this compound to accumulate in the membrane[21].

This inhibition occurs at low concentrations of bacitracin (of the same order as those required to inhibit the growth of *M. lysodeikticus*). However, bacitracin had no effect on the enzymatic dephosphorylation of *p*-nitrophenyl phosphate by a non-specific phosphatase from *E. coli*, although bacitracin inhibited the dephosphorylation of C_{55}-isoprenyl pyrophosphate by the same phosphatase. This suggested that the inhibition resulted from an interaction of the bacitracin with the substrate rather than with the enzyme. Bacitracin A is a cyclic dodecapeptide with a six-membered amino acid ring and an acyclic pentapeptide tail (Figure 8.4). The side chain has a peculiarity in its structure which is probably related to the antimicrobial action of bacitracin. The terminal two amino acids are L-cysteine and L-isoleucine and the —SH group of the cysteine reacts with the isoleucine carbonyl group to form a

Table 8.1 Structures, growth-inhibitory levels, and association constants of bacitracin and C_{55}-isoprenyl pyrophosphate[23]

Bacitracin peptide	*Structure differing from Bacitracin A*	*Minimal concentration for inhibition of growth*	K_a	K_a *in presence of EDTA*
Bacitracin A		1×10^{-7}	1.1×10^6	5.2×10^3
Mixture of bacitracin A and bacitracin B		2×10^{-7}	1×10^6	
DNP-Bacitracin	δ-Amino group of ornithine substituted with DNP	3×10^{-7}	2.8×10^4	
Bacitracin F	Deaminated in thiazoline ring*	5×10^{-5}	9.8×10^3	6.3×10^3
Desamido bacitracin		7×10^{-4}	1.4×10^3	2×10^3

* Thiazoline ring structure is therefore destroyed

thiazoline ring. Formation of a thiazoline ring brings the N of the cysteine close enough to the N of the isoleucine that a metal ion can be chelated between these two atoms. In addition the pyrophosphate can also fit in a pocket in the molecule (shown by the structure in Figure 8.4).

Divalent metal ions were known to potentiate the action of bacitracin on intact microbial cells. A divalent metal ion was also shown to be required for the inhibition of the phosphatase by bacitracin (although metal ions are not in fact required for the enzymatic reaction itself). The efficiency of various metals varies widely; Zn^{2+} is far more effective than Ca^{2+}, Mg^{2+}, Ni^{2+}

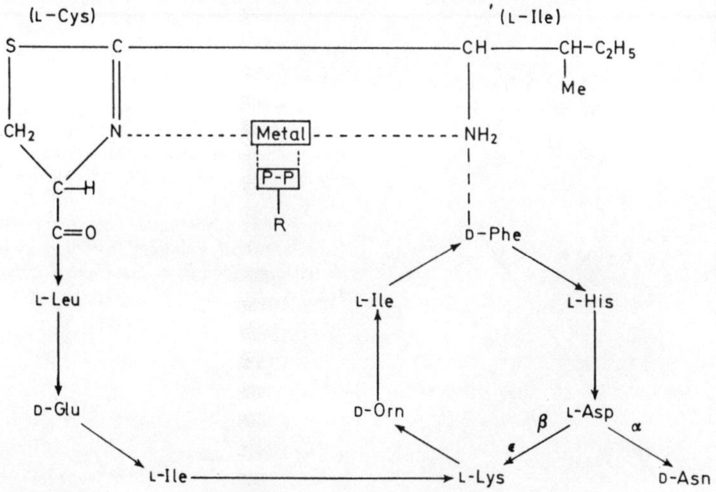

Figure 8.4 Proposed structure of bacitracin A. Arrows signify C—N bonds. The structure of the L-cysteine and L-isoleucine residues at the 'tail' of the acyclic side chain are shown in detail to illustrate the thiazoline ring. Dotted lines indicate where the metal and pyrophosphate may be chelated in the complex

Co^{2+} or Cd^{2+}. The addition of EDTA or other metal chelating agents to the system reversed the inhibition produced by bacitracin in the presence of metal, and this reversal was overcome by the addition of higher concentrations of metal ions. These facts suggested that bacitracin was in some way interfering with the capacity of the substrate to interact with the enzyme rather than with the basic catalytic activity of the enzyme. It was possible then to demonstrate the formation of a tight one-to-one complex of bacitracin with the substrate in the presence of a divalent cation[22]. The pyrophosphate group of the lipid intermediate appears to fit into a pocket in the bacitracin molecule and the metal ion may be the ligand between the antibiotic and this polar end of the substrate. This complex thus removes an essential intermediate from the reaction cycle of cell wall synthesis, and in this way the process is arrested. The proposal for the precise nature of the complex which is formed[22] requires some refinement. It does not take account of n.m.r. studies which have shown that the histidine residue of bacitracin is involved in the binding of metal. The binding of bacitracin and C_{55}-isoprenyl pyrophosphate in the absence of enzyme is pH-dependent, with an optimum at about pH 7 (Figure 8.5(a))[23]. The pK_a (*ca.* 8) of the descending limb suggests that an

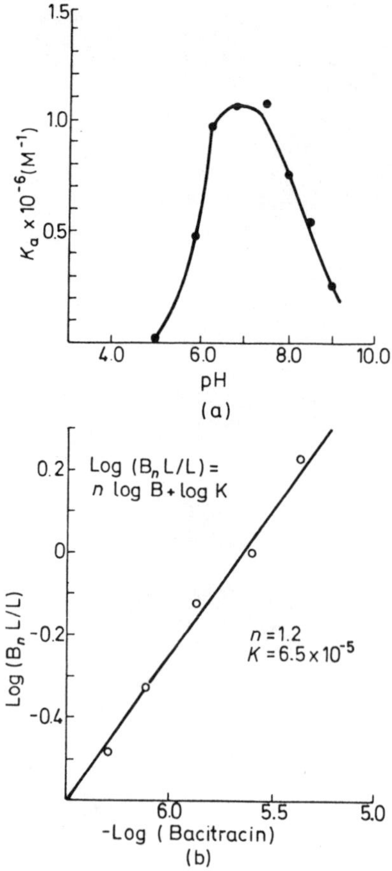

Figure 8.5 (a) Dependence of binding of bacitracin to C_{55}-isoprenyl pyrophosphate on pH[23]. Optimum is about 7.
(b) Measurement of the association constant (K_a) of bacitracin and C_{55}-isoprenyl pyrophosphate (see text). (From Storm and Strominger[23], by courtesy of the American Society of Biological Chemists Inc.)

α-amino group in bacitracin must be protonated for binding to occur. The only α-amino group which could be involved is that of the terminal isoleucine residue (since the δ-amino group of ornithine can be dinitrophenylated without greatly disturbing the formation of the complex or antimicrobial activity). This finding is not compatible with the model previously suggested, since a protonated amino group would be a poor metal chelator. The ascending limb of the pH curve corresponds to the pK_a of the pyrophosphate residue and suggests that this group must be fully ionised for optimum complex formation. It is also near the pK_a of histidine.

The association constant of bacitracin to the substrate could be measured by varying the amount of bacitracin (B) while keeping the amount of lipid pyrophosphate (L) constant[23]. By plotting the value log (BnL/L) against log-[bacitracin], a straight line was obtained, the slope of which provided the value n, the number of bacitracin molecules in the complex (Figure 8.5(b)). When the value of BnL/L is one, its log is 0 and the association constant (K_a) is estimated from the appropriate value on the abscissa of (Figure 8.5(b)). C_{55}-Isoprenyl pyrophosphate and C_{15}-farnesyl pyrophosphate have extremely high K_a values, whereas the association between bacitracin and C_5-isopentenyl pyrophosphate is much weaker (Table 8.1). Inorganic pyrophosphate itself is also very weakly bound. There is evidently a very important interaction between lipid and the bacitracin peptide in addition to the chelation reaction. The strength of this interaction can be measured by the K_a values obtained in the presence of EDTA.

Bacitracin is well known to be highly toxic particularly to the kidney when given parenterally. The binding of C_{15}-farnesyl pyrophosphate to bacitracin suggested that bacitracin might also be an inhibitor of the synthesis of sterols in animals. This was readily demonstrated with rat liver preparations using several different precursors (Figure 8.6)[24]. For example, the conversions of

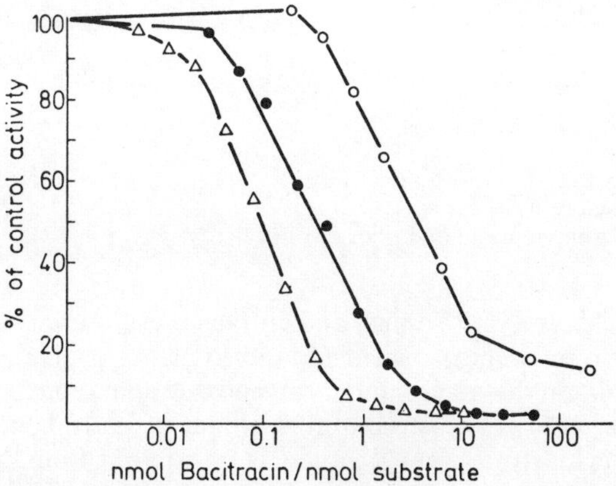

Figure 8.6 Effect of bacitracin on the utilisation of various precursors for sterol synthesis with rat liver preparations[24] MVA = mevalonic acid; IPP = C_5-isopentenyl pyrophosphate; FPP = C_{15}-farnesyl pyrophosphate. (From Storm and Strominger[24], by courtesy of the American Society of Biological Chemists Inc.)

isopentenyl pyrophosphate to geranyl pyrophosphate and farnesyl pyrophosphate are actively inhibited by low concentrations of bacitracin. Conceivably, at least part of the toxicity of bacitracin could be accounted for by its extreme effectiveness as an inhibitor of sterol biosynthesis. Could bacitracin also be useful as a hypocholestremic agent?

8.4 CYCLOSERINE AND PHOSPHONOMYCIN

The mode of action of bacitracin which forms a complex with the *substrate* of an enzymatic reaction may be contrasted with that of two other antibiotics that interfere with the earlier stages of cell wall synthesis, namely cycloserine and phosphonomycin. As described earlier, the last step in the synthesis of UDP–acetylmuramyl-pentapeptide is the addition of the dipeptide, D-alanyl–D-alanine. This dipeptide is formed by the initial conversion of L-alanine to D-alanine catalysed by a racemase, followed by the joining of two D-alanine molecules, catalysed by D-alanyl–D-alanine synthetase, to produce the dipeptide. Cycloserine is an inhibitor of both of these reactions[8, 25]. The structures of D-alanine and D-cycloserine (Figure 8.7(a)), obtained from

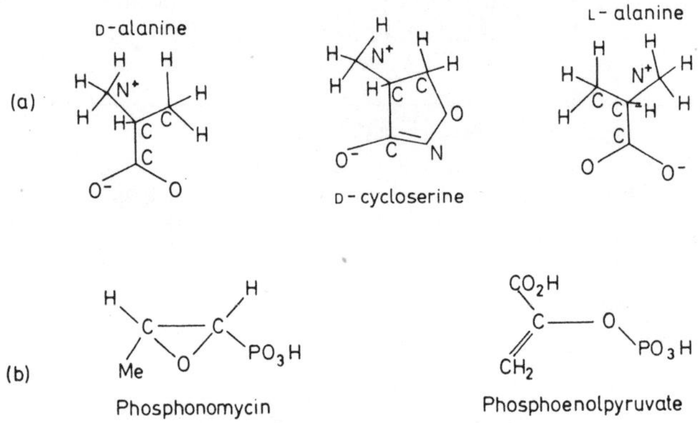

Figure 8.7 (a) Zwitterion forms of D- and L-alanine compared with cycloserine.
(b) Phosphonomycin and phosphoenol pyruvate

molecular models) are very similar and in fact D-cycloserine is a true competitive inhibitor of alanine racemase and D-alanyl–D-alanine synthetase. Phosphonomycin, on the other hand, is an epoxide containing a phosphonate group (Figure 8.7(b)). It is a structural analogue of phosphoenolpyruvate[26]. As indicated previously, a pyruvyl moiety is transferred from this compound to UDP-*N*-acetylglucosamine to form the 3-*O*-pyruvate-enol ether of the nucleotide, which is then reduced by NADPH to UDP-*N*-acetylmuramic acid. During the reaction the pyruvyl residue becomes substituted on the enzyme as a thioether and is then transferred to the nucleotide. The enzyme which catalyses this transfer is attacked by phosphonomycin at the analogous carbon atom. The epoxide opens, but phosphonate cannot be expelled. A stable C—S bond is formed and the enzyme becomes irreversibly inactivated.

Thus these three antibiotics each act on different enzymatic reactions by various modes of action: cycloserine by competitive and phosphonomycin by irreversible inhibition of an enzyme, and bacitracin by complexation with a substrate.

8.5 PENICILLIN-SENSITIVE REACTIONS

All but one of the series of reactions of cell wall synthesis has been considered to this point. None of these was found to be penicillin-sensitive. Only one more reaction remains — the final cross linking of the peptidoglycan chains to form the gigantic macromolecule. Suffice it to say that this reaction is indeed the one sensitive to penicillins and cephalosporins[27, 28].

The cross linking of peptidoglycan strands takes place outside the cell membrane at the site of the pre-existing wall. Since there is no ATP or other obvious energy source available at this site, a mechanism independent of any external energy source has apparently been evolved for this reaction. The

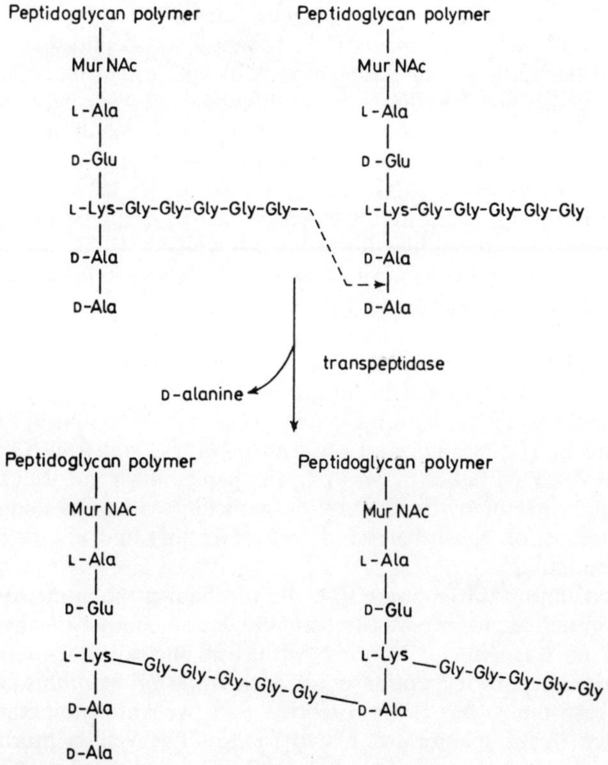

Figure 8.8 The third stage of cell wall synthesis in *S. aureus:* cross linking of peptidoglycan polymers by transpeptidation with the elimination of D-alanine

reaction is a transpeptidation in which the terminal amino end of an open crossbridge attacks the terminal peptide bond in an adjacent strand to form a cross link. The terminal alanine residue from the strand which becomes cross linked is thus eliminated (Figure 8.8). It is this reaction that is sensitive to penicillins and cephalosporins. It has been proposed that penicillin inhibits the reaction by acting as a structural analogue of the terminal D-alanyl–D-alanine residue of the peptidoglycan strand[28]. The similarity between the conformation of the penicillin molecule and one of the conformations of the dipeptide, D-alanyl–D-alanine, is shown in Figure 8.9. It was suggested, therefore, by analogy with similar reactions, that the transpeptidase would first react with the substrate to form an acyl enzyme intermediate, with the elimination of D-alanine, and that this active intermediate would then react with another strand to form the cross link and regenerate the enzyme. If penicillin were an analogue of alanyl-alanine, it should fit the substrate-binding site with the highly reactive —CO—N— bond in the β-lactam ring in the same position as the bond involved in the transpeptidation. It might therefore acylate the enzyme, forming a penicilloyl enzyme, and thereby inactivate it (Figure 8.10).

The initial studies were hampered by the inability to find a cell-free enzyme system which would catalyse the transpeptidation. After several years, this was finally found in *Escherichia coli*[29-31]. In fact it was found that there were two terminal reactions in cell wall synthesis in this organism, and each was inhibited by penicillin. The first was catalysed by the transpeptidase which cross links the two strands with the elimination of one D-alanine residue. The second enzyme is a D-alanine carboxypeptidase the precise function of which is not known. It may act to catalyse the removal of a D-alanine residue from the second strand, thus limiting the extent of cross linking by preventing the formation of further cross links. Since the cell walls of *E. coli* are such more loosely cross linked than those of *S. aureus*, the carboxypeptidase may be involved in this phenomenon.

Although both the transpeptidase and the carboxypeptidase were sensitive to penicillins, the carboxypeptidase was sensitive to extremely low concentrations, far below those required for inhibition of growth of the cells. The sensitivity of the transpeptidase, however, was very similar to that of the growth of the organism. The inhibition of the transpeptidase was irreversible, i.e. it could not be reversed either by washing the penicillin out of the particulate enzyme preparation or by destroying the penicillin with penicillinase[31]. The carboxypeptidase, on the other hand, was reversibly and competitively inhibited by penicillin[32].

It was next important to prove that the mechanism of inactivation of an irreversibly inhibited enzyme was by penicilloylation. For various reasons the *E. coli* enzyme was difficult to work with, and instead a then newly discovered D-alanine carboxypeptidase present in *Bacillus subtilis* was studied. It differed from the *E. coli* carboxypeptidase in two important respects: (a) it was located in the membrane; and (b) it was irreversibly inactivated by penicillin[33]. The fact that this enzyme might be uncoupled transpeptidase (Figure 8.10) provided a rationale for its study. However, this enzyme is now known not to be essential for cell viability and it is unlikely to be the transpeptidase.

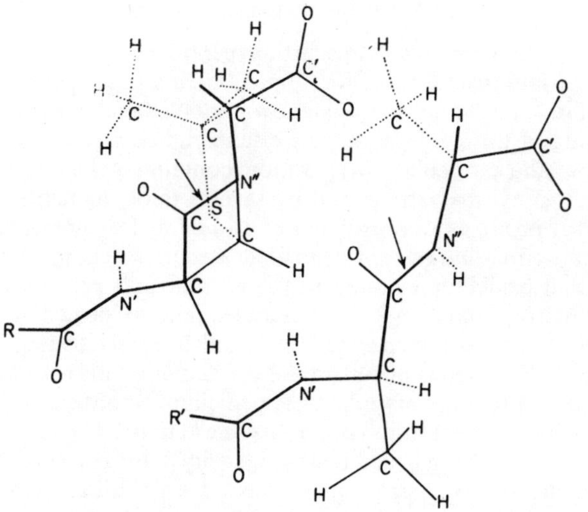

Figure 8.9 Dreiding stereomodels of penicillin (left) and of the D-alanyl–D-alanine end of the peptidoglycan strand (right)[28]. Arrows indicate the position of the CO—N bond in the β-lactam ring of penicillin and of the CO—N bond in D-alanyl–D-alanine at the end of the peptidoglycan strand (From Tipper and Strominger[28], by courtesy of the National Academy of Science, U.S.)

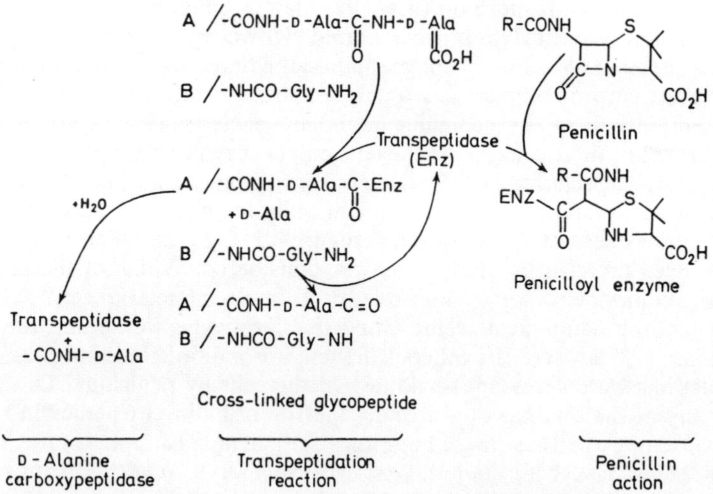

Figure 8.10 Proposed mechanism of inhibition of transpeptidation by penicillins[28]. A represents the end of the main peptide chain of the glycan strand. B represents the end of the pentaglycine substituent from an adjacent strand. If the acyl enzyme intermediate can react with water instead of the acceptor (left), the enzyme would be regenerated and the substrate released. The overall reaction would be the hydrolysis of the terminal D-alanine residue of the substrate (D-alanine carboxypeptidase activity). Such a reaction could be an 'uncoupled transpeptidation' reaction, but it also seems likely that carboxypeptidation (left) and transpeptidation (centre) reactions can be catalysed by separate proteins, perhaps using similar mechanisms

The particles that contained the carboxypeptidase also bound radioactive penicillin[34]. It had been known for some 30 years that bacterial cells could bind penicillin[35], but very little was known about the chemical nature of the actual binding of this compound. The bound radioactive penicillin could be released from the particulate preparation containing the carboxypeptidase both by neutral hydroxylamine and by several thiols including ethanethiol. These reversal products were isolated and identified as penicilloyl hydroxamate and an α-ethyl-thio ester of penicilloic acid, thus lending credence to the hypothesis that penicillin is bound to the enzyme as a penicilloyl derivative.

Neutral hydroxylamine not only released bound penicillin, but it also restored the activity of the enzyme which had been inactivated by penicillin G. A study of the kinetics of restoration of activity with neutral hydroxylamine showed that the rate of removal of penicillin from the carboxypeptidase paralleled the rate of restoration of the enzyme activity. The carboxypeptidase has been solubilised and purified to homogeneity[36]. It has been shown that the purified enzyme shows a single protein band on sodium dodecyl sulphate (SDS) gels which corresponds exactly to the band of bound radioactive penicillin, indicating that the radioactive penicillin was not released after denaturation with SDS. The enzyme, which requires 1 mole of penicillin per mole of enzyme to be completely inactivated, has four sulphydryl groups, one of which disappears as it is titrated with penicillin. The penicillin is therefore bound as a penicilloyl thioester. Further studies on the kinetics of inactivation of the enzyme suggest that the efficacy of different penicillins in inhibiting it depends more on their efficiency in binding to the enzyme than on their ability to acylate it, once bound. However, an important question had to be asked. Was the carboxypeptidase the only protein component in the particulate enzyme preparation which bound radioactive penicillin?

Essentially all of the penicillin bound by bacterial cells is bound to the membranes of these cells, i.e. to the particulate enzyme preparation. However, contrary to expectation, all the bacterial membranes that have been studied were found to contain multiple penicillin-binding sites[37]. *B. subtilis* has five, *B. cereus* has at least six major ones (Figure 8.11); *E. coli* perhaps as many as ten or even more, if the minor ones are considered. Which of the penicillin binding components was responsible for killing the bacterial cell? Which of these binding components represented the carboxypeptidase and the transpeptidase? What were the others? If there are multiple binding sites, which interactions were necessary for killing of the cells by penicillin? Or, indeed, were any of the binding sites actually the site of killing by penicillin?

To attempt to answer these questions more detailed studies on the binding components have been undertaken, especially in *B. subtilis*[38-40]. In this organism, the major penicillin-binding component is the D-alanine carboxypeptidase (Figure 8.12). It accounts for 70% of the bound penicillin, but its inactivation is not lethal for the micro-organisms. For example, cephalothin kills *B. subtilis* at concentrations which do not inactivate carboxypeptidase. By contrast, 6-aminopenicillanic acid has no effect on viability at a concentration which totally inactivates the carboxypeptidase[38]. Thus, the penicillin binding components of *B. subtilis* have different reactivities with different β-lactam antibiotics. A comparison of their rates of inactivation for different antibiotics (i.e., $\mu g\ ml^{-1}$ of antibiotic $\times$ time), with the killing rates for intact

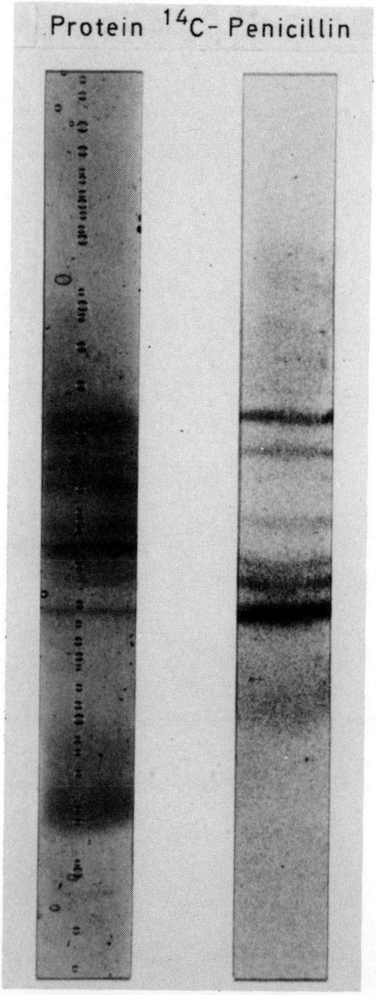

Figure 8.11 The penicillin binding components of *Bacillus cereus*[37]. The technique used was isoelectric focusing of solubilised membrane proteins containing bound ^{14}C-penicillin G in buffer containing detergent and urea. Duplicate gels were stained for protein or exposed to x-ray film

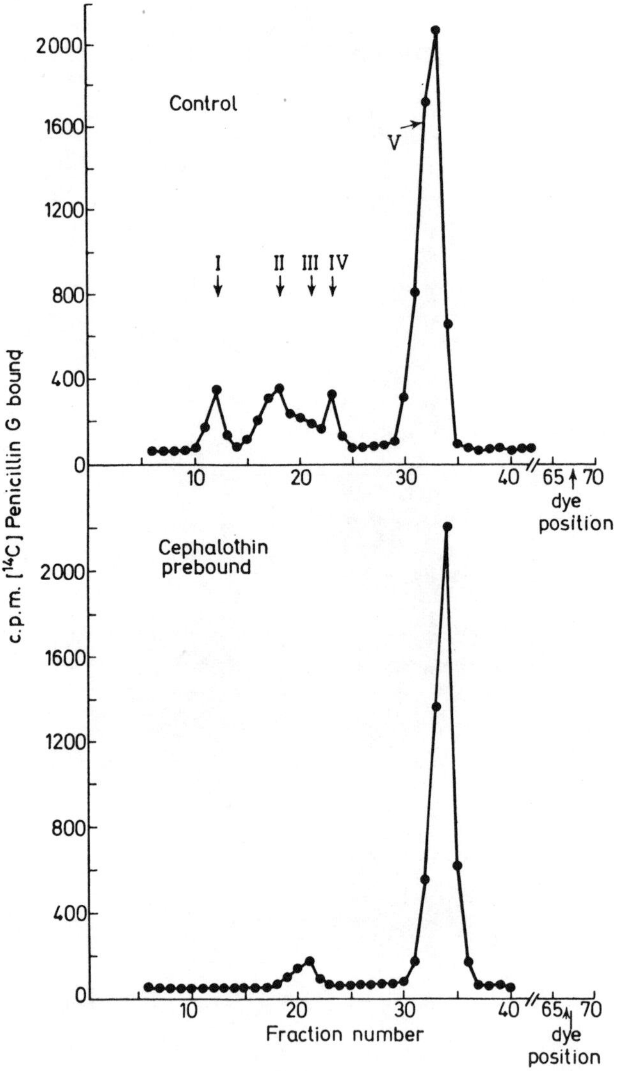

Figure 8.12 Penicillin-binding components of *Bacillus subtilis*[39]. The technique used in this case was SDS gel electrophoresis of solubilised membranes containing bound ^{14}C-penicillin G. Radioactivity in 1 mm gel slices is shown. Component V is the *D*-alanine carboxypeptidase. The failure of components III and V to react with cephalothin under the conditions used is shown below; prior treatment with cephalothin does not affect labelling of these components with ^{14}C-penicillin G. Components I, II and IV could be killing sites

cells suggests that the minor components I, II and IV (Figure 8.12) could be killing sites[39].

The five binding components of *B. subtilis* have now been isolated by covalent affinity chromatography (Figure 8.13)[40]. They are therefore now available in relatively substantial amounts, rather than as radioactive traces, and it should be possible to investigate in more detail the nature of their reactions with penicillins and cephalosporins and their possible functions. By further modification of the procedure, the carboxypeptidase, which makes up 1% of the total membrane protein, has been obtained in pure form in a single step with 50–80% overall recovery of enzymatic activity.

Further studies in other bacteria have yielded additional information which may complicate the story more. In *Bacillus megaterium* three formally different transpeptidation reactions have been identified (Figure 8.14)[41]. Studies with different penicillins suggest that these reactions maybe catalysed by different transpeptidases[42]. However, these transpeptidases appear to be *reversibly* inhibited by penicillins, i.e. most of the activity could be recovered by treatment with penicillinase. Thus, it is possible that in some organisms acylation by penicillin is not a necessary component of the killing effect.

Thus, we are left with many questions that still need to be answered. Although there are multiple penicillin-binding sites in bacteria, it is not yet known which of these is the site at which penicillin exhibits its lethal action, and whether or not there may be more than one killing site. Do these sites represent multiple transpeptidases in microbial cells, or, if not, what are their functions? Although the same type of cell wall is laid down during both septation and elongation of the cell, the enzymes which participate in these two processes might be different. Indeed, it was observed a long time ago that at sublethal concentrations penicillin induces filamentation of *E. coli*[43]. This simple observation appears to represent clear evidence that the system for cell septation is somehow different from that for cell elongation and more sensitive to penicillin. Are the corners of rod-shaped bacteria different from the rest of the structure and does their synthesis require a special enzyme? If so, then the observation that the rod-shaped *Bacillus proteus* grows in the presence of penicillin as a sphere with a wall of normal structure and particularly with normal cross linking could be explained[44]. Perhaps in this organism the septation and elongation systems are penicillin resistant (so the organism can grow and divide) but the 'corner synthesising' system is penicillin sensitive (so it grows as a sphere). The discovery that the transpeptidase in *B. megaterium* is *reversibly* inhibited by penicillin adds an additional dimension to the problem. One other phenomenon deserves comment—that of the step-wise development of resistance to penicillin first described more than 30 years ago by Demerec[45], one of the early microbial geneticists. Demerec observed that if increasing amounts of penicillin were added to a culture of *S. aureus no* resistant survivors were found at high penicillin concentrations, in contrast to the situation with streptomycin, for example, where highly resistant mutants could be obtained in a single selection step. To obtain organisms highly resistant to penicillin Demerec had to repeat the selection many times. A relatively small increase in resistance was found at each step. One explanation for this step wise development of high level resistance to penicillin could be the presence of multiple killing sites with

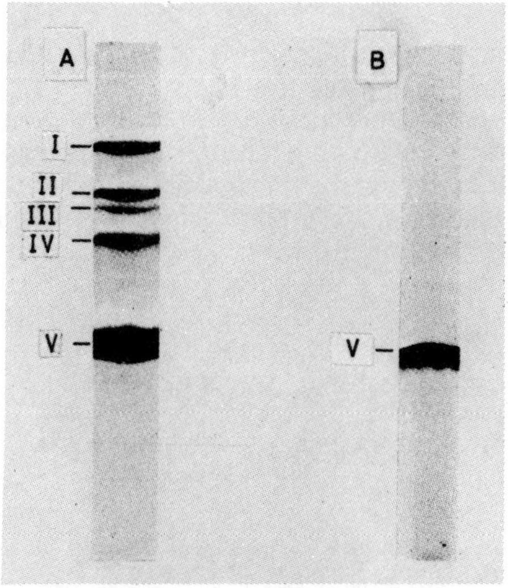

Figure 8.13 Penicillin-binding components in *Bacillus subtilis* separated by covalent affinity chromatography[40]. Succinyl diaminopropylamino Sepharose 4B is first prepared, and 6-aminopenicillanic acid is coupled to it by the use of 1-ethyl-3-(3-dimethylaminopropyl) carbodiimide. The substituted sepharose is then permitted to react with membranes (or whole cells) solubilised with detergents. The unbound proteins are removed by washing the Sepharose column with high salt buffer, and the penicillin binding components are released and eluted from the column with hydroxylamine. After dialysis and concentration on a DEAE-cellulose column, SDS gels are prepared and are shown in A. The bands are stained with Coomassie blue. They should be compared to the radioactive trace shown in Figure 8.12. B, right, shows band V, D-alanine carboxypeptidase on SDS gel, purified from membranes in one step by a modification of the above method involving treatment of the membranes with cephalothin before adding the substituted sepharose. Cephalothin binds more readily with bands I, II and IV than with V, and therefore effectively removes them from the sphere of action, and permits band V to be obtained in pure form after subsequent treatment as described above. Component III is eluted with difficulty by hydroxylamine and under the conditions used in B remained largely bound to the column. (From Blumberg and Strominger[40], by courtesy of the National Academy of Science, U.S.)

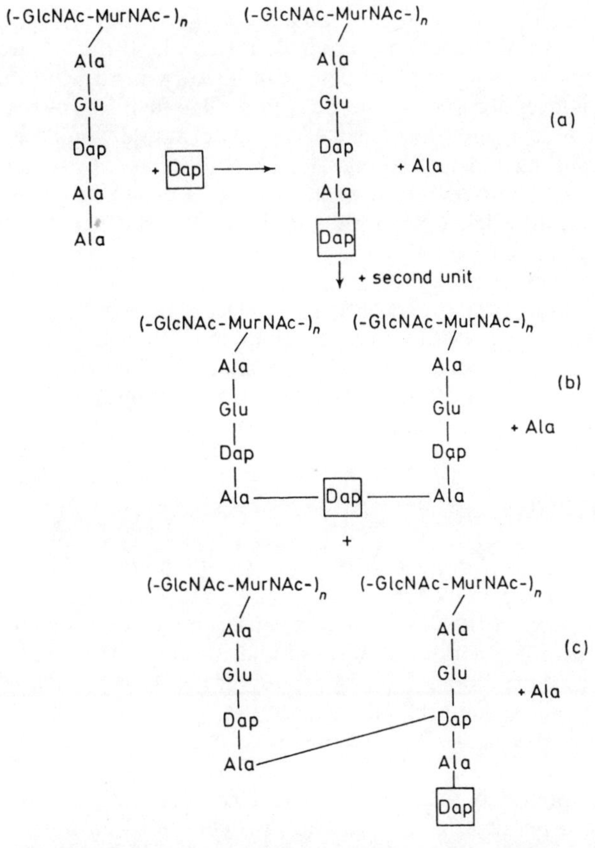

Figure 8.14 Transpeptidation reactions in particulate enzymes from *B. megaterium*[41]. Three separate transpeptidation reactions can occur.
(a) Diaminopimelic acid (Dap) is incorporated at the end of the peptide subunit replacing D-alanine. Either *meso*- or (D),(D)-diaminopimelic acid can be utilised (but not the (L),(L) form).
(b) If (D),(D)-diaminopimelic acid is the substrate in reaction 1, a second subunit can interact with the first one, with the two amino groups of Dap forming a bridge between the two peptidoglycan units.
(c) If *meso*-diaminopimelic acid is the substrate in reaction 1, a different transpeptidation results in a normal cross link between the two strands and a diaminopimelic acid terminal residue

different sensitivities. For instance, a hypothetical wild type organism could have three killing sites, sensitive to 1, 5, and 10 $\mu g\ ml^{-1}$ of penicillin, respectively. If the killing sites are independent of each other, then the organism would be killed by 1 $\mu g\ ml^{-1}$ of penicillin, the sensitivity of the most sensitive site. A mutant of this organism capable of survival at a higher penicillin concentration, would have to be mutated, at this site. If this site were mutated to a sensitivity of 15 $\mu g\ ml^{-1}$ of penicillin, the organism would nevertheless be killed at a level of 5 $\mu g\ ml^{-1}$ of penicillin, the sensitivity of the second site of killing. The next mutation selected in this example would be a mutation at the third killing site, and so on. There would be no way to acquire a high level of resistance in one step. In each case the most sensitive site would need to be mutated in order to reach the next level of resistance. So the existence of multiple penicillin killing sites would provide an explanation for the multistep development of resistance.

And so, although the broad outlines are clear, there is still a great deal to learn about the mechanism by which penicillin kills bacterial cells. Perhaps it would not be pretentious to steal a line from the English writer, Leonard Woolf, who entitled the last volume of his autobiography finished shortly before his death, '*The Journey Not The Arrival Matters.*'

ACKNOWLEDGEMENT

This manuscript was derived from a transcript of the Charles E. Dohme Memorial Lectures delivered at the Johns Hopkins Medical School on November 13 and 14, 1972. A similar manuscript will also be published in the Bulletin of the Johns Hopkins Hospital and is being published here with the permission of the Editors. I am extremely grateful to Dr. Pamela Samuels Talalay who edited the original transcript.

I wish to acknowledge my indebtedness to Drs. P. M. Blumberg, D. R. Storm, J. Umbreit and G. Wickus and to a host of students and postdoctoral fellows who preceded them in the laboratory for the devotion and effort which made it possible for me to write this chapter. I have been generously supported in this work by research grants from the U.S. Public Health Service (AI-09152 and AM-13230) and the National Science Foundation (GB–29747X and GB–30690X).

References

1. Lederberg (1957). *J. Bacteriol.*, **73**, 166
2. Park, J. T. and Johnson, M. (1949). *J. Biol. Chem.*, **170**, 585
3. Park, J. T. and Strominger, J. L. (1957). *Science*, **120**, 119
4. Strominger, J. L. and Ghuysen, J.-M. (1967). *Science*, **156**, 213
5. Ghuysen, J.-M. (1968). *Bacteriol. Rev.*, **32**, 425
6. Kandler, O. (1972). *Bacteriol. Rev.*, **36**, 407
7. Salton, M. R. J. (1964). *The Bacterial Cell Wall* (Amsterdam: Elsevier)
8. Strominger, J. L. (1970). *Harvey Lectures, series 64*, 179 (New York: Academic Press)
9. Ito, E. and Strominger, J. L. (1973). *J. Biol. Chem.*, **248**
10. Chatterjee, A. N. and Park, J. T. (1964). *Proc. Nat. Acad. Sci. U.S.*, **51**, 9
11. Meadow, P. M., Anderson, J. S. and Strominger, J. L. (1964). *Biochem. Biophys. Res. Commun.*, **14**, 382

12. Anderson, J. S., Matsuhashi, M., Haskin, M. A. and Strominger, J. L. (1965). *Proc. Nat. Acad. Sci. U.S.*, **53,** 881
13. Anderson, J. S., Meadow, P. M., Haskin, M. A. and Strominger, J. L. (1966). *Arch. Biochem. Biophys.*, **116,** 487
14. Anderson, J. L., Matsuhashi, M., Haskin, M. A. and Strominger, J. L. (1967). *J. Biol. Chem.*, **242,** 3180
15. Higashi, Y., Strominger, J. L. and Sweeley, C. C. (1970). *J. Biol. Chem.*, **245,** 3697
16. Wright, A., Dankert, M., Fennessey, P. and Robbins, P. W. (1967). *Proc. Nat. Acad. Sci. U.S.*, **57,** 1798
17. Scher, M., Lennarz, W. J. and Sweeley, C. C. (1968). *Proc. Nat. Acad. Sci. U.S.*, **58,** 1313
18. Evans, P. J. and Hemming, F. W. (1973). *FEBS Lett.*, **31,** 335
19. Behrens, N. H., Parodi, A. J., Leloir, L. F. and Krisman, C. R. (1971). *Proc. Nat. Acad. Sci. U.S.*, **68,** 2857 and (1971). *Arch. Biochem. Biophys.*, **143,** 375
20. Wedgwood, J. F. and Warren, C. D. (1973). *Federation Proceedings*, **32,** 1709; see also Wedgwood, J. F., Warren, C. D. and Strominger, J. L. (1974). *J. Biol. Chem.*, 249, in press
21. Siewert, G. and Strominger, J. L. (1967). *Proc. Nat. Acad. Sci. U.S.*, **57,** 767
22. Stone, K. J. and Strominger, J. L. (1971). *Proc. Nat. Acad. Sci. U.S.*, **68,** 3223
23. Storm, D. R. and Strominger, J. L. (1973). *J. Biol. Chem.*, **248,** 3940
24. Stone, K. J. and Strominger, J. L. (1972). *J. Biol. Chem.*, **247,** 5107
25. Neuhaus, F. C. (1967). *Antibiotics* Vol. 1, 40 (D. Gottlieb and P. Shaw, editors) (New York: Springer)
26. Cassidy, P. J. and Kahan, F. M. (1973). *Biochemistry*, **12,** 1364
27. Wise, E. M. and Park, J. T. (1965). *Proc. Nat. Acad. Sci. U.S.*, **54,** 75
28. Tipper, D. J. and Strominger, J. L. (1965). *Proc. Nat. Acad. Sci. U.S.*, **54,** 1133
29. Izaki, K., Matsuhashi, M. and Strominger, J. L. (1966). *Proc. Nat. Acad. Sci. U.S.*, **55,** 656
30. Araki, Y., Shimida, A. and Ito, E. (1966). *Biochem. Biophys. Res. Commun.*, **23,** 518
31. Izaki, K., Matsuhashi, M. and Strominger, J. L. (1968). *J. Biol. Chem.*, **243,** 3180
32. Izaki, K. and Strominger, J. L. (1968). *J. Biol. Chem.*, **243,** 3198
33. Lawrence, P. J. and Strominger, J. L. (1970). *J. Biol. Chem.*, **254,** 3660
34. Lawrence, P. J. and Strominger, J. L. (1970). *J. Biol. Chem.*, **245,** 3653
35. Cooper, P. D. (1956). *Bact. Rev.*, **20,** 28
36. Umbreit, J. and Strominger, J. L. (1973). *J. Biol. Chem.*, **248,** 6759, 6767
37. Suginaka, H., Blumberg, P. M. and Strominger, J. L. (1972). *J. Biol. Chem.*, **247,** 5279
38. Blumberg, P. M. and Strominger, J. L. (1971). *Proc. Nat. Acad. Sci. U.S.*, **68,** 2814
39. Blumberg, P. M. and Strominger, J. L. (1972). *J. Biol. Chem.*, **247,** 8107
40. Blumberg, P. M. and Strominger, J. L. (1972). *Proc. Nat. Acad. Sci. U.S.*, **69,** 3751
41. Wickus, G. and Strominger, J. L. (1972). *J. Biol. Chem.*, **247,** 5297
42. Wickus, G. and Strominger, J. L. (1972). *J. Biol. Chem.*, **247,** 5307
43. Gardner, A. D. (1940). *Nature (London)*, **146,** 837
44. Katz, W. and Martin, H. H. (1970). *Biochem. Biophys. Res. Commun.*, **39,** 774
45. Demerec, M. (1948). *J. Bacteriol.*, **56,** 63

9
Turnover of Membrane Proteins in Animal Tissues

R. T. SCHIMKE
Stanford University

9.1 INTRODUCTION

Continual flux amidst apparent constancy is a concept that dates back to Greek philosophers, and is embodied in the statement attributed to Heracleitus in 500 B.C.: 'all things forever flow and change . . . even in the stillest matter there is unseen flux and movement'. The biological concept that cellular constituents are undergoing continual replacement, i.e. turnover, has been developed extensively in the past 20 years since the publication of the classical work of Schoenheimer and his associates in 'The Dynamic State of Body Constituents'[1].

In the past 3 years it has become increasingly evident that cellular membranes are also continually turning over, and that the constancy of structure occurs within a dynamic process of continual replacement. The properties of this turnover, and the associated questions of how membranes are generated and replaced will be the subject of this review. Clearly, there will be far more questions raised than there are answers available, and it is hoped that a discussion of concepts, techniques, and problems will serve as a stimulus to persons interested in entering this important field.

9.2 GENERAL PROPERTIES OF TURNOVER OF PROTEINS

As a basis for understanding the properties of turnover of membrane constituents, it would be useful first to enumerate some general properties of turnover of proteins in animal tissues. The most extensive studies have been undertaken with rat liver. However, these properties appear to be common to essentially all tissues and organisms studied, including unicellular eukaryotes, prokaryotes, as well as mammals.

9.2.1 Turnover is extensive

Studies of Swick[2], Buchanan[3] and Schimke[4] have indicated that essentially all proteins of rat liver take part in the continual replacement process. These studies have used the general technique of continuous administration of an isotope of known specific activity, and subsequent comparison of the specific activity of isotope isolated from defined protein with that of the administered isotope[5,6]. For example, in studies using an algal diet of constant ^{14}C specific activity, Buchanan estimated that approximately 70% of rat liver protein was replaced every 4–5 days from the dietary source[3].

9.2.2 Turnover is largely intracellular

Turnover of proteins can represent synthesis and secretion of secretory proteins, replacement of cells, i.e. cell turnover, or intracellular synthesis and degradation. The life-span of hepatic cells is of the order of 160–400 days[3,7,8], and hence the extensive turnover occurring in 4–5 days precludes cell replacement as the explanation for the turnover observed in liver. In addition, the extensive replacement of protein cannot represent synthesis and secretion of serum proteins (albumin), since the steady state level of such proteins in liver is of the order of only 1–2% of total liver protein[9]. Hence, it can be concluded that the majority of turnover is intracellular synthesis and degradation.

9.2.3 There is a marked heterogeneity of rates of replacement of different proteins (enzymes)

Table 9.1 indicates a representative list of rates of degradation (expressed as half-lives) for various specific enzymes of rat liver, as well as the methodology employed for measurement of such rates. More extensive lists are given by Schimke and Doyle[6]. The wide range of half-lives for these specific proteins is remarkable, ranging from 11 min for ornithine decarboxylase to 16 days for LDH_5. In addition, there is no necessary correlation between half-lives and metabolic functions of the enzymes. For instance, glucokinase and LDH_5, both involved in carbohydrate metabolism, have markedly different half-lives (30 h *v.* 16 days), as do tyrosine aminotransferase and arginase (1.5 h *v.* 4–5 days), both of which are involved in amino acid catabolism. There is no correlation between the rate of turnover and the cell fraction from which the enzyme is isolated. This is shown for δ-aminolevulinate synthetase and ornithine aminotransferase, both of which are associated with mitochondria (half-lives of 60 min and 1 day) compared with total mitochondrial protein (half-life of 4–5 days), as well as the 'cytoplasmic' enzymes mentioned above.

In addition, the rates at which specific proteins are degraded can vary with the physiological state of the animal. For instance, Schimke *et al.*[10] have shown that the administration of tryptophan to animals results in an accumulation of tryptophan oxygenase (pyrrolase), an effect caused by continued synthesis of enzyme with a cessation of the normally occurring degradation of the enzyme (half-life normally of 2 h). Schimke[11] has also demonstrated that during starvation, arginase accumulates in rat liver as a result of decreased degradation in the presence of continued synthesis. In contrast, Majerus and Kilburn[12] have shown that starvation increases the rate of degradation of acetyl CoA carboxylase. Thus, for the soluble proteins of rat liver, there is a continual process in which the total complement of any given enzyme is being replaced at different rates, and those rates can be altered by various physiological conditions.

9.2.4 The degradation of an enzyme molecule, once synthesised, is a random process

This conclusion is based on the essentially universal finding that the decay of an isotopically labelled (specific) enzyme following a single (pulse) administration of isotope, follows first-order kinetics[6,13]. Although there are several interpretations of such data, the most likely is that once a given enzyme molecule is synthesised and enters a pool of like molecules, its chance of being degraded (or otherwise removed from the pool of like molecules) is a random process. Thus, there is no evidence for an accumulation of damage, i.e. ageing, to explain why any given enzyme molecule is degraded.

Table 9.1 Half-lives of specific enzymes and subcellular fractions of rat liver

	Half-life	*Reference*	*Method of measure*
Enzymes			
Ornithine decarboxylase (soluble)	11 min	84	Loss of activity after puromycin
δ-aminolevulinate synthetase (mitochondria)	70 min	81	Loss of activity after puromycin
Alanine-aminotransferase (mitochondria)	0.7–1.0 days	73	Time course of enzyme change
Catalase (peroxisomal)	1.4 days	83	Recovery of activity after irreversible inhibition of activity
Tyrosine aminotransferase (soluble)	1.5 h	80	Isotope decay*
Tryptophan oxygenase (soluble)	2 h	10	Isotope decay*
Glucokinase (soluble)	1.25 days	82	Change of enzyme activity
Arginase (soluble)	4–5 days	11	Isotope uptake and decay*
Glutamic-alanine transaminase	2–3 days	86	Time course of enzyme change
Lactate dehydrogenase isozyme-5	16 days	78	Isotope uptake*
Acetyl CoA carboxylase (soluble)	2 days	12	Isotope decay*
Cell fractions			
Nuclear	5.1 days	15	Isotope decay
Supernatant	5.1 days	15	Isotope decay
Endoplasmic reticulum	2.1 days	15	Isotope decay
Plasma membrane	2.1 days	15	Isotope decay
Ribosomes	5.0 days	15	Isotope decay
Mitochondria	4–5 days	79	Isotope decay

* Denotes use of immunoprecipitation techniques

9.2.5 There is an apparent correlation between the size of proteins and their rates of degradation

Dehlinger and Schimke[14] first showed that large proteins have greater relative rates of degradation than small proteins, as measured by the double-isotope method of Arias, Doyle and Schimke[15]. This same correlation holds for proteins of other organelles, including ribosomes[16], proteins associated with chromatin[17] and membranes[18]. This last point will be discussed in more detail in a subsequent section. In addition Dice, Dehlinger and Schimke[19] have shown that those soluble proteins that turn over most rapidly *in vivo* are also more sensitive to proteolytic cleavage by proteases including trypsin, chymotrypsin, and pronase. Large proteins were also shown to be more rapidly degraded by these same proteases than small proteins. Such studies have led these workers to propose that the correlation of size and rate of degradation is based on the overall greater change of a larger protein when 'hit' by a protease, producing an initial rate limiting peptide bond cleavage, with subsequent unfolding and rapid degradation to amino acids. Their surprising finding that intracellular organelles do not turn over as units also led them to propose that there is a continual association–dissociation of assembled macromolecular structures, and that, only in the dissociated state were the proteins degraded[19]. This suggestion has been further supported by studies of Tweto, Dehlinger and Larrabee[20] in showing that the dissimilar subunits of the fatty acid synthetase complex of rat liver are turning over at different rates, again with large subunits turning over more rapidly than smaller subunit components.

9.3 PROPERTIES OF TURNOVER OF MEMBRANE PROTEINS

Studies on the characteristics of turnover of membrane constituents are far more limited, in large part because of the difficulties in obtaining pure preparations of specific membrane fractions and highly purified specific membrane-associated proteins. The reported studies indicate that the general properties found for protein turnover also hold for the turnover of membrane-associated proteins. The majority of these studies have been performed also in rat liver. These properties include:

(1) There is a continual synthesis and degradation of membrane proteins. Studies from various groups have estimated that the average half-life of total proteins of the plasma membrane and both rough and smooth endoplasmic reticulum fractions are approximately 30–60 h[15, 21-23]. Such differences in estimated mean half-lives may appear to be very large. However, the estimates were made by different techniques and different isotopic labels were used. Because of the multiple problems involved in obtaining accurate or true measures of half-lives[3], such differences in estimates should not be considered as indicating different rates of turnover. For instance, Arias *et al.*[15] estimated the half-life of total smooth endoplasmic reticulum to be 48 h when the decay of radioactivity was determined following administration of guanidino-^{14}C-L-arginine, whereas a value of 60 h was obtained following single administration of uniformly labelled L-leucine. This difference can be ascribed to the

Table 9.2 Half-lives of membrane-associated enzymes

	Half-life	*Reference*	*Method of measure*
Nucleoside diphosphatase	1.5 days	66	Isotope decay
Cytochrome c reductase (endoplasmic reticulum)	60–80 h	32	Isotope decay
Cytochrome b_5 (endoplasmic reticulum)	100–120 h	32	Isotope decay
NAD glucohydrolase (endoplasmic reticulum)	16 days	25	Isotope decay
Hydroxymethylglutaryl CoA reductase (endoplasmic reticulum)	2–3 h	29, 30	Activity decay after cycloheximide and isotope decay

differing degrees of re-utilisation of the two isotopic amino acids. Thus guanidino-^{14}C-L-arginine has minimal re-utilisation because of the high activity of arginase in rat liver[24]. It should also be emphasised that in other tissues where arginase activity is low (which includes most tissues, including cultured cells[11]), use of this isotopic form of arginine does not obviate the problem of isotope reutilisation.

(2) The rates of turnover of membrane proteins are heterogeneous. Table 9.2 provides a representative list of specific enzymes isolated from the endoplasmic reticulum, and their estimated apparent half-lives, as well as the methods employed in obtaining such estimates. Half-lives vary widely, just as with soluble enzymes, with a range of from 2 h for hydroxymethylglutaryl CoA reductase, to 16 days for NAD glycohydrolase. Such estimates have been done using techniques that give mean estimates of mean half-lives for total membrane proteins for 60 h, and hence the differences truly indicate heterogeneity of turnover rates. It is also interesting to note from the studies of Bock *et al.*[25] with NAD glycohydrolase, that the half-life of this enzyme was the same in plasma membrane and endoplasmic reticulum fractions that were otherwise different by the criterion of enzyme markers.

The heterogeneous nature of the turnover of membrane proteins is further indicated in studies of Dehlinger and Schimke[18], using the double isotope method developed by Arias, Doyle, and Schimke[15]. This method permits a measure of the relative rate constants of degradation, and is illustrated schematically in Figure 9.1. Here is depicted the theoretical labelling and decay patterns of two hypothetical proteins which are present in the same steady-state concentrations, but one of which (A) is synthesised and degraded at twice the rate of the other (B). In this method, one isotopic form of an amino acid (^{14}C) is administered initially to the animal and allowed to decay a

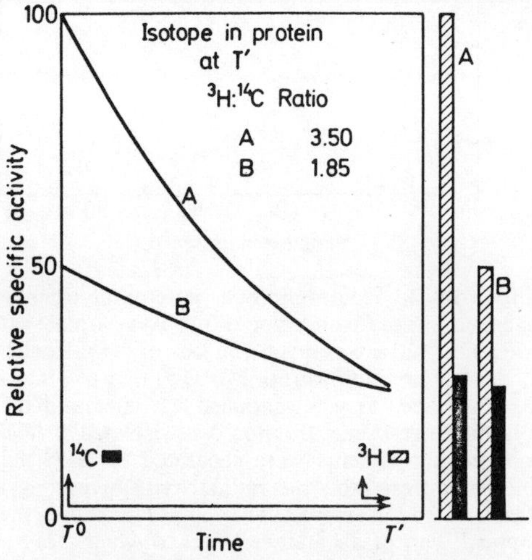

Figure 9.1 Theoretical labelling of two proteins with a two fold difference in degradation rates by a double-isotope technique. Proteins A and B are present at the same steady-state concentration, with A turning over at twice the rate of B[15]

specified time, thereby defining a point on the exponential decay curve. The second isotopic form (^{3}H) of the amino acid is administered to the same animal to establish an initial time point on the decay curve. Thus, a protein which is turning over more rapidly will lose a greater relative amount of its incorporated ^{14}C-amino acid, and will also incorporate more of the ^{3}H amino acid. Therefore, proteins which are turning over more rapidly will have greater ^{3}H:^{14}C ratios, irrespective of what the absolute specific radioactivity of the protein is[15, 26].

This experimental approach has been used to determine the relative rates of degradation of proteins in a rat liver plasma membrane fraction and in endoplasmic reticulum, as separated by SDS gel electrophoresis. A study of

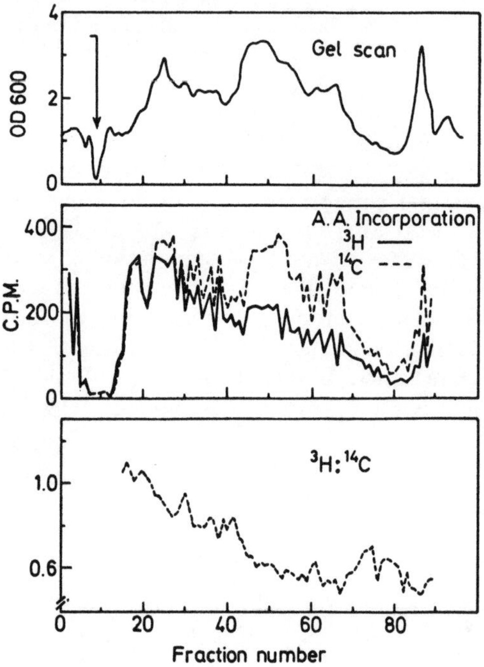

Figure 9.2 Electrophoretic patterns of double-labelled plasma membrane of rat liver. A male rat weighing 120 g was given 250 μCi of ^{3}H-L-leucine 4 days after administration of 100 μCi of ^{14}C-L-leucine. The rat was sacrificed 4 h later and the plasma membrane fraction was isolated[18]. The protein constituents were separated by SDS gel electrophoresis and the radioactivity of each gel fraction was measured. The *upper box* shows the optical scan of an analytical gel to which 225 μg of protein were applied, and subsequently stained with Acid-fast Green. The *middle box* indicates the actual radioactivity measured when 5 mg of protein were applied to a 19 × 75 mm gel. The *lower box* indicates the calculated ^{3}H:^{14}C ratios obtained

the relative degradation rates of plasma membrane proteins is shown in Figure 9.2. As indicated by the variation in the $^3H:^{14}C$ ratios, there is a marked heterogeneity in the rates of degradation of the proteins. Furthermore, there is a correlation between the size of the protein or subunit, and its relative rate of degradation. The larger proteins and subunits have higher $^3H:^{14}C$ ratios than the smaller ones, indicating a more rapid turnover rate for higher molecular weight proteins. In a control experiment, where both isotopic labels were administered at the same time, no systematic variation in $^3H:^{14}C$ ratios was observed.

A similar finding was also made for the endoplasmic reticulum. In these experiments, sufficient time was allowed for labelled plasma proteins to be secreted from the liver before membrane fractions were isolated. Figure 9.3

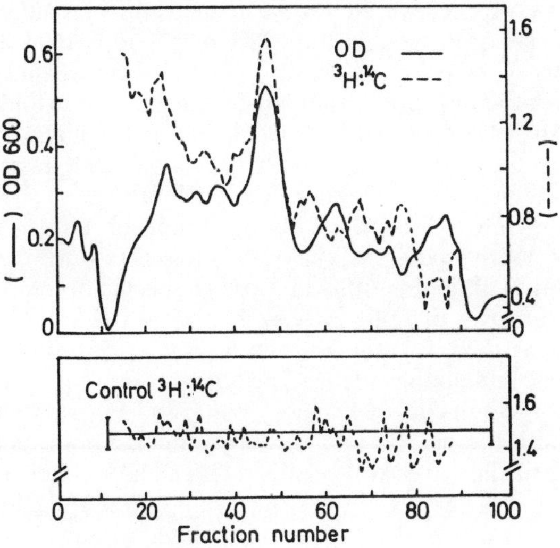

Figure 9.3 Relative turnover of Triton X-100-insoluble proteins of rat liver smooth endoplasmic reticulum. The labelling schedules were similar to those described in Figure 9.2. The smooth endoplasmic reticulum was isolated and the protein fraction insoluble in 1% Triton X-100 was examined[18], as described in Figure 9.2. The *upper box* shows the calculated $^3H:^{14}C$ ratios compared with the optical scan of an acid-fast Green-stained gel. The *lower box* shows the result of a control experiment in which a comparable animal was given 50 μCi of ^{14}C-L-leucine and 250 μCi of ^3H-L-leucine simultaneously and sacrificed 4 h later

compares the $^3H:^{14}C$ ratios observed for those membrane proteins which are insoluble in 1% Triton X-100 with a control in which both isotopic labels were administered simultaneously. Again, one observes a general trend in which larger proteins turn over more rapidly. An exception to the molecular size/turnover correlation is observed for the endoplasmic reticulum membrane proteins migrating in the 50 000 molecular weight range. These proteins, which appear to be major constituents of the endoplasmic reticulum are turning over at a relatively rapid rate. Essentially, the same results are also observed for those membrane proteins which are soluble in 1% Triton X-100[18].

Just as with cytoplasmic proteins[6], the rates of both synthesis and degradation of different proteins can be altered by physiological states of the organism. For instance, the synthesis of HMG-CoA reductase is cyclic throughout the day, with a peak of synthesis at approximately 9–12 a.m., and essentially no synthesis during the remainder of the 24 h period[29]. Since this enzyme turns over so rapidly (half-life of 2 h), there is a remarkable cyclic variation in the enzyme activity[29,30]. This is another example of the fact that enzymes with rapid rates of turnover will manifest rapid changes in levels, when either the rate of synthesis or the rate of degradation is altered[31]. Another example of selective alteration in the rates of synthesis of specific membrane-associated proteins is the effect of phenobarbital administration on the synthesis of cytochrome b_5 and NADPH cytochrome c reductase. Synthesis of the latter enzyme is stimulated some three–fourfold, whereas that of cytochrome b_5 is not altered[15,32]. Dehlinger and Schimke have also presented evidence for selective effects of phenobarbital administration on synthesis of proteins of the endoplasmic reticulum[33].

Effects on the rate of turnover of membrane proteins have also been observed. Thus, Kuriyama *et al.*[32] have shown that following administration of phenobarbital to animals results in a rather specific decrease in the normally occurring turnover of both cytochrome b_5 and NADPH cytochrome c reductase, whereas there is little diminution in the mean rate of turnover of total endoplasmic reticulum proteins. The turnover of total membrane protein has also been studied in cultured cells by Glick and Warren[34]. They provided evidence that both the protein and carbohydrate components of the surface membrane were turning over at the same rate, and more importantly, that the rate of turnover was far less in cells growing logarithmically than in cells grown in stationary phase. Both of the above cited instances of alterations in rates of turnover of proteins but these findings are subject to a certain degree of question because of the problem of isotopic re-utilisation. If there were a far greater degree of re-utilisation of isotope in the growing state, or following phenobarbital administration to intact rats, an apparent decrease in the rate of turnover of protein constituents would be found. Clearly, the effect of physiological variables on the rate of turnover of membrane components has not been the subject of many investigations, and more are needed.

In summary, then, the turnover of membrane-associated proteins has many properties in common with those of proteins generally in cells, including heterogeneity of rates of turnover, alterations in rates of turnover under different physiological states, and a correlation between the size of the protein and its relative rate of degradation.

9.4 TURNOVER OF LIPID CONSTITUENTS OF MEMBRANES

Just as protein components of membranes are turning over at relatively rapid and heterogeneous rates, lipid components turn over rapidly and at apparently heterogeneous rates[21, 22]. The overall rate of turnover of lipid components of the endoplasmic reticulum is approximately the same as that of the protein components. This may suggest that the protein and lipid components are degraded as units. However this conclusion, plus the observed differences in heterogeneity of the various lipid components, may be questioned as a result of the work of Wirtz and Zilversmit[27] who have demonstrated a rapid exchange of phospholipid components between endoplasmic reticulum and mitochondrial membranes. Furthermore, Zilversmit has shown that the exchange is facilitated by a specific cytoplasmic protein from rat liver[28]. Presumably such exchange also occurs in the intact tissue, although a direct demonstration of such exchange in intact cells has not been provided. Thus the turnover of phospholipids (and other lipids) may not reflect degradation of a protein-lipid complex, i.e. membrane, but rather the exchange phenomenon. Furthermore, the heterogeneity of rates of turnover of different lipid components may only indicate differences in rates of metabolic reactions of the various lipids once they have exchanged out of the membrane environment.

9.5 ON THE MECHANISMS OF GENESIS AND TURNOVER OF MEMBRANES IN ANIMAL TISSUES

The following discussion and speculation is based on the general properties of turnover of membrane proteins discussed previously.

(1) The rates of turnover of different proteins is heterogeneous, and such rates can be differently altered. This includes rates of both synthesis and degradation of membrane proteins.

(2) There is an apparent correlation between the rate of turnover and the molecular size for essentially all proteins, whether cytoplasmic or associated with RNA (ribosomal), DNA (chromosomal), or phospholipid (membrane).

(3) Pulse labelling-decay studies of membrane proteins follow exponential kinetics, indicating a random loss of proteins once synthesised and assembled into a membrane.

The fundamental question to be posed given the above characteristics of turnover as studied in animal cells is: *How do proteins get into membranes, and how are they removed*? There are currently no adequate answers to such questions, and this discussion will be highly speculative with a selective but by no means exhaustive review of potentially relevant papers.

Let us consider two extreme models for the synthesis and turnover of membranes. Figure 9.4 depicts a model in which membrane genesis occurs at discrete sites in a membrane, and in which degradation (removal) of membrane occurs at other discrete sites. Experimental verification of such a model might come from membrane proteins with the demonstration of isotope incorporation into proteins with subsequent retention of label over a time interval required for the membrane to flow from the site of synthesis to the point at which it is removed (for instance by entrapment in a lysosome, or transfer to another membrane fraction). Such a theoretical labelling pattern, together with actual labelling patterns, is shown in Figure 9.4. The consistent finding of exponential decay of isotope in total membrane proteins, as well as specific proteins, does not hold with such a simplistic model, since the exponential decay suggests a random process. The model, however, cannot be discounted on the basis of the demonstration of exponential decay kinetics, since the random process implied by such kinetics may not be the degradation itself but rather a random lateral diffusion of newly synthesised protein molecules to a discrete site of removal[35].

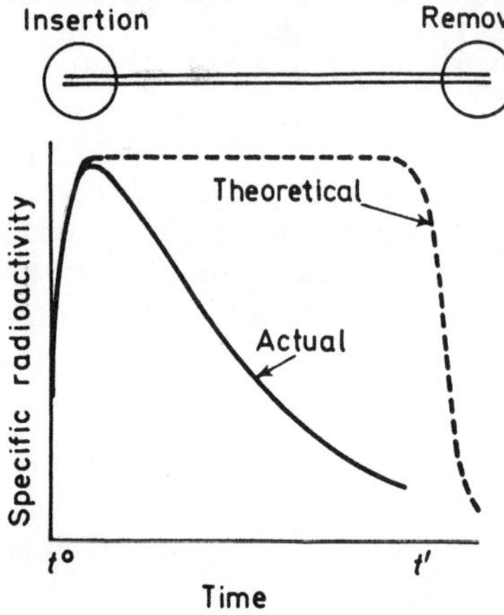

Figure 9.4 Theoretical labelling pattern of pulse-chase experiment. Membrane is assembled at discrete points at time t_0, and removed at discrete points at time t_1 (- - - -). Actual experimental data idealised for specific proteins[25, 32, 66]

The model described in Figure 9.4 is consistent with the proposal of Palade[36] that membrane components are synthesised on rough endoplasmic reticulum and subsequently appear in the smooth fraction. Some experimental corroboration of such a model comes from studies of Omura and Kuriyama[37] who have shown that following pulse labelling of rat liver proteins in intact rats with guanidino-^{14}C-arginine, a higher specific radioactivity of purified NADPH cytochrome c reductase and cytochrome b_5 occurs in rough endoplasmic reticulum within the first 15 minutes than in smooth endoplasmic reticulum. This differential is followed in time by an equilibrium of specific radioactivity of the purified proteins in the two membrane fractions. Nevertheless, these same workers observed that there was no such pattern of apparent transfer of higher specific radioactivity protein from rough to smooth membrane fractions when total membrane protein was analysed. Such a finding suggested to these workers that the rough to smooth progression does not hold for membrane proteins generally. Either the interpretation of Omura and Kuriyama is correct, or it may be that membrane proteins are synthesised on ribosomes associated with membrane, and that the lateral movement of membrane proteins generally (but not for the two enzymes studied) is far too rapid within the time period studied (15 min) to observe the differential labelling pattern.

The model of Palade has recently received support from studies by Hirano *et al.*[38] showing ferritin-conjugated conconavalin directed against specific saccharide moieties on glycoproteins. These workers have shown binding to the cisternal, but not the ribosomal side of endoplasmic reticulum fractions of myeloma cells and that the outer surface of the plasma membrane contains similar antigenic sites. They propose that the cisternal surface of the endoplasmic reticulum becomes the outer side of the plasma membrane by fusion of membrane packets with the plasma membrane. However, such a model is not entirely compatible with the fact that the surface (plasma) membrane of hepatocytes does not contain enzymes (including cytochrome P-450) similar to those of the endoplasmic reticulum[25], neither does it have comparable SDS-acrylamide gel protein profiles[18]. Thus, if such a model for the genesis of surface membranes is correct, then it is also necessary to invoke the specific loss of many of the proteins which are characteristically directed in their functional groups to the 'inside' of the cell, as opposed to the cisternal side. This will be discussed more fully with respect to the removal (turnover) of proteins from the membrane.

The model (Figure 9.4) of specific sites for genesis of membranes implies a mechanism for the recognition of sites of insertion of proteins. That there are specific insertion sites for certain membrane proteins is suggested from studies with influenza and parainfluenza viruses[39] in which specific viral coat proteins appear in the plasma membrane of the host cells at discrete sites. It is not known whether this involves synthesis of membrane-bound ribosomes, whether the viral coat proteins are inserted randomly into the membrane and subsequently undergo association following rapid lateral movement, or whether insertion of the first coat protein(s) forms an association point at which subsequent coat proteins are inserted.

The simplest means of visualising the existence of such discrete insertion sites would be for the membrane-associated proteins to be synthesised as a group, i.e. a polycistronic mRNA in which the mRNA is bound to a membrane (i.e. rough endoplasmic reticulum) with the growing nascent chains inserted directly into the phospholipid matrix. The first problem with this mechanism concerns the existence of polycistronic mRNAs in animals. At the present time, there is no evidence for such messenger RNAs[40-44]. One intriguing possibility would be an analogy to the mRNA for coat proteins of polio virus, which are synthesised as a long, single polypeptide and are divided into specific proteins by endopeptidase attack[45]. However, such a possibility is not likely since this hypothesis would require a constant stoichiometry of membrane protein synthesis, whereas studies have already been cited indicating preferential stimulation of NADPH cytochrome c reductase following phenobarbital administration to rats[15, 32].

Of fundamental importance is the question of whether membrane-associated proteins are synthesised on membrane-bound ribosomes. There is essentially no adequate answer to this question, and the few studies directed at this question have provided incomplete and equivocable results[46]. If *all* membrane proteins are so synthesised, then there are intriguing questions as to the mechanism for such specificity whereby certain mRNAs and ribosomes, i.e. polysomes, recognise the membrane. Certainly the ultimate specificity must reside with a specific mRNA. Is there a region on the mRNA that directly recognises a membrane protein, or is there a specific mRNA-associated protein[47] that recognises by some fashion a membrane? Alternative to such a proposal is one in which specific 60S subunits reside on the membrane[48, 49]. This proposal then requires some recognition process whereby the mRNA-40S ribosomal subunit complex binds specifically with certain membrane-bound 60S subunits. Such a mechanism therefore requires specific recognition sites on both the 60S and mRNA-40S complexes. Perhaps simpler would be a mechanism in which the specificity resides in the growing nascent chain which would associate with a phospholipid matrix, and hence fix the ribosome-mRNA complex to the membrane. Although the latter proposal is attractive, it would require, in its simplest form, that the *N*-terminal region of the membrane protein be the hydrophobic region of the molecule. However, recent studies of Ozols[50], indicate that the hydrophobic region of cytochrome b_5 is actually the *C*-terminal end.

It would not, however, appear that all membrane proteins are synthesised on membrane-bound ribosomes, based on knowledge of the regulation of the M protein (permease) of the lac operon[51]. This structural gene is directly continguous to the structural gene for the cytoplasmic protein, beta galactosidase[52]. Inasmuch as translation of the two genes occurs at the same time as the mRNA is being transcribed from DNA[53,54], it is difficult to conceive of a part of the mRNA being membrane bound, i.e. the part synthesising M protein, whereas directly continguous mRNA in this polycistronic mRNA is directing synthesis of a cytoplasmic protein.

Let us now turn to that aspect of this model (Figure 9.4) which suggests that membrane proteins are degraded at finite sites within the cell on an all or none basis. This model is not consistent with the remarkable heterogeneity of rates of turnover of known membrane proteins (see Table 9.2). How can such heterogeneity be explained? Firstly, and perhaps most obviously, would be the possibility that the membrane fractions isolated are extremely heterogeneous, and that domains of membrane are, in fact, degraded as units. Although such an explanation may hold for certain of the enzymes listed, it would not appear to be tenable for the differences in turnover rates of NADPH-cytochrome C reductase and cytochrome b_5. Both proteins appear to be involved in the same coupled electron transport reactions of the mixed function oxygenase system[55,56], and also can be shown to undergo association in resolved protein fractions[57]. Hence they are presumably associated in the same functional membrane unit, yet they differ significantly in half-lives.

In summary, then, one extreme of the model (Figure 9.3) in which membranes are assembled as units at finite growing points, and degraded as units at finite sites, has not been proved or disproved conclusively. The studies on this problem are extremely limited, and the results can be interpreted in various ways. However, certain evidence appears to be sufficiently strong to exclude such a mechanism as functioning for *all* membrane proteins.

Let us, then, look at another extreme model of membrane genesis and turnover, depicted in Figure 9.5. This model has proteins inserted randomly into membranes and randomly removed from membranes. The insertion process can be one that involves either membrane-bound or free ribosomes. Certain aspects of these different possibilities have been discussed previously. If some or many of the membrane-associated proteins are synthesised on cytoplasmic ribosomes, then the important question arises of how such proteins, synthesised in a presumably aqueous environment, become associated with a lipid environment. Among possibilities to be considered are modifications of the protein by covalent reactions, or by association with a limited number of phospholipid molecules such that conformational changes occur leading to exposure of hydrophobic regions of the protein and thereby allowing it to become imbedded in the membrane. In addition, cleavage of a portion of the protein, with subsequent conformational alteration should be considered. It is interesting to note the recent demonstrations that immunoglobin light chains are synthesised as a precursor molecule, with subsequent cleavage of some 12–25 amino acid residues at the *N*-terminal end[58,59]. This finding was possible only because of the ability to isolate specific mRNA, and synthesise the precursor in a cell-free protein synthesising system.

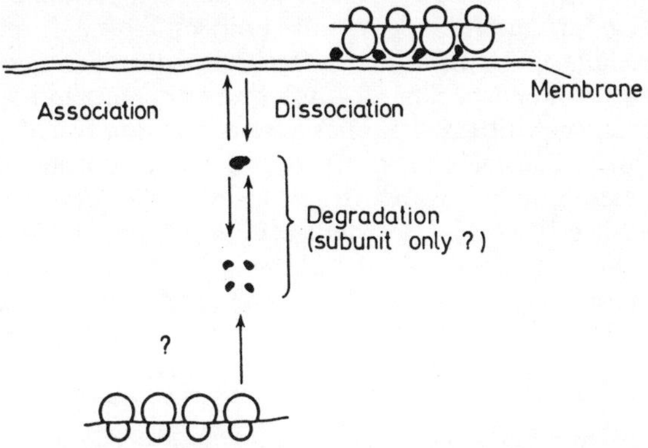

Figure 9.5 *Proposed schematic model of membrane genesis and degradation.* Some constituent membrane proteins are synthesised on membrane-bound polysomes and inserted directly into the membrane matrix, while others may be made on free polysomes and subsequently associate with the matrix by virtue of their physico-chemical properties. A finite free cytoplasmic pool exists for membrane proteins, with degradation occurring randomly in the dissociated state. The degradation rate of an individual membrane protein is determined by its properties as a substrate for the same protease system(s) acting on most or all intracellular proteins

The questions raised above are important in understanding the genesis of membranes, and are no longer beyond the capacity of current techniques. Answers to such questions require the ability to identify polysomes synthesising a specific protein. Recent developments utilising antibodies that can react with growing nascent chains, and which can be labelled with ^{125}I, provide the necessary tool[60,61,44] to localise the synthesis of membrane proteins to free or membrane-bound ribosomes. In addition, the binding of antibody to specific polysomes allows for the immunologic precipitation of the specific polysomes and isolation of specific mRNA[61,62]. Definitive answers will come only when specific and pure membrane proteins are studied, and in those instances when an antibody to the protein will react with nascent chains.

Random insertion of proteins into membranes is consistent with studies of Leskes *et al.*[63], who showed by histochemical techniques that glucose-6-phosphatase, a membrane-associated protein, appears randomly throughout the endoplasmic reticulum in the early postnatal period, when enzyme activity is accumulating rapidly. Again, as in all such studies that appear to localise specific membrane proteins to a given area, any conclusion can be questioned if rapid lateral diffusion of proteins in the plane of the membrane is a common or universal phenomenon[35].

The model presented in Figure 9.5 also proposes that proteins turn over, i.e. leave the membrane, at independent rates. The nature of the 'leaving' or degradation process is totally unknown. One of the major problems in all studies involved in the turnover of proteins is that once the protein leaves the identifiable fraction, or loses enzymatic activity or immunologic reactivity, the 'degradation' product(s) are generally unidentifiable. Although it is clear that for intracellular proteins in tissues generally, turnover involves degradation to the level of amino acids[1], for any given protein, and specifically for membrane protein constituents, this has not been demonstrated. Three general possibilities can be considered for the process whereby proteins are removed from specific membranes. The first is that membrane proteins are released to the outside of the cell. This is consistent with observations on the shedding of glycoproteins into the medium of cultured cells[34,64,65]. Thus, one could consider that there is a flow of membrane from one fraction to another, to the surface membrane[38,66], and finally shedding of proteins at the surface. One could propose a mechanism for the shifting populations of proteins as they progress to the surface of a cell if the membrane proteins were exposed to both the inner and outer surface of the membrane, and as they become exposed to the cell surface, a process such as protease attack or glycolysation (or saccharide hydrolysis) might alter the ability of a protein to remain inserted into the membrane and hence be shed. Clearly, more work is necessary regarding this possibility.

The second general mechanism for the degradation (release) of membrane proteins would be that they are subject to proteolytic attack while resident within the membrane. A resident protease activity has been demonstrated in red blood cells[67], but has not been observed with endoplasmic reticulum fractions[68]. An alternative to a resident protease would be attack by a cytoplasmic protease activity. Such a proposal would require that the proteins that are turning over more rapidly be those that are in a more peripheral position on (in) the membrane. A number of lines of evidence do not support this possibility. Thus, HMG-CoA reductase, whose half-life is 2h (Table 9.2) is clearly an 'integral' membrane protein, and is extremely refractory to solubilisation by detergent, as well as ionic extractants[69]. Hence, it does not appear to be a peripheral membrane protein. In addition, trypsin digestion of identifiable proteins of endoplasmic reticulum preferentially releases cytochrome b_5 (half-life of 80 h) compared with NADPH cytochrome c reductase (half-life of 50 h)[70]. If proteins that turn over rapidly were more accessible to exogenous proteases, just the opposite would be expected. Lastly, in experiments where membrane proteins were double-labelled to reveal heterogeneity of turnover rates (see Figure 9.1) and then subjected to digestion by trypsin or pronase, there was no preferential release of soluble radioactivity with a high ^3H:^{14}C ratio[71].

The last general mechanism for the turnover of membrane proteins would have proteins dissociate from the membrane, be released into the cytoplasm, and then be degraded (or re-enter the membrane). The ability to reconstitute purified membrane proteins with phospholipid mono- or bi-layers with functional properties[72-74], indeed suggests that association can occur. Theoretically, the dissociation can also occur. Whether, in fact, dissociation and association occur in the living cell remains to be demonstrated. One could, therefore, visualise the 'leaving' process as a part of a thermodynamic equilibrium, presumably far to the side of the membrane-associated state. Such an equilibrium could be perturbed by the membrane environment, including the nature of the lipids or the associated proteins. In this respect, Negishi and Omura[75] have recently reported that there is normally in rat liver an initial rapid turnover of NADPH cytochrome c reductase with a half-life of 1–2 h, followed by the slow turnover characteristic of the 50 h value (Table 9.2). This rapid phase of the turnover is not observed following phenobarbital administration to rats. They make the interesting suggestion that the membrane has been altered such that the newly inserted enzyme does not undergo the rapid removal process.

One can also envisage, rather than an equilibrium phenomenon, a process whereby release of a protein from a membrane-bound state requires specific intervention of a chemical process, such as covalent modification of the protein by phosphorylation, acetylation, etc. by specific enzymes, or by peroxidative reactions involving the protein or its immediately adjacent lipid environment[76, 77].

9.6 CONCLUDING REMARKS

What should be clear from this discussion is that very little is known about the manner in which membranes are synthesised, assembled, and removed. It is highly likely, just as with many biological processes, that no single mechanism will be found for all specific proteins and all membranes. However, with the development of more sophisticated means of separating and isolating membrane proteins, and new developments which allow for identification of specific polysomes, specific mRNAs, many of the questions posed here give promise of being answered.

References

1. Schoenheimer, R. (1942). in *The Dynamic State of Body Constituents*, (Cambridge, Mass: Harvard University Press)
2. Swick, R. W. (1957). *J. Biol. Chem.*, **231,** 751
3. Buchanan, D, L. (1961). *Arch. Biochem. Biophys.*, **94,** 500
4. Schimke, R. T. (1964). *J. Biol. Chem.*, **239,** 3808
5. Schimke, R. T. (1974). *Methods in Enzymology*, (in the press)
6. Schimke, R. T. and Doyle, D. (1970). *Ann. Rev. Biochem.*, **39,** 929
7. MacDonald, R. A. (1961). *Arch. Internal Med.*, **107,** 335
8. Swick, R. W., Koch, A. L. and Handa, D. T. (1956). *Arch. Biochem.*, **63,** 226
9. Campbell, P. N. and Stone, N. E. (1957). *Biochem. J.*, **66,** 669
10. Schimke, R. T., Sweeney, E. W. and Berlin, C. M. (1965). *J. Biol. Chem.*, **240,** 135
11. Schimke, R. T. (1964). *J. Biol. Chem.*, **239,** 136
12. Majerus, P. W. and Kilburn, E. (1969). *J. Biol Chem.*, **244,** 6254
13. Schimke, R. T. (1970). in *Mammalian Protein Metabolism*, 177 (H. N. Munro, editor) (New York: Academic Press)
14. Dehlinger, P. J. and Schimke, R. T. (1970). *Biochem. Biophys. Res. Commun.*, **40,** 1473

15. Arias, I. M., Doyle, D. and Schimke, R. T. (1969). *J. Biol. Chem.*, **244,** 3303
16. Dice, J. F. and Schimke, R. T. (1972). *J. Biol. Chem.*, **247,** 98
17. Dice, J. F. and Schimke, R. T. (1973). *Arch. Biochem. Biophys.*, **158,** 97
18. Dehlinger, P. J. and Schimke, R. T. (1971). *J. Biol. Chem.*, **246,** 2574
19. Dice, J. F., Dehlinger, P. J. and Schimke, R. T. (1973). *J. Biol. Chem.*, **248,** 4220
20. Tweto, J., Dehlinger, P. J. and Larrabee, A. R. (1972). *Biochem. Biophys. Res. Commun.* **48,** 1371
21. Holtzman, J. L. and Gillette, J. R. (1968). *J. Biol. Chem.*, **243,** 3020
22. Omura, T., Siekevitz, P. and Palade, G. E. (1967). *J. Biol. Chem.*, **242,** 2289
23. Schimke, R. T., Ganschow, R., Doyle, D. and Arias, I. M. (1968). *Fed. Proc.*, **27,** 1223
24. Swick, R. W. and Handa, D. T. (1956). *J. Biol. Chem.*, **218,** 557
25. Bock, K. W., Siekevitz, P. and Palade, G. E. (1971). *J. Biol. Chem.*, **246,** 188
26. Glass, R. D. and Doyle, D. (1972). *J. Biol. Chem.*, **247,** 5234
27. Wirtz, K. W. A. and Zilversmit, D. B. (1968). *J. Biol. Chem.*, **243,** 3596
28. Zilversmit, D. B. (1971). *J. Biol. Chem.*, **246,** 2645
29. Higgins, M., Kawachi, T. and Rudney, H. (1971). *Biochem. Biophys Res. Commun.*, **45,** 138
30. Shapiro, D. J. and Rodwell, V. W. (1971). *J. Biol. Chem.*, **246,** 3210
31. Berlin, C. M. and Schimke, R. T. (1965). *Mol. Pharmacol.*, **1,** 149
32. Kuriyama, Y., Omura, T., Siekevitz, P. and Palade, G. E. (1969). *J. Biol. Chem.*, **244,** 2017
33. Dehlinger, P. J. and Schimke, R. T. (1972). *J. Biol. Chem.*, **247,** 1257
34. Glick, M. C. and Warren, L. (1968). *J. Cell Biol.*, **37,** 729
35. Singer, S. J. and Nicolson, G. (1972). *Science*, **175,** 1222
36. Palade, G. (1959). in *Subcellular Particles*, 64 (T. Hayashi, editor) (New York: Ronald Press)
37. Omura, T. and Kuriyama, Y. (1971). *J. Biochem.* (*Tokyo*), **69,** 651
38. Hirano, H., Parkhouse, B., Nicolson, G., Lennox, E. S. and Singer, S. J. (1972). *Proc. Nat. Acad. Sci.*, **69,** 2945
39. Compans, R. W. and Choppin, P. W. (1971), in *Comparative Virology*, 407 (K. Maramorosch and F. Kurstak, editors) (New York: Academic Press)
40. Heywood, S. M., Dowben, R. and Rich, A. (1967). *Proc. Nat. Acad. Sci.*, **57,** 1002
41. Lockard, R. E. and Lingrel, J. B. (1969). *Biochem. Biophys. Res. Commun.*, **37,** 204
42. Rhoads, R., McKnight, S. and Schimke, R. T. (1971). *J. Biol. Chem.*, **246,** 7407
43. Stavenezer, J. and Huang, R. C. C. (1971). *Nature*, **230,** 172
44. Taylor, J. M. and Schimke, R. T. *J. Biol. Chem.* (in the press)
45. Jacobson, M. F., Asso, J. and Baltimore, D. (1970). *J. Mol. Biol.*, **49,** 657
46. Ragnotti, G., Lawford, G. R. and Campbell, P. N. (1969). *Biochem. J.*, **112,** 139
47. Kwan, S. W. and Brawerman, G. (1972). *Proc. Nat. Acad. Sci.*, **69,** 3247
48. Ragland, W. L., Shires, J. K. and Pitot, H. C. (1971). *Biochem. J.*, **121,** 271
49. Sunshine, G. H., Williams, D. J. and Rabin, B. R. (1971). *Nature New Biology*, **230,** 133
50. Ozols, J. (1972). *J. Biol. Chem.*, **247,** 2242
51. Fox, C. F. and Kennedy, E. P. (1965). *Proc. Nat. Acad. Sci.*, **54,** 891
52. Beckwith, J. R. (1967). *Science*, **156,** 597
53. Miller, O. L., Jr., Hamkalo, B. A. and Thomas, C. A., Jr. (1970). *Science*, **169,** 392
54. Zubay, G. and Chambers, D. H. (1969). *Cold Spring Harbor Symp. Quant. Biol.*, **34,** 753
55. Estabrook, R. W., Franklin, M., Baron, J., Shigematsu, A. and Hildebrandt, A. (1971). In *Drugs and Cell Regulation*, 228 (E. Mihich, editor) (New York: Academic Press)
56. Hara, T. and Minakami, S. (1971). *J. Biochem* (*Tokyo*), **69,** 317
57. Okuda, T., Mihara, K. and Sato, R. (1972). *J. Biochem.* (*Tokyo*), **72,** 987
58. Mach, B., Faust, C. and Vassalli, P. (1973). *Proc. Nat. Acad. Sci.*, **70,** 451
59. Milstein, C., Brownlee, G. G., Harrison, T. R., and Matthews, M. B. (1972). *Nature New Biology*, **239,** 117
60. Palacios, R., Palmiter, R. D. and Schimke, R. T. (1972). *J. Biol. Chem.*, **247,** 2316
61. Palacios, R., Sullivan, D., Summers, N. M., Kiely, M. L. and Schimke, R. T. (1973). *J. Biol. Chem.*, **248,** 540
62. Sullivan, D., Palacios, R., Taylor, J. M., Faras, A. J., Kiely, M. L., Summers, N. M., Bishop, J. M. and Schimke, R. T. *J. Biol. Chem.* (in the press)

63. Leskes, A., Siekevitz, P. and Palade, G. E. (1971). *J. Cell Biol.*, **49,** 264
64. Eidam, C. R. and Merchant, D. J. (1965). *Exptl. Cell Res.*, **37,** 147
65. Kornfield, S. and Ginsburg, V. (1966). *Exptl. Cell Res.*, **41,** 592
66. Kuriyama, Y. (1972). *J. Biol. Chem.*, **247,** 2979
67. Morrison, W. L. and Neurath, H. A. (1953). *J. Biol. Chem.*, **200,** 39
68. Dehlinger, P. J. and Schimke, R. T. unpublished observations
69. Kawachi, T. and Rudney, H. (1970). *Biochemistry*, **9,** 1700
70. Omura, T. and Sato, R. (1964). *J. Biol. Chem.*, **239,** 2370
71. Taylor, J. M., Dehlinger, P. J., Dice, J. F. and Schimke, R. T. (1973). *Drug Metab. Disp.*, **1,** 84
72. Hong, K. and Hubbell, W. L. (1972). *Proc. Nat. Acad. Sci.*, **69,** 2617
73. Rothfield, L., Romeo, D. and Hinckley, A. (1972). *Fed. Proc.*, **31,** 12
74. Strittmatter, P., Rogers, M. J. and Spatz, L. (1972). *J. Biol. Chem.*, **247,** 7811
75. Negishi, M. and Omura, T. (1972). *J. Biochem.* (*Tokyo*), **72,** 1407
76. Hatefi, Y. and Hanstein, W. G. (1970). *Arch. Biochem. Biophys.*, **138,** 73
77. Tappel, A. L. and Zalken, H. (1960). *Nature*, **185,** 35
78. Fritz, P. J., Vesell, E. S., White, E. L. and Pruitt, K. M. (1969). *Proc. Nat. Acad. Sci.* (*Wash*), **62,** 558
79. Swick, R. W., Rexroth, A. K. and Stange, J. L. (1968). *J. Biol. Chem.*, **243,** 3581
80. Kenney, F. T. (1967). *Science*, **156,** 525
81. Marver, H. S., Collins, A., Tschudy, D. P. and Rechcigl, M., Jr. (1966). *J. Biol. Chem.*, **241,** 4323
82. Niemeyer, H. (1966). *Nat. Cancer Inst. Monogr.*, **27,** 29
83. Price, V. E., Sterling, W. R., Tarantola, V. A., Hartley, R. W. Jr. and Rechcigl, M., Jr. (1962). *J. Biol. Chem.*, **237,** 3468
84. Russell, D. and Snyder, S. H. (1968). *Proc. Nat. Acad. Sci.* (*Wash.*), **60,** 1420
85. Schimke, R. T., Sweeney, E. W. and Berlin, C. M. (1964). *Biochem. Biophys. Res. Commun.*, **15,** 214
86. Segal, H. L. and Kim, Y. S. (1963). *Proc. Nat. Acad. Sci.* (*Wash.*), **50,** 912

10
Membrane Transport

WOLFGANG EPSTEIN
University of Chicago

10.1 INTRODUCTION

A survey of the field of membrane transport presents a pleasant challenge because of the large amount of work in the field and the rapid pace of discoveries in recent years. The field has become so large that comprehensive coverage is not possible in a mere chapter, perhaps not even in a book. I will examine in detail a few examples of transport in which considerable progress has been made in recent years. The choice reflects the author's interest in transport systems with a high degree of specificity for the transported substrate. Many such systems are capable of concentrative uptake. The term substrate will be used to refer to the molecule transported. The analogy with enzymes is intentional since transport systems share many common features with enzymes even though most transport systems are not enzymes since they do not produce a chemical change in the substrate.

10.2 CARRIERS IN MEMBRANE TRANSPORT

Some fundamental concepts about membranes and transport have been with us for many decades and are still applicable even though the ideas have evolved to accommodate new knowledge. It is generally accepted that the primary permeability barrier of membranes is a lipid bilayer arranged with the polar portions of lipids facing the external solution while the hydrocarbon regions extend into the bilayer forming a hydrophobic phase. The basic model for membranes first proposed explicitly by Gorter and Grendel[1] almost half a century ago remains the basis of most membrane models[2,3]. This view provides a rather good explanation for the fact that the permeation of membranes by a variety of small molecules can be estimated from their molecular weight and lipid solubility, two major factors affecting diffusion through a lipid bilayer[4].

The other idea, that of carriers, was proposed to reconcile the view of membranes as a lipid bilayer with the fact that many larger molecules which are virtually insoluble in lipids can cross membranes readily by a path which is saturable and highly specific for the substrate[5]. Carriers were assumed to be lipid soluble molecules that bind substrate on one side of the membrane, diffuse through the lipid bilayer and then dissociate on the other side of the membrane releasing free substrate. The binding site accounts for specificity, and saturation is due to diffusion by a limited number of carrier-substrate complexes through the lipid bilayer being the rate limiting step. If a carrier has no particular orientation in the membrane nor interacts specifically with particular membrane lipids it is described as a truly mobile carrier. This type of mobile carrier probably characterises few biological transport systems, but the concept of a carrier is useful to describe membrane components which bind substrate specifically and have hydrophobic interactions with the membrane. Carriers are no longer an attractive hypothesis—they have become facts. Fox and Kennedy[6] in 1965 were the first to identify a carrier, the M protein which is the substrate binding site of the lactose transport system of *Escherichia coli*. Since then quite a number of molecules for which a carrier role is postulated have been identified; some of these will be discussed here.

The effect of temperature on membranes and transport processes has been especially revealing because biological membranes undergo thermal transitions at temperatures determined largely by the fatty acid composition of the phospholipids in the membrane[7-15]. The transition temperatures are normally below the physiological temperature range for a cell but by incorporating fatty acids with high melting temperatures, such as elaidic acid, transitions at temperatures approaching 37°C and higher can be achieved[9,10,13-16]. In general, the two transition temperatures can be defined: above the upper one all lipids are in a fluid state, below the lower one the lipids are in an ordered quasicrystalline state, while between the temperatures both liquid and quasicrystalline phases are present[16-18]. The effect of temperature on transport rates has been used to reveal the role of the state of the lipids in bacterial transport systems[11-16]. This technique is of less value for animal cells which have cholesterol in their membranes, because sterols tend to suppress thermal phase transitions[19, 20].

10.2.1 Ionophores

Ionophores are small polymers of molecular weight ranging from 500 to 3000 which bind ions and mediate their movement through both natural and artificial membranes. Chemically they are rather diverse, including some with only peptide linkages, others with both peptide and ester linkages, and poly ethers[21, 22]. Like the peptide antibiotics to which class some ionophores belong, the naturally occurring ionophores usually have both the unnatural D- as well as the natural L-isomer amino acids. Ionophores are lipid soluble and very poorly soluble in water. When added to the aqueous phase of a membrane system these compounds are rapidly taken up by the membranes with little remaining in aqueous solution. The ion selectivity of ionophores is discussed in Chapter 2 of this volume.

10.2.1.1 Valinomycin

Most ionophores appear to bind ions not by ionic bonds, but primarily by ion-dipole interactions between the ion and polar groups such as carbonyl oxygen and amide nitrogen atoms. The structural basis of ion binding is exemplified by binding of K^+ by valinomycin. Valinomycin is a cyclic depsipeptide containing the following sequence repeated three times: D-valine—L-lactate—L-valine—D-α-hydroxyisovalerate. The molecule has a doughnut shape with hydrophobic residues extending outward while polar residues including six important carbonyl oxygen residues face the central cavity[23, 24]. The carbonyl oxygen atoms coordinate with the ion in the central cavity, the net effect being that valinomycin replaces the hydration shell of the ion. The marked preference of valinomycin to bind K^+ and Rb^+ is apparently due to the fact that the K—O distance in the valinomycin—K^+ complex is very close to the K—O distance in the hydrated ion, whereas for other ions (except Rb^+) the fit is not as good[23].

Two different mechanisms underlie ionophore action. Valinomycin is representative of the group of ionophores that appear to act as truly mobile carriers. Three arguments for such a mechanism are that valinomycin mediates uptake of K^+ into bulk lipid phases as long as a lipid soluble anion is present to prevent charge separation[25], valinomycin is active in thick membranes[26], and changes in the state of the membrane lipids drastically alter valinomycin-mediated ion movements[27]. Valinomycin mediated K^+ conductance in artificial membranes drops at least four orders of magnitude when the membrane is cooled to below the temperature at which the lipids undergo a transition to a quasi-crystalline state[27]. Diffusion of a large molecule such as valinomycin would be expected to be severely restricted by the ordered hydrocarbon array present in membranes below the transition temperature.

The ion specificities of ionophores determine not only which ions can move, but also whether the movement will be electrogenic resulting in an uncompensated movement of charge or whether it will be electrically neutral. Since only K^+, of the cations normally present in and around cells, is efficiently transported by valinomycin, K^+ movement will be electrogenic and no net movement of K^+ will occur unless there are other paths to allow for compensating movement of cations or anions. Nigericin, another ionophore which appears to act as a mobile carrier, can also transport protons. This property allows nigericin to perform an electrically neutral exchange of K^+ for H^+ [25].

10.2.1.2 Gramicidin

Gramicidin is a linear polypeptide containing 15 amino acids of both D and L configuration. The *N*-terminus is formylated, while ethanolamine is at the *C*-terminus. The commercial preparation is a mixture of several types differing only in the aromatic amino acid at position 11[28]. Gramicidin-mediated ion movements exhibit three features none of which has been observed with mobile carrier compounds such as valinomycin: (1) the peptide is active only in bilayers and not in thick membranes[29]; (2) the conductance shows gating, meaning a voltage dependence of the conductance[30]; and (3) the conductance does not vary continuously but changes by discrete small steps each of which is close to an integral multiple of a common value[30, 31]. This behaviour suggests that gramicidin forms units of conductance which exist either in a conducting or non-conducting configuration. These properties are readily explained as those of ion-conducting channels which span the membrane. The stepwise changes in conductance are due to the addition of one or more channels to the group that is in the conducting configuration, while gating is due to sensitivity of the channel structure to the electrical gradient in the membrane. Channel formation is supported by the small effect on gramicidin-mediated conductances in an artificial membrane when the membrane is cooled to below its transition temperature[27]. The ability of ions to move through a channel formed by a protein should not be very sensitive to the physical state of lipids around the protein forming the channel.

The structure of channel-forming peptides in the membrane is thought to be helical with hydrophobic residues facing outward and hydrophilic ones inward. Urry and co-workers have suggested a number of helical structures for such peptides[32, 33]. His group has synthesised peptides containing repeated sequences of just three amino acids which were predicted to behave as channel-forming ionophores which show gating. Some of these peptides behave as expected[34], indicating that the basic ideas of how these peptides function are probably correct. In what is the most probable configuration for gramicidin the length of the molecule down the axis is less than 20 Å[31, 32], which is far too small to span a membrane. However, two molecules end to end should suffice, and this dimer configuration is consistent with the observation that conductance varies approximately as the square of the gramicidin concentration[35].

10.2.1.3 Ionophores in biological transport systems

There is good reason to believe that channel-type structures similar to those of some ionophores participate in at least some types of biological ion transport. The properties of the energetically passive movements of Na^+ and K^+ accompanying nerve and muscle excitation have long been considered to occur via channels which are gated and show considerable ion discrimination[36]. Preparations with ionophore activity have been obtained from bacterial and animal cells. An as-yet poorly characterised bacterial material called EIM produces large conductance channels in artificial membranes[37-39]. It seems that the form of this material active in artificial membranes has little relation to ion transport in the bacteria from which it is obtained since EIM shows virtually no discrimination between different monovalent cations[40]. Extraction of brain or electroplax membranes with methanol or dimethylsulphoxide has resulted in preparations which appear to exhibit channel-type conductance specific for either K^+ or for Na^+ [41]. Tryptic digestion of a partially purified preparation of the Na^+—K^+ ATPase leads to material with Na^+ ionophore activity which may be of the channel type[42]. The anion transport system of human erythrocytes seems to be formed by a large protein which spans the membrane. This conclusion is suggested by studies with certain stilbene derivatives which react covalently with proteins in intact erythrocytes and irreversibly inhibit anion transport[43]. When a radioactive stilbene derivative is used, label appears almost exclusively in a protein band of 95 000 molecular weight[43]. It is known that the major protein in this band spans the erythrocyte membrane[44].

10.2.2 Bacterial binding proteins

A class of proteins that has some of the properties expected of carriers is the binding proteins. The most extensively studied are those of gram-negative bacteria. Pardee and co-workers were the first to describe such a protein, the sulphate-binding protein of *Salmonella typhimurium*[45]; shortly thereafter Piperno and Oxender reported one in *E. coli* for branched-chain amino acids[46]. At present a large number of such proteins has been identified and characterised. There are even binding proteins for some vitamins[47, 48]. The topic has been discussed in a number of recent reviews[49-51].

The binding proteins are typical soluble proteins. Their molecular weights range from 20 000 to 42 000. They are monomeric proteins with a single high-affinity binding site per monomer*. They are found in the periplasmic space, the region between the inner and the outer membranes of the cell.

* The galactose binding protein exhibits unusual behaviour not fitting this generalisation. Each mole of the protein binds considerably less than one mole of sugar with high affinity, but can bind a total of 2 moles of sugar at high sugar concentration[52]. The amount of binding with high affinity varies with the experimental conditions[52, 53].

This type of double membrane system is characteristic of Gram negative bacteria. The inner membrane is a typical plasma membrane, containing many enzymatic and transport activities and representing the principal permeability barrier to small molecules. The outer membrane has virtually no biochemical activities. It restricts movement of proteins and intermediate-sized molecules down to those with molecular weights in the range of 1000. A periplasmic location for the binding proteins is based on three types of evidence: (1) weakening of the outer envelope layers by lysozyme and EDTA or similar procedures results in loss of the binding proteins and other periplasmic proteins, (2) a similar loss is produced by cold osmotic shock, a technique developed by Neu and Heppel[54], and (3) binding proteins *in vivo* are not accessible to antibodies which cannot penetrate the outer bacterial membrane but can be inactivated by chemicals which penetrate the outer membrane but not the inner membrane[55,56]. There is no reason to doubt the periplasmic location of the binding proteins although for most this location is based solely on their removal by osmotic shock. This test can be misleading since some intracellular proteins are removed by osmotic shock[57].

It is now well established that binding proteins participate in the transport of substances they bind. Kinetic analysis relating uptake rate to substrate concentration for many transport systems shows that two or more components of transport can be distinguished with differing affinities for the substrate[58-64]. When the binding protein is absent, one of these components of transport, usually the one with the highest affinity for the substrate, is lost. Loss of the binding protein markedly reduces transport at low concentrations of substrate but generally produces less impairment at high substrate concentrations. Concomitant loss of a component of transport and of a binding protein has been demonstrated by a number of techniques. The most common is to remove the binding protein by osmotic shock. Osmotic shock must be performed with caution, since under some conditions impairment of non-binding protein transport systems and even cell death can occur[46]. Confirmation of the correlation inferred from osmotic shock studies comes from work with changes in the level of binding protein produced by induction or repression[61-66], and with mutants which either make no functional binding protein[59,61-64,67,68], a functionally altered binding protein[61,63,69,70], or supra-normal levels of the binding protein[59,63,71,72]. In each case there is a good quantitative correlation between the amount of the binding protein and the rate of uptake via the binding protein associated component of transport. Yet another line of evidence is the fact that the binding constant of a binding protein is usually the same within experimental error as the apparent affinity (K_m) of the binding protein component of uptake.

It should be possible to restore transport in cells lacking a binding protein if the protein can gain access to the periplasmic space where it presumably acts. Since the external membrane normally retains periplasmic proteins, proteins added to the medium should not be able to gain access to the periplasmic space unless the outer membrane is damaged in some way. Osmotic shock treatment with distilled water may be a suitable way of damaging the external membrane, since such a treatment allowed Medveczky and Rosenberg[68] to demonstrate restoration of PO_4 uptake in *E. coli*. These workers showed that the initial rate of $^{32}PO_4$ uptake was stimulated, that uptake into the cell was occurring since there was an increase in the rate of appearance of esterified $^{32}PO_4$, and that antibody to purified binding protein blocked restoration. Restoration occurred in a mutant lacking the binding protein, but not in a mutant defective in transport but which had normal levels of binding protein. Most other attempts to restore transport with binding protein have not been successful. The reason for these negative results is not known. It may be that success requires the right pre-treatment of the cells and may be successful only with certain bacterial strains. Partial restoration of leucine and galactose transport was reported some years ago, but the primary effect was an increase in plateau level of accumulation with little effect on the initial rate of uptake[73]. Other workers do not appear to have confirmed these results[49].

What is the molecular role of binding proteins in transport? They cannot be a general feature of bacterial transport systems since they have not been found in Gram-positive bacteria and quite a number of transport systems in Gram-negative bacteria do not have an associated binding protein. They have been considered candidates for a carrier role because they bind substrates tightly and very specifically and have no known enzymatic activity. However, their other properties argue against a role as a transmembrane carrier of the substrates they bind. They are soluble proteins and do not seem to have any special proclivity to associate with membranes. Their location in the periplasmic space is not evidence for a carrier role. They are found in this space because the cell secretes them to the outside of the inner membrane, and they cannot normally leave this space. Many enzymes are found in the periplasmic space yet for none is there evidence to suggest a specific association with the membranes.

A number of lines of evidence indicate that one or more specific gene products in addition to the binding protein are required to permit the normal functioning of a transport system. Loss of the binding protein usually abolishes only one component of transport, whereas loss of other gene products will abolish several components of transport including that associated with the binding protein[59, 63, 67, 68, 74]. For example, histidine transport in *S. typhimurium* has a number of components, one with very high affinity for substrate (K_m of 3×10^{-8} M) being associated with the *hisJ* binding protein[59]. Mutants in another gene, *hisP*, abolish both the component associated with the *hisJ* protein as well as a component with a K_m of 2×10^{-7} M which is probably not due to a binding protein since such a component is found in membrane vesicles of the closely related species *E. coli* and such vesicles lack binding proteins[75]. The simplest explanation of these findings is that the *hisJ* protein works in conjunction with the *hisP* transport system; in the absence of the *hisP* system the *hisJ* protein cannot function in transport[59]. Comparison of uptake by whole cells with uptake by membrane vesicles which are virtually devoid of binding and other soluble proteins shows that in most cases uptake is intact, only that the component associated with a binding protein is absent[75]. This shows that most transport systems function well, albeit with lessened apparent affinity for substrate, in the absence of binding proteins.

There are some data suggesting that binding-protein-associated transport differs from that by transport systems which have no binding protein.

Glutamine uptake in *E. coli* is markedly dependent on a binding protein; in the absence of the binding protein uptake is virtually abolished[71]. Membrane vesicles of *E. coli* do not take up significant amounts of glutamine[75]. By the use of extensively starved cells Berger[76] showed that glutamine uptake was much more dependent on cell pools of ATP than on the availability of oxidative energy while the opposite is true of most bacterial transport systems (see below). A dependence on ATP is not the rule for transport associated with binding proteins. Uptake via the galactose binding protein system is fuelled by oxidative energy and only indirectly by ATP[77]. The result with glutamine may be due in part to stimulation by ATP of subsequent metabolism of glutamine rather than uptake, since glutamine is rapidly metabolised by a number of aminotransferases all of which are ATP dependent.

The role for binding proteins that this reviewer favours is that of an unusual scavenger for substrate, binding it with high affinity and then transferring it in a special way to a membrane-bound transport system so that the K_m for the overall transport process is essentially the same as that for binding of substrate by the protein. This type of role has been suggested by a number of workers and is consistent with most of the published data. This model implies that the binding proteins interact in a specific way with a transport system. Attempts to demonstrate such an interaction in membrane vesicles, either by showing binding of protein to the vesicles or an alteration of transport kinetics by addition of the protein have not been successful[77].

10.2.3 Other binding proteins

Proteins similar in properties to the binding proteins of Gram-negative bacteria have been found in animals. A Ca^{2+} binding protein occurs in intestine, kidney and uterus of a number of animal species[78]. There is a reasonably good correlation between levels of the binding protein in the intestine and the rate of Ca^{2+} uptake under a number of conditions (see References 78 and 79 for review), suggesting that the binding protein is part of the intestinal Ca^{2+} transport system. However, the correlation is far from perfect, since in kinetic studies *in vivo* uptake is stimulated by vitamin D before there is a detectable rise in binding protein[80], while in organ culture a rise in binding protein can be detected before transport activity is stimulated[81]. The lack of correlation in these studies may be due to the presence of several Ca^{2+} uptake systems only one of which is dependent on the binding protein. A binding protein for vitamin B_{12} has been known for a long time as the intrinsic factor of gastric juice absent in individuals with pernicious anaemia. This protein has been purified and shown to be a glycoprotein which binds vitamin B_{12} with high affinity[82].

A function for these animal binding proteins in membrane transport is likely in view of the general correlation between presence of the protein and enhanced transport in the intestinal tract. However, the lack of perfect correlation in the kinetic studies with Ca^{2+} indicate that, as in bacteria, the situation is probably complex and that a satisfactory understanding of the role of the binding proteins will require much more knowledge of the nature and function of other components of intestinal transport systems.

10.3 ACTIVE TRANSPORT

Whenever a substance is transported against its electrochemical energy gradient the process is commonly referred to as active transport. This definition is not ideal, since it is often difficult if not impossible to measure with reasonable accuracy electrical potential differences across cell membranes and activities of solutes in cells. None the less, in most cases it is not difficult to distinguish such transport from those where solute transport occurs without coupling to energy. Study of a wide variety of transport systems has revealed at least three rather different mechanisms of energy coupling: (1) group translocation where a covalent change in the transported substrate is part and parcel of the transport system, the energy coming from the energy of formation of the derivative of the substrate, (2) a process often called *primary* active transport where a chemical reaction involving the transport system provides energy by covalent bond formation resulting in conformational changes, and (3) where transport of one solute is coupled to flow of a second solute down its electrochemical energy gradient. Below, examples of each of these are discussed, as well as the still-unresolved mechanism of transport closely coupled to oxidative processes seen in bacteria and mitochondria.

10.3.1 Group translocation

Group translocation has recently emerged as an important transport mechanism. This model has been proposed over the years by a number of workers, apparently first by Wilbrandt in 1933[83], but the phosphoenolpyruvate-dependent phosphotransferase system (PTS) of bacteria is the first well-documented example of this transport mechanism[84,85]. It is important to distinguish between true group translocation in which transport and chemical alteration (in this case phosphorylation) are mechanistically linked, and a process usually called *trapping* in which transport is closely followed by covalent alteration but where there is no direct coupling between these two. Where cell pools of unaltered substrate are low and the subsequent chemical reaction fast, it may be impossible to distinguish group translocation from trapping by kinetic measurements alone.

The PTS is responsible for the phosphorylation of many sugars and related compounds in most but not all bacteria[86-92]. Common to all systems are two soluble proteins, enzyme I and Hpr. Enzyme I transfers phosphate from phosphoenolpyruvate (PEP) to Hpr, a small, rather heat stable protein which is phosphorylated at the N–1 position of a histidine residue. The next steps differ in different types of organisms. In *E. coli* and related Gram-negative bacteria phosphate is transferred to sugar by the mediation of a membrane bound and substrate-specific activity called an enzyme II. Each enzyme II consists of two distinct proteins called IIa and IIb[93]. In *Staphylococcus aureus* phosphate is first transferred from phospho-Hpr to a soluble sugar-specific protein called a factor III, and finally transferred to the sugar by a membrane-bound sugar-specific enzyme II[94,95]. There is some evidence suggesting that the underlying mechanisms of the enzyme II-catalysed steps in *S. aureus* and *E. coli* may be quite similar, and that the function of factor

III of the former is similar to that of the IIa protein in the latter[87]. Some variations on the above pattern have been reported. In *E. coli* a soluble factor III involved in glucose phosphorylation has been identified[93]. In the phosphorylation of fructose by *Aerobacter aerogenes* the role normally filled by Hpr is replaced by a fructose-induced soluble protein[96]. Since this protein accepts phosphate directly from PEP in a reaction catalysed by enzyme I, this protein is different from the factor III proteins found in *S. aureus*. That the function of Hpr is readily replaceable by other proteins in Gram negative bacteria such as *E. coli* is suggested by the observation that all mutants defective solely in Hpr are able to utilise PTS sugars at a low rate, and many of the revertants able to grow rapidly on PTS sugars are found to retain the defect in Hpr (M. Saier, Jr., personal communication).

Many of the proteins of the PTS have been purified, several to homogeneity[87,93,97-99]. Studies with the purified soluble proteins have shown that the intermediates in the reaction, phospho–Hpr and for *S. aureus* phospho-III_{lac}, are all so-called high energy phosphates whose free energy of hydrolysis is close to that of PEP[95]. Thus the free energy available in the sugar phosphorylating step is large, sufficient not only to form the low energy sugar–phosphate bond but for accumulation of the sugar phosphate to high levels. Three different IIa proteins from *E. coli*, each specific for a different sugar, have been purified to homogeneity[93]. The most intriguing component in terms of understanding the function of the PTS, the membrane-bound enzyme II of *S. aureus* or IIb of *E. coli*, have stubbornly resisted most attempts at solubilisation. The purest active preparations reported to date are enriched for enzyme II activity but are probably far from homogeneous[93,98].

Two types of observation indicate that the PTS performs coupled transport and phosphorylation: (1) the first form of the sugar to appear inside the cell is not unchanged substrate, but the phosphorylated derivative, and (2) mutants defective in either sugar-specific or common components of the PTS neither phosphorylate nor transport the affected sugars. The kinetic evidence is most clear cut for analogues whose metabolism does not proceed beyond the point of phosphorylation, such as α-methylglucoside which is taken up by an enzyme II in *E. coli*[93]. The first form of this analogue to appear in the cells is the phosphate; only later does the free glycoside accumulate in the cell[100]. There are also data from inhibitor studies suggesting that intracellular free glycoside is derived from the phosphorylated form, and not vice versa[101].

Mutants defective in one of the sugar-specific components of the PTS, either a factor III or a component of an enzyme II, are defective in both phosphorylation and uptake of sugars with affinity for that system[102,103]. Loss of function of the enzyme II, the component that appears to carry the substrate binding site, would be expected to abolish both functions. The fact that loss of factor III which probably does not directly recognise the substrate also abolishes both activities is less expected. The inference is that the series of conformational changes required to effect transport of sugar across the cell membrane cannot occur at a significant rate in the absence of factor III, since even uptake to the point of equilibration with the external medium is extremely slow in the factor III mutants[102]. Mutants lacking one of the common components of the PTS, enzyme I and Hpr, fail to transport or phosphorylate substrates of the PTS[85,102,104]. However, these mutants also fail to transport or metabolise a number of sugars and related compounds whose metabolism is independent of the PTS. This property of the mutants initially suggested that the PTS is needed for transport even of substances which are not substrates of the PTS[86]. This view has been revised as it became clear that mutations in enzyme I or Hpr had other effects which are not yet completely understood[105]. One of these effects is an impairment in enzyme induction in such strains; failure to utilise certain substances was traced to lack of induction of the needed enzymes or transport activities[106, 107]. In *E. coli* this effect is partly relieved by the addition of cyclic AMP to the culture medium, suggesting that the enzyme I mutations alter cyclic AMP metabolism. Cyclic AMP is known to be necessary for induction of many catabolic enzymes in Gram negative organisms[108]. Failure of induction could also be due in part to impaired transport of inducer into the cell[109]. This factor is plausible for most sugars, but could hardly explain the failure of the mutants to grow on glycerol since this compound readily enters cells when present in high concentrations in the medium[110]. Other possible effects include an alteration of glycolysis, with changes in cell pools of intermediates and enhanced feed-back inhibition of some steps. Whatever the mechanisms underlying the various effects of mutations in the common components of the PTS, it is clear that the properties of these mutants are consistent with the idea that the PTS performs coupled transport and phosphorylation.

The major question about the PTS is how do the membrane-bound enzymes function. Most workers view the system as having a binding site for sugar on the outer face of the membrane, and that interaction with the phosphorylating component (phospho–Hpr, phospho–III or whatever) on the inside results in a conformational change such that the sugar is brought into the cell and phosphorylated as a single operational step. This mechanism implies that transport and phosphorylation are not separable functions under ordinary circumstances; i.e. that each alone should not be performed by the enzyme II. Evidence on such separation of transport from phosphorylation is contradictory. There is very good evidence that phosphorylation without transport does occur in *E. coli* in the case of thiomethylgalactoside[111], and suggestive evidence that α-methylglucoside undergoes some preliminary interaction with the cell, possibly mediated entry, prior to phosphorylation[100]. The possibility that transport can occur without phosphorylation is suggested by the finding that mannitol liberated inside *E. coli* from a disaccharide exits more rapidly in strains induced by growth on mannitol than in uninduced strains[112]. The inference is that enzyme II is the component being induced and it accounts for the more rapid exit of mannitol. On the other hand, there is also strong evidence indicating that glucose cannot enter most enzyme I mutants at significant rates, even though such mutants have supra-normal levels of the glucose-specific enzymes II. In *S. typhimurium* it was shown that such mutants will grow on glucose, presumably utilising the ATP-dependent glucokinase present constitutively in Gram negative bacteria, if they carry a mutation to constitutive expression of a non-PTS linked transport system which can mediate glucose movement into the cell[113]. Kinetic studies on various mutants in both *S. typhimurium* and *S. aureus* suggest that the presence of enzyme II in the absence of the ability to phosphorylate does not allow transport of substrate at anything but very low rates[102, 104].

The discovery of the PTS has stimulated interest in this type of mechanism. While kinetic evidence consistent with group translocation has been found in other cases, no other well-documented example of a group translocation system has been reported. Adenine uptake in *E. coli* has been proposed as mediated by group translocation, the coupling reaction being conversion to AMP through the action of adenine phosphoribosyltransferase (PRT)[114]. The kinetic data are consistent with group translocation, but also fit a trapping system in which only low concentrations of free adenine accumulate in the cell. A major problem for the proposed mechanism is that the coupling enzyme is a typical soluble protein without any evidence suggesting that it is membrane-bound or membrane-associated. Membrane vesicles retain a fraction of the PRT activity of whole cells[114], but this seems to be due to the small contamination of vesicles with soluble cell proteins since the fraction of total PRT retained is similar to the fraction of total soluble proteins found in vesicle preparations[115]. The partial loss of PRT and many other enzymes of nucleoside metabolism had been interpreted as evidence for a periplasmic location of these enzymes[116]. A recent investigation of this question, comparing loss in osmotic shock with loss upon conversion to spheroplasts and inactivation of the enzymes by a diazo reagent which cannot penetrate the cell, showed that these enzymes are probably intracellular, and unusual in being partly removed by osmotic shock[57].

Group translocation in the renal transport of amino acids has been proposed as the biological role for a cyclic series of reactions leading to the synthesis and breakdown of γ-glutamyl amino acids[117]. The enzyme catalysing formation of γ-glutamyl amino acids from glutathione and free amino acids is membrane-bound, and thus could serve in transport. Very high rates of excretion of an intermediate of the cycle in a patient with an unusual genetic defect in amino acid metabolism are consistent with a transport role of the cycle[118], but there is no direct evidence in support of such a role. The renal transport of a number of amino acids is Na^+-dependent[119, 120] suggesting that the energy is provided by the Na^+ gradient as is the case for most intestinal transport (see below). Such a mechanism is energetically much more efficient than the γ-glutamyl cycle. If there is a 1:1 stoichiometry with Na^+ movement, one mole of ATP can transport three moles of amino acid, while in the γ-glutamyl cycle it takes three moles of ATP to transport one mole of amino acid, an efficiency nine-fold lower. The high levels of the γ-glutamyl system found in the kidney imply that this system plays some important biological role, but it is not clear what role this system has in transport.

10.3.2 ATP-driven transport

The one transport system which has been most extensively studied and is probably best understood is that for Na^+ and K^+ in animal cells, here referred to as the Na/K system. The properties of this system have been reviewed frequently[121-125]. This system is present in the plasma membranes of higher animal cells and plays a central role in maintaining cell ionic and osmotic homeostasis as well as serving other roles such as indirectly providing energy for the transport of other substances linked to Na^+ movement (see below). The system transports Na^+ out of the cell, K^+ into the cell and consumes ATP in the process. A few other ion transport systems, notably those for Ca^{2+} in the sarcoplasmic reticulum of muscle[126] and in other membranes[127], seem to be similar in their basic mechanism to that of the Na/K system.

Recent advances owe much to the discovery of an ATPase activity associated with this transport system. In 1957 Skou observed that crab nerve microsomes had an ATPase that was stimulated when both Na^+ and K^+ were present[128]. He inferred that the component stimulated by the combined presence of the two cations represented the activity of the transport system responsible for extrusion of Na^+ from the cell. This view has been amply substantiated in subsequent work which has shown an impressive correlation between this ATPase and the transport of Na^+ and K^+. The rates of these two are comparable in tissues ranging from those with sluggish activity such as human erythrocytes to the very high values of kidney and electroplax tissue[129, 130]. A useful property of the Na/K system is its selective inhibition by ouabain and other cardiotonic steroid glycosides[131]. Measurement of the glycoside inhibited component of ATPase is a convenient way of assaying activity of the Na/K system in the presence of other ATPases. However, it is not always correct to equate the glycoside-inhibitable portion of an activity with the function of the Na/K system, since some of the partial reactions of the system persist to varying degrees in the presence of saturating concentrations of glycosides[132-137]. The Na/K system appears to be found only in animal cells. Reports suggesting the existence of this system in bacteria[138] and plants[139] are not very convincing, since the properties of the activities measured are rather different from those of the Na/K system of animal cells.

The Na/K system presents very different functional aspects to the two surfaces of the cell membrane. On the inside are binding sites for the uptake of Na^+, for ATP, and sites for the release of K^+ transported in and the ADP and Pi produced in the reaction. On the external surface is the binding site for K^+ uptake, the site of release of transported Na^+, and the site where glycosides bind to inhibit. Measurements of the stoichiometry indicate that approximately three moles of Na^+ are extruded per mole of ATP utilised[140-148]. In erythrocytes approximately two moles of K^+ are taken up per three moles of Na^+ extruded[140, 142, 143].

A detailed analysis of the reaction sequence of the ATPase became possible with the detection of a phosphoenzyme intermediate[144, 145]. The formation of the intermediate from ATP is dependent on Na^+, while K^+ leads to a rapid decay of the intermediate[144, 146]. The intermediate has the stability properties of an acyl phosphate. Phosphate appears to be linked to an aspartyl residue, based on studies of a phosphorylated tripeptide isolated after pronase digestion of phosphorylated enzyme[147].

The most widely held current view of the Na/K system is that transport is accomplished by a cycle of alternating phosphorylation and dephosphorylation, the former associated with Na^+ transport and the latter with K^+ transport. A simplified scheme of the reaction sequence is shown in Figure 10.1, where the system has been drawn as if the enzyme moved to different positions in the cycle. The various positions are used to represent changes in

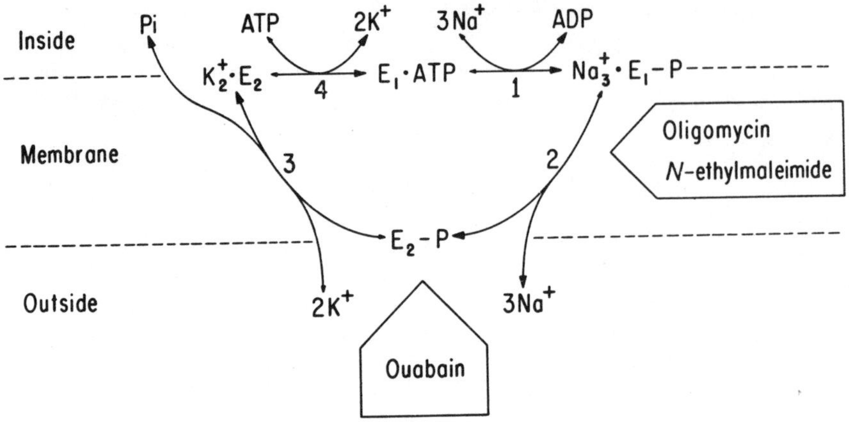

Figure 10.1 A schematic representation of the operation of the Na/K transport system of animal cells. Dashed lines represent the two surfaces of the cell membrane. The four steps described in the text are represented by numbers 1 through 4. For each step the forward, physiological direction is shown by the large arrowhead, while small arrowheads indicate the reverse direction which can be observed under suitable conditions. E_1 and E_2 refer to two conformationally different states of the enzyme inferred from its kinetic properties. The subscripts after K^+ and Na^+ designate the stoichiometry of the system. Compounds written in large arrows inhibit the steps indicated

affinities for substrates as well as changes in the accessibility of the cation binding sites from the inside or outside of the cell. Movement is used here to designate conformational change; no actual movement of the whole enzyme is implied. The four steps shown correspond to four forms participating in the normal function of the enzyme and which are sufficiently stable under suitable conditions that their properties can be studied directly.

In step 1 phosphoenzyme E_1–P is formed from ATP in the presence of Na^+. This step requires only low concentrations of ATP and Mg^{2+} in the range of 10^{-6} M, is stimulated by Na^+ but no other cations, and is reversible[146]. Reversibility is readily demonstrated when step 2 is blocked, a very active Na^+-dependent ATP–ADP exchange then being evident[148, 149]. In step 2 the phosphoenzyme undergoes a spontaneous change in configuration to E_2–P. This change does not result in a change in the covalent bonding of P to the enzyme[150], but alters the reactivity of the phosphoenzyme so that it becomes sensitive to discharge by K^+. This is the step in the cycle selectively inhibited by low Mg^{2+} concentrations, by oligomycin, and by *N*-ethylmaleimide[148-150]. These inhibitors do not affect formation of phosphoenzyme from ATP nor the ATP–ADP exchange reaction, but they block discharge of phosphoenzyme by K^+. In the absence of inhibitors step 2 is rapid compared to step 1 so that most of the phosphoenzyme is in the E_2–P form and is readily discharged by K^+ but not by ADP[150]. Step 2 probably represents a change in configuration associated with transport of Na^+ to the outside of the cell. In step 3 K^+ discharges P from the phosphoenzyme. Specificity for the cation is not very great; Rb^+ and Tl^+ appear to be even better than K^+, and all other alkali metal cations and NH_4^+ (but not Na^+) will serve[146, 151].

The product of step 3 is a relatively unreactive form of dephosphoenzyme designated E_2. This form, unlike E_1, does not rapidly form phosphoenzyme in the presence of ATP and Na^+ [151, 152]. Since the rate of conversion of E_2 to the reactive E_1 is dependent on the cation used in step 3, E_2 must be in a relatively stable complex with K^+ or other cation used in step 3[151]. The stability of the $K^+.E_2$ complex is partly an artefact of the *in vitro* conditions used in kinetic experiments where only low concentrations of ATP are present. When higher concentrations of ATP are added, the $K^+.E_2$ complex is much more rapidly converted to E_1[151]. This fact is the basis for drawing step 4 as a reaction in which ATP acts to liberate free K^+ and produce an E_1.ATP complex ready for step 1. There is good evidence that E_1 forms a complex with ATP in the absence of Na^+ and that prebound ATP is used to phosphorylate the enzyme when Na^+ is added[150, 153, 154].

A requirement for high concentrations of ATP in step 4 can account for the paradox that whereas step 1 requires only low concentrations of ATP and is not highly specific for ATP, the overall reaction of the system measured either as nucleotide triphosphate hydrolysis or transport of Na^+ and K^+ requires much higher ATP concentrations to saturate and is more specific for ATP[146, 155-157]. It is possible that other steps in the reaction sequence are also stimulated by adenine nucleotides.

Cardiac glycosides bind slowly to the Na/K system to produce a form which cannot carry out most of the reactions of the system. It has been known for a long time that K^+ partly antagonises the action of glycosides, and more recently it has become clear that all of the ligands of the system affect ouabain inhibition, and that the degree of inhibition correlates with the extent of glycoside binding[158-162]. The data are compatible with the view that the E_2–P form is the one that reacts with ouabain; ligands that tend to increase the concentration of this form increase sensitivity to and binding of ouabain, and ligands that reduce the amount of E_2–P reduce ouabain sensitivity and binding[161]. The ouabain enzyme is readily phosphorylated by Pi to form a phosphoenzyme in which the bonding of P appears to be identical to that formed from ATP[163, 164]. Glycosides can be seen as reacting with the E_2–P form and making this so stable as to virtually block all reactions of the system. Glycoside binding has been useful as a measure of the number of Na/K transport sites[165]. In tissues with a low density of transport sites, the high affinity specific binding must be distinguished from binding to sites not related to the Na/K system[166].

The ATP–ADP exchange reaction, and the formation of phosphoenzyme from Pi in the presence of ouabain are but two of a number of partial reactions of the system which have been useful in determining the function and properties of the different intermediate forms and steps involved. Ouabain-sensitive Na^+–Na^+ exchange transport in human erythrocytes, observed in K^+-free media, requires both ATP and ADP inside, is not coupled to net ATP hydrolysis, but is sensitive to oligomycin[143, 157, 167]. These facts show that both steps 1 and 2 are needed for Na^+ transport, the dependence on ATP and ADP reflecting the requirements of step 1, while oligomycin sensitivity testifies to the participation of step 2. An ouabain-sensitive K^+–K^+ exchange can also be observed in erythrocytes if internal Na^+ is depleted[135, 168, 169]. This exchange requires both Pi and ATP. The K_m for ATP is 60 μM, much higher than the ATP requirement in step 1, but just as for the Na^+–Na^+ exchange no net ATP hydrolysis accompanies the K^+–K^+ exchange. These observations support the idea that both steps 3 and 4 are necessary for K^+ transport, the ATP requirement being due to participation of the nucleotide in step 4. Pi is needed for the reversal of step 3. Such reversal implies the formation of phosphoenzyme from Pi even in the absence of ouabain.

Two reactions have been described which probably reflect activity of step 3 alone. The most recent to be noted is a K^+-dependent ^{18}O exchange between Pi and water[170]. This reaction is inhibited by Na^+ as well as by ouabain, and can be seen as the result of sequential phosphorylation of the enzyme by Pi and then hydrolysis to liberate Pi once again. An interesting finding of the study is that concentrations of deoxycholate that do not inhibit ATPase, but in fact stimulate it, totally inhibit Pi–H_2O exchange. This suggests that some detergents may alter the enzyme slightly to make step 3 virtually irreversible, and may explain why Lindenmayer *et al.* observed efficient labelling of protein from Pi in a crude enzyme preparation even in the absence of ouabain, but only in the presence of ouabain with a preparation purified with deoxycholate[163].

Another activity best explained as an expression of step 3 is the K^+-dependent acyl phosphatase of the Na/K system[171]. Acetylphosphate is the best substrate[172] but in most work *p*-nitrophenylphosphate is used because of its greater stability and ease of assay for its hydrolysis. Recent work by Garrahan and colleagues has shown that the substrate site for *p*NPP is inside the cell, while ligands affecting phosphatase activity (K^+, Na^+, ATP) act on the same side of the membrane as they do in the normal operation of the cycle[173]. Energy-dependent ion movements are not driven by *p*NPP[174]. Acyl phosphates lead to formation of the phosphorylated intermediate too[175]. The phosphatase can be seen as a manifestation of step 3 in which an acyl phosphate rather than Pi is the phosphate donor in the reverse operation of this step, while in the subsequent forward mode of operation Pi is released. A prediction of this interpretation is that *p*NPP should be able to replace Pi in supporting ouabain-sensitive K^+-K^+ exchange.

The ultimate in 'partial' reactions, namely the full reversal of the whole pump cycle, has been demonstrated quite convincingly by Glynn and Lew[176]. In this work erythrocytes were depleted of Na^+ and loaded with ^{32}Pi, then during the experimental period the cells were incubated in K^+-free solutions. Incorporation of label into ATP and ADP, the latter presumably arising from ATP through the action of adenylate kinase, was stimulated by external Na^+, and abolished by ouabain or addition of K^+ to the medium. There was good quantitative agreement between the rate of ATP synthesis and the rate of ion movements going through the pump in the reverse direction[135]. Only approximately 50% of the total incorporation into ATP was related to the action of the pump. The ouabain-sensitive increment in labelling probably represents synthesis of ATP and not a Pi–ATP exchange activity, since the latter has never been demonstrated to be part of the Na/K system. Estimates of ATP synthesis corrected for breakdown during the experimental period showed that approximately one mole of ATP was synthesised per two moles of K^+ which flowed out of the cells.

The model of this system presented here and outlined in Figure 10.1 is undoubtedly somewhat simplified but contains sufficient detail to fit most of the known facts of the system. An important aspect of this type of model is that interference by one ligand of the effects of another is not necessarily a sign of direct competition between them. Instead the interfering ligand is seen as shifting the equilibrium of the system so that more of the enzyme is in a form which does not readily react with the other ligand. The interaction of ouabain is very well explained by this view of ligand effects[161]. The stimulation of the phosphatase activity by Na^+ plus ATP seen in the presence of sub-optimal concentrations of K^+ is readily understood as due to shuttling enzyme from the E_1 to the E_2 form via steps 1 and 2, thereby increasing fraction of enzyme reactive with the acyl phosphate. Such an interpretation explains why the stimulation by Na^+ plus ATP is sensitive to oligomycin although the basic K^+-stimulated phosphatase is not[175,177].

Efforts to purify the Na/K ATPase have been rewarded by reports of what can be considered nearly pure preparations of active enzyme from several laboratories. Kyte purified the enzyme from canine kidneys[178]; Jørgensen worked with a rabbit kidney enzyme[179,180], while Hokin utilised rectal glands of the dogfish shark as starting material[181]. None of the preparations is pure in the usual sense because lipids needed for activity are present. Neither the composition of the lipids nor their amount in the preparations has been carefully studied. In terms of their protein content, however, the preparations are close to pure. The major protein is a large polypeptide of molecular weight near 135 000. This peptide carries the residue which is phosphorylated[181,182]. The preparations also contain a smaller peptide of molecular weight 35 000 containing 13% carbohydrate[181,183]. These molecular weight figures are obtained by gel filtration[183]; somewhat different and probably less reliable values were obtained by sodium dodecylsulphate acrylamide electrophoresis[180,181,183]. All of the preparations retained a high degree of activity[179,181,184].

It is not clear whether the smaller peptide is part of the enzyme or a contaminant. Reconstruction experiments utilising separated small and large peptides should resolve this question. A role for the smaller peptide is inferred from the fact that the two purify together and the cross-linking of the two in a 1:1 molar ratio by dimethyl suberimidate[183]. Using the gel filtration molecular weight figures the molar ratio of small to large peptide is approximately 1.7 in all of the preparations. It is clear that the small peptide is not an adventitious breakdown product of the larger one, since only the smaller one has covalently bound carbohydrate.

10.3.3 Transport coupled to oxidative reactions

The energy for many bacterial transport systems and probably for many of those in mitochondria is closely coupled to oxidative reactions in the membranes of these entities and does not depend on soluble high energy compounds such as ATP or phosphoenolpyruvate. A great deal of recent work has addressed itself to this problem, with the goal of defining in molecular terms how energy liberated in oxidative reactions in the membrane drives transport.

The most direct data implying a close link between transport and oxidative reactions have come from studies of bacterial membrane vesicles developed and extensively exploited by Kaback and co-workers. These vesicles are prepared by osmotic lysis of spheroplasts, followed by washing to remove soluble proteins[115]. The resulting small vesicles are quite stable to frozen storage, are capable of high rates of transport, and retain less than 1 % of the soluble proteins of the cells. Depending on the strain used, the vesicles may be predominantly single layered or have multiple concentric layers of membranes. The vesicles transport substances at a low rate utilising the modest remaining endogenous energy sources. Transport is dramatically stimulated by several compounds all of which are substrates for membrane-bound flavin-linked dehydrogenases. D-lactate is the best natural substrate for vesicles prepared from glucose-grown *E. coli*, but succinate, L-lactate, and NADH are also stimulatory[185]. The most effective substrate is the ascorbate-phenazine methosulphate combination[186]. Extensive metabolism of the substrate is not required; D-lactate is converted to pyruvate most of which is not further metabolised[185]. A large number of bacterial transport systems are specifically stimulated by these oxidative substrates: almost all systems for amino acids and many of those for sugars. The only major exceptions are sugars taken up via the phosphotransferase system.

The way oxidative reactions drive transport is the subject of some controversy. The pre-eminent effectiveness of D-lactate led Kaback to suggest that this dehydrogenase has a special role in transport. In Kaback's model successive oxidation and reduction of a carrier produces cyclic conformational changes resulting in transmembrane movement of the substrate of the carrier[187]. The idea that D-lactate dehydrogenase has a special role has not been generally accepted. Mutants lacking this activity are perfectly capable of transport when other oxidative substrates are added[188, 189]. A requirement for some component in addition to a specific dehydrogenase is demonstrated in mutant strains defective in electron transport. These strains are capable of oxidising D-lactate and other oxidative substrates and have the ATPase of oxidative phosphorylation, but oxidative substrates drive neither transport nor presumably ATP synthesis[189]. The relatively poor effectiveness of succinate and NADH in driving vesicle transport is readily explained as a permeability problem. The vesicles are not as impermeable as intact cells since phosphoenolpyruvate is able to enter the vesicles[190], but the vesicles are far from freely permeable since they are able to create and maintain concentration gradients of transport substances. Large and/or multiply-charged molecules such as succinate and NADH should not cross membranes readily, and it can be assumed that entry is needed since the normal site for oxidation of these compounds is at the inner face of the membrane. Lactic acid seems to cross membranes quite readily. The D-isomer is much more effective because the dehydrogenase for D-lactate is present at high levels in glucose-grown cells. The L-lactate dehydrogenase is inducible and present at low levels in glucose grown cells. The observation that the vesicles oxidise succinate and NADH quite well even though unable to utilise these substances for rapid rates of transport can be attributed to the presence of leaky and/or inside-out vesicles[191]. All substrates have ready access to dehydrogenases in such altered vesicles but no transport can be observed.

Work both with vesicles and intact cells has ruled out a requirement for ATP in transport driven by oxidative reactions, but ATP can drive this type of transport. Since the vesicles lack virtually all low molecular weight solutes including nucleotides, transport in the vesicles cannot require ATP formation. Bacterial mutants lacking the ATPase associated with oxidative phosphorylation (here referred to as the OP ATPase to avoid confusion with other ATPases) cannot synthesise ATP from oxidative substrates and therefore cannot grow on such substrates, but they can use oxidative substrates to drive transport[191, 192]. Reduced transport in some of the ATPase mutants[188] is not due to any requirement for ATP synthesis but seems to be due to an uncoupling effect of some of the mutations since an inhibitor of the OP ATPase, dicyclohexylcarbodiimide, restores transport driven by oxidative reactions[191]. Studies of the effects of inhibitors of glycolysis and respiration strongly suggest that transport can be driven either by oxidative substrates without the mediation of ATP, or by glycolytically produced ATP[193]. Addition of ATP to vesicle preparations usually produces no stimulation of transport[185], but a modest stimulation is seen when special efforts are made to obtain ATP entry into the vesicles[191]. The poor effectiveness of ATP is probably due to slow uptake and rapid degradation inside the vesicles.

An absolute requirement for ATP is seen in cation transport in *Streptococcus fecalis*. This organism is unusual in that it cannot perform oxidative phosphorylation to any significant extent and lacks an oxidative chain. In this organism the OP ATPase inhibitor dicyclohexylcarbodi-imide blocks transport[194], and this effect is specific since a mutant resistant to the transport effects of the inhibitor has an OP ATPase that is resistant *in vitro*[195]. Thus *S. fecalis* appears to act just as *E. coli* would anaerobically or when oxidative reactions are blocked in requiring ATP in order to couple energy into the membrane reactions of transport.

The pattern that has emerged of this type of bacterial transport is strongly reminiscent of the situation in mitochondria where a number of different types of energy linked reactions—ATP synthesis, oxidation, transport, and a transhydrogenase—all appear to be linked through a high-energy intermediate of unknown nature. Of the various suggestions that have been made for the nature of this intermediate, the chemiosmotic theory of Mitchell must be given special consideration. This theory specifies the nature of the intermediate, namely an energy difference for protons (the proton-motive force, abbreviated PMF), and thereby leads directly to a number of testable predictions. Since a number of these predictions have been confirmed, the theory merits serious consideration.

The chemiosmotic theory was developed by Mitchell as a logical extension and development of earlier ideas that electrical potential differences across cell membranes could be created by oxidative reactions which would liberate protons on one side of the membrane while releasing the electrons on the other side[196-198]. Mitchell has single-handedly taken this idea, fleshed it out in specific terms, and suggested that it applies to a very large range of membrane-mediated energy transformations. Since it was first proposed[199] in 1961, the details of the theory have been altered in keeping with experimental facts[200, 201]. A number of reviews have dealt with the theory[202, 203]. Here we will examine its relevance to bacterial transport.

The chemiosmotic theory proposes that energy-yielding reactions in the membrane act to liberate protons and electrons on opposite sides of the membrane, resulting in both a pH difference and an electrical potential difference across the membrane. Together these two constitute the energy of the PMF. Protons can move down their electrochemical gradient at high rates only on specific proton carriers each of which is linked to some energy utilising reaction such as transport or ATP synthesis. These latter reactions as well as the oxidative reactions which create the PMF are assumed to be reversible, so that under suitable conditions ATP hydrolysis or energetically downhill movement of solutes can generate a PMF and this in turn can reverse the normal flow of electron transport. All of the energy transducing reactions of the membrane whether creating or dissipating the PMF must span the membrane if they are to move protons across the membrane. While it is obvious that transport systems must span the membrane, this is not so obvious for the ATP-synthesising enzymes nor for those of electron transport. There is evidence in mitochondria that at least parts of the electron transport chain are arranged in a manner consistent with the expectations of the chemiosmotic theory[204].

The chemiosmotic theory makes a number of predictions, all of which are, in theory, testable: (1) both energy generating and energy utilising reactions show stoichiometric coupling to proton movement, (2) energy transduction is dependent on a low dissipative (non-energy linked) proton permeability of the membrane, (3) energy-linked processes can be driven by a proton gradient no matter how the gradient is generated, and (4) the free energy released in proton movement is large enough to drive the reactions to which it is presumed to be coupled. There is considerable support for the first three of these predictions. Galactoside uptake in both *E. coli*[205] and *Streptococcus lactis*[206] is accompanied by proton movements; in *E. coli* the stoichiometry is one proton per mole of galactoside[206]. In an *E. coli* mutant with impaired coupling of energy to galactoside transport, the stoichiometry of protons: galactoside is reduced to well below one[207]. A requirement for proton impermeability is seen in the ability of a wide variety of proton conductors, of which dinitrophenol is a classical example, to uncouple transport from energy yielding reactions. Strong evidence for the role of an electrical potential difference comes from studies in intact cells of two species of *Streptococcus*[206, 208] and in *E. coli* vesicles[209]. When loaded with K^+ and suspended in K^+-free media, addition of valinomycin leads to high specific electrogenic K^+ permeability, resulting in a large transient diffusion potential, interior negative, which is dissipated as other ion movements allow K^+ to leave. A transient accumulation of amino acids or galactoside was seen, coincident with the duration of the electrical potential differences as measured by accumulation of a lipid soluble cation, dimethyldibenzyl ammonium (DDA)[209]. In *S. lactis* imposition of a sudden pH gradient led to a transient accumulation of galactosides, again suggesting that a PMF can drive transport[206]. The latter result is reminiscent of the generation of ATP in chloroplasts subjected to a large pH gradient[210].

The prediction most difficult to test is the last one, that the magnitude of the PMF is always large enough to drive reactions coupled in this way. The major part of the PMF must be contributed by an electrical potential difference since the interior of the cell will not tolerate a pH far removed from neutrality. In *S. fecalis* the interior is approximately one pH unit more alkaline when glucose is the energy source, but there is no measurable difference when arginine supplies energy[211]. Lower values for the difference in pH have been reported for *E. coli*[212, 213], but these measurements were made after washing and in the absence of substrate and so may not reflect the condition in cells which are actively metabolising substrates. The electrical potential difference cannot be measured directly in bacteria, but can be estimated by the distribution of a non-metabolised lipid soluble cation. Dimethyldibenzyl ammonium ion (DDA) has been used among others for this purpose. In the presence of small amounts of a lipid-soluble anion, such as tetraphenyl boron, DDA is taken up by bacteria and by vesicles[209, 214]. In Na^+ loaded cells and vesicles large electrical potential differences are indicated by the uptake of DDA; the values are approximately −100 mV in the vesicles[209], and −200 mV in the cells[214]. These, even without any pH difference, are sufficient to account for rather large accumulations of substrates whose transport is linked to protons. While indirect measures such as those with DDA may not be accurate reflections of electrical differences, it is clear that micro-organisms can generate large potential differences. In *Neurospora crassa*, values as high as −240 mV have been measured with microelectrodes[215].

There is one condition which provides some problems for a chemiosmotic explanation, and this is the case for cells and vesicles which contain the normal intracellular cation K^+. Vesicles loaded with K^+ rather than Na^+ do not take up much DDA, suggesting that the electrical potential is low. However, vesicles are defective in K^+ transport compared to intact cells, and it has been suggested that the low uptake of DDA is an artefact due to inability of intracellular K^+ to leave and liberate anionic sites for DDA[209]. This explanation cannot explain why Na^+-loaded cells of *S. fecalis* which concentrate DDA approximately 600-fold release most of the DDA to the medium when provided with K^+ to take up. Here there can be no shortage of anionic sites, but the cells seem to prefer K^+ to DDA. If DDA is an accurate measure of the electrical potential differences, then this difference is small in cells containing K^+ and therefore cannot account for large accumulations of either K^+ or other substances accumulated by K^+-rich cells. The existence of a rather small membrane potential in growing *E. coli* is suggested by measures of Cl^- distribution[216]. It is a reasonable but untested assumption that Cl^- is passively distributed in these cells and therefore a measure of the membrane potential.

It is not yet possible to reach a conclusion as to the role of the PMF in bacterial transport. It takes only one contradictory piece of evidence to eliminate the chemiosmotic theory as a general mechanism for energy coupling, and the data with K^+ loaded cells and vesicles appear to be inconsistent with the theory. However, the indirect means used to measure membrane potentials with DDA and other lipid soluble cations may be misleading, so that the interpretation of such data must be approached with caution. It can be concluded that proton movements are important in bacterial transport; any proposed mechanism of energy coupling will have to take account of this fact. The chemiosmotic explanation takes account of it in a most complete and simple manner. Alternatives halfway between classical theories of a chemical intermediate and Mitchell's view have been proposed, namely that protons do provide energy but are not normally released to generate a difference across the membrane[217]. Rather the view is that protons are liberated within the membrane to drive transport, so that coupling is similar to that proposed by Mitchell except that the protons have to be delivered directly to the coupling reaction. One of the attractive features of Mitchell's model is that coupling is via a difference across the membrane so that all systems that span the membrane can couple to the PMF and no direct interaction between PMF generating and utilising systems is needed. In summary it is fair to say that Mitchell has come close to the truth, but whether he has hit the mark in explaining bacterial transport remains to be determined.

10.3.4 Na^+-linked transport

Observations that many active transport systems in animals were Na^+ dependent led several groups of investigators[218-220] to postulate that such transport was mediated by a carrier which coupled Na^+ movement to the movement of substrates of transport. Since cell Na^+ is kept low through the action of the Na/K transport system, there is a driving force for Na^+ entry which, through coupling by the common carrier, is translated into a force leading to accumulation of substrate in the cell. Assuming a molar stoichiometry between Na^+ and substrate of 1:1 and ignoring (for the moment) electrical effects, maintenance of cell Na^+ concentration at 1/10th that outside can lead in the limit to a tenfold concentration of the substrate in the cell. In the years since this type of model was first proposed a great deal of work has appeared, most of it supporting this type of energy coupling which is generally referred to as Na^+-linked or Na^+-gradient transport. For a critical and comprehensive discussion of work up to 1970 the reader is referred to the review by Schultz and Curran[221]. This topic is also the subject of a recent book edited by Heinz[222] and a review by Kimmich restricted to intestinal sugar transport[223]. Here the current status of energy coupling by Na^+-linked transport systems will be examined.

The adequacy of the driving force for Na^+ as the sole source of energy for such transport systems has been questioned by observations in which either the extent of substrate accumulation was greater than expected on the basis of the Na^+ concentration ratio[224, 225], where metabolic inhibitors reduced the extent of accumulation even though the Na^+ gradient was but little altered[226], or where reversal of the Na^+ gradient did not lead to a net efflux of the substrate[227, 228]. To explain these apparent discrepancies it has been suggested that energy is also provided by other ion gradients such as that for K^+ [224, 226] or through direct coupling to energy such as by an ATP-driven step[223, 225, 228]. However, careful examination indicates that there are large uncertainties as to the magnitude of the energy available from Na^+ movements. It now appears that these may be adequate to account for the observations referred to above.

The amount of energy available per mole of Na^+ transported is given by the sum of (1) the energy available in the chemical (concentration) gradient determined by the ratio of external to cytoplasmic Na^+ ion activity plus, if the transport is electrogenic, (2) the electrical energy set by the magnitude of the electrical potential difference across the cell membrane. One major uncertainty is in the correct value for cytoplasmic Na^+ activity. The usual estimate for this figure, the average cell concentration of Na^+, certainly overestimates the activity, since there is good evidence that the cell nucleus contains a disproportionately high fraction of cell Na^+ and a number of different techniques suggest that the activity coefficient in the cell is probably less than 0.5 and may be even lower (see summary of data in Terry and Vidaver[229]). It seems that cytoplasmic Na^+ activity is in the range of 25% to 50% of the average cell Na^+ concentration, a correction which would allow a given Na^+ gradient to give a two- to four-fold higher substrate concentration in the cell than expected from calculations based on average cell Na^+ concentration.

The importance of the electrical potential as contributing to the driving force for Na^+ received little attention in most of the earlier work. Transport by a carrier which moves Na^+ and a neutral substrate such as a sugar or neutral amino acid will be electrogenic unless movement of the carrier is coupled to a compensatory ion movement. Compensatory ion movements must occur to maintain electroneutrality, but the essential point is that unless these are *coupled* to entry of Na^+ and substrate the Na^+-linked transport process will be electrogenic and thus energetically sensitive to the electrical potential difference across the cell membrane. Efflux of K^+ partially compensating Na^+ influx with glycine was noted by Eddy[230] and originally considered linked to the Na^+-dependent glycine transport system, but more recent work makes this interpretation dubious[231].

The electrical potential has been carefully measured with microelectrodes in two of the cell types also used extensively in studies of Na^+-linked transport. In rabbit ileum an average potential difference of -36 mV is found[232], while the value for Ehrlich ascites tumour cells is -26 mV[233]. The latter value is the same as that obtained from the Cl^- distribution in ascites tumour cells[233]. Electrical potentials of this magnitude represent an extra driving force for Na^+ entry allowing for an extra two- to four-fold concentration of solute above that based solely on energy in the chemical gradient for Na^+ movement.

The stoichiometry of coupling betweeen Na^+ and substrate must also be known. Sugar transport in the intestine is coupled by a 1:1 ratio to Na^+ movement[234]. Alanine transport in the intestine shows variable coupling ranging from low values when external Na^+ is low to approach one at high external Na^+ concentrations[235]. Glycine uptake by ascites tumour cells also shows a 1:1 coupling with Na^+ movement[230]. Amino acid uptake in erythrocytes may be driven by the movement of more than one Na^+ ion per molecule of substrate since stoichiometries well above one have been seen[236-238]. Whether the stoichiometries significantly higher than one represent the coupling ratio of the transport system is not clear since there are large unavoidable errors in such measurements. Only part of the Na^+ influx is associated with any particular transport system, and if addition of substrate stimulates Na^+ influx not only via the transport system under study but also through indirect effects such as changes in cell volume the measured increase in Na^+ influx will overestimate Na^+ influx linked to substrate transport.

When the corrections discussed here for cytoplasmic Na^+ activity and for the influence of the electrical potential difference are considered, the extent of substrate concentration which can be driven by Na^+ alone is in the range of three to twelve times higher than calculated from the Na^+ concentration ratio alone. These calculations assume a 1:1 stoichiometry between Na^+ and substrate movement. If systems with higher stoichiometries exist, as may be the case for avian erythrocytes, even higher substrate concentration gradients can be achieved. For a system with a 2:1 stoichiometry the concentration ratio will be equal to the square of that achieved by systems with a 1:1 stoichiometry. Thus the apparently anomalous substrate concentration ratios noted above[224, 225] can actually be accounted for by the driving force for Na^+. A contribution by the electrical gradient explains the diminution of accumulation by metabolic poisons in the face of a high Na^+ gradient[226] as well as the persistence of accumulation in the face of a reversed Na^+ gradient of moderate magnitude[227, 228].

There is direct experimental evidence suggesting that transport via Na^+-linked systems is electrogenic, thereby justifying including an electrical term in the calculations. White and Armstrong[239] and Rose and Schultz[232] observed that addition of transport substrates to intestinal preparations produced a transient drop in the electrical potential difference across the mucosal surface which is the surface where Na^+-linked transport occurs. This finding is most readily explained by assuming that the transport is electrogenic and thus tends to discharge the membrane potential. Gibb and Eddy[231] showed that in poisoned cells with a negligibly small Na^+ gradient the imposition of a large K^+-determined membrane potential by the addition of valinomycin resulted in a large accumulation of methionine, a finding also suggesting that the membrane potential contributes to the driving force for Na^+-linked transport.

It is clear that there is considerably more energy in Na^+ gradients than had earlier been suspected, after making due allowances for cytoplasmic Na^+ activity and the membrane potential. It is not necessary at the present time to assume that a source of energy in addition to that for Na^+ movement contributes to any Na^+-linked transport system, and thus we are back to an energetic model identical to that first proposed[218-220]. This conclusion is supported by recent work[229, 240]. Although there seems to be no need for coupling to other energy sources, this does not exclude the possibility that some Na^+-linked transport systems may be coupled to additional sources of energy.

10.4 SUMMARY

The extent of diversity in biological transport systems is being reduced as understanding of transport systems increases. At the present time there is firm evidence for three basic mechanisms for active transport: (1) group translocation exemplified by the bacterial phosphotransferase system, (2) direct coupling to energy as in the ATP-driven Na/K transport system, and (3) coupling transport of one substance to the energetically downhill movement of another small molecule seen in the Na^+-linked transport systems.

There is a good possibility that the transport systems of bacteria and mitochondria which are closely associated with oxidative reactions are another example of the last mechanism but where protons provide the driving force, but this mechanism has not yet been established to the satisfaction of all.

A major challenge in this field is the creation of a three-dimensional picture of transport. A major contribution in this area have been the channel ionophores which now give us a picture of a system which can move ions down their electrochemical energy gradient. Whether this type of structure is a good general model for transport systems, and how energy can be coupled to transport via such structures, are just two of the many questions which remain to be answered in the field of membrane transport.

Note added in Proof

Recent studies of energy coupling in bacterial transport tend to confirm Berger's finding on glutamine transport[76] that ATP, or a high-energy phosphate compound which can be formed from ATP, drives many binding protein associated transport systems. Transport of several other amino acids[241], a dipeptide[242], and a sugar[243], for all of which a periplasmic binding protein has been detected, seems to be driven by high-energy phosphate compounds and not by oxidative reactions. Our studies suggest high-energy phosphate coupling for two major K^+ uptake systems of *E. coli* (the Kdp and TrkA systems[244]) because both are very sensitive to arsenate inhibition (D. B. Rhoads, unpublished observations). These studies strongly suggest that a considerable number of bacterial active transport systems, most of them associated with periplasmic binding proteins, obtain energy from high-energy phosphate compounds. This type of energy coupling has not previously been demonstrated to exist in bacteria.

Uptake of solutes via transport systems which appear to derive their energy from high-energy phosphate compounds has not been demonstrable in Kaback vesicles[75,245,246]. Failure to demonstrate transport via ATP-dependent systems may be due to an inability to provide energy for them in the vesicles. The vesicles as made in the usual way are depleted of low molecular weight solutes and soluble proteins, and are nearly impermeable to ATP and ADP.

References

1. Gorter, E. and Grendel, F. (1925). *J. Exp. Med.*, **41,** 439
2. Danielli, J. F. and Davson, H. (1934). *J. Cellular Comp. Physiol.*, **5,** 495
3. Singer, S. J. (1971). *Structure and Function of Biological Membranes*, 146 (L. I. Rothfield, editor) (New York: Academic Press)
4. Collander, R. (1949). *Physiol. Plantarum*, **2,** 300
5. Osterhout, W. J. V. (1935). *Proc. Nat. Acad. Sci. USA*, **21,** 125
6. Fox, C. F. and Kennedy, E. P. (1965). *Proc. Nat. Acad. Sci. USA*, **54,** 891
7. Steim, J. M., Tourtellotte, M. E., Reinert, J. C., McElhaney, R. N. and Rader, R. L. (1969). *Proc. Nat. Acad. Sci. USA*, **63,** 104
8. Engelman, D. M. (1970). *J. Mol. Biol.*, **47,** 115
9. Esfahani, M., Barnes, E. M. Jr. and Wakil, S. J. (1969). *Proc. Nat. Acad. Sci. USA*, **64,** 1057

10. Esfahani, M., Limbrick, A. R., Knutton, S., Oka, T. and Wakil, S. J. (1971). *Proc. Nat. Acad. Sci. USA*, **68**, 3180
11. Schairer, H. U. and Overath, P. (1969). *J. Mol. Biol.*, **44**, 209
12. Wilson, G., Rose, S. and Fox, C. F. (1970). *Biochem. Biophys. Res. Commun.*, **38**, 617
13. Overath, P., Schairer, H. U. and Stoffel, W. (1970). *Proc. Nat. Acad. Sci. USA*, **67**, 606
14. Overath, P., Hill, F. F. and Lamnek-Hirsch, I. (1971). *Nature New Biol.*, **234**, 264
15. Tsukagoshi, N. and Fox, C. F. (1973). *Biochemistry*, **12**, 2816
16. Linden, C. D., Wright, K. L., McConnell, H. M. and Fox, C. F. (1973). *Proc. Nat. Acad. Sci. USA*, **70**, 2271
17. Phillips, M. C., Ladbrooke, B. D. and Chapman, D. (1970). *Biochim. Biophys. Acta*, **196**, 35
18. Shimshick, E. J. and McConnell, H. M. (1973). *Biochemistry*, **12**, 2351
19. Ladbrooke, B. D., Williams, R. M. and Chapman, D. (1968). *Biochim. Biophys. Acta*, **150**, 333
20. Shimshick, E. J. and McConnell, H. M. (1973). *Biochem. Biophys. Res. Commun.*, **53**, 446
21. Mueller, P. and Rudin, D. O. (1969). *Current Topics in Bioenergetics*, **3**, 157 (D. R. Sanadi, editor) (New York: Academic Press)
22. Pedersen, C. J. (1968). *Fed. Proc.*, **27**, 1305
23. Shemyakin, M. M., Ovchinnikov, Y. A., Ivanov, V. T., Antonov, V. K., Vinogradova, E. I., Shkrob, A. M., Malenkov, G. G., Evstratov, A. V., Laine, I. A., Melnik, E. I. and Ryabova, I. D. (1969). *J. Membrane Biol.*, **1**, 402
24. Pinkerton, M., Steinrauf, L. K. and Dawkins, P. (1969). *Biochem. Biophys. Res. Commun.*, **35**, 512
25. Pressman, B. C., Harris, E. J., Jagger, W. S. and Johnston, J. H. (1967). *Proc. Nat. Acad. Sci. USA*, **58**, 1949
26. Mueller, P. and Rudin, D. O. (1967). *Biochem. Biophys. Res. Commun.*, **26**, 398
27. Krasne, S., Eisenman, G. and Szabo, G. (1971). *Science*, **174**, 412
28. Sarges, R. and Witkop, B. (1965). *Biochemistry*, **4**, 2491
29. Goodall, M. C. (1971). *Arch. Biochem. Biophys.*, **147**, 129
30. Hladky, S. B. and Haydon, D. A. (1970). *Nature (London)*, **225**, 451
31. Hladky, S. B. and Haydon, D. A. (1972). *Biochim. Biophys. Acta*, **274**, 294
32. Urry, D. W. (1971). *Proc. Nat. Acad. Sci. USA*, **68**, 672
33. Urry, D. W. (1972). *Proc. Nat. Acad. Sci. USA*, **69**, 1610
34. Goodall, M. C. and Urry, D. W. (1973). *Biochim. Biophys. Acta*, **291**, 317
35. Tosteson, D. C., Andreoli, T. E., Tieffenberg, M. and Cook, P. (1968). *J. Gen. Physiol.*, **51**, 373s
36. Hodgkin, A. L. and Huxley, A. F. (1952). *J. Physiol.*, **117**, 500
37. Mueller, P. and Rudin, D. O. (1963). *J. Theor. Biol.*, **4**, 268
38. Bean, R. C., Shepherd, W. C., Chan, H. and Eichner, J. (1969). *J. Gen. Physiol.*, **53**, 741
39. Ehrenstein, G., Lecar, H. and Nossal, R. (1970). *J. Gen. Physiol.*, **55**, 119
40. Latorre, R., Ehrenstein, G. and Lecar, H. (1972). *J. Gen. Physiol.*, **60**, 72
41. Goodall, M. C. and Sachs, S. (1972). *Nature New Biol.*, **237**, 252
42. Shamoo, A. E. and Albers, R. W. (1973). *Proc. Nat. Acad. Sci. USA*, **70**, 1191
43. Cabantchik, Z. I. and Rothstein, A. (1974). *J. Memb. Biol.*, **15**, 207
44. Steck, T. L. (1972). *Membrane Research*, 71 (C. F. Fox, editor) (New York: Academic Press)
45. Pardee, A. B. and Prestidge, L. S. (1966). *Proc. Nat. Acad. Sci. USA*, **55**, 189
46. Piperno, J. R. and Oxender, D. L. (1966). *J. Biol. Chem.*, **241**, 5732
47. Taylor, R. T., Norrell, S. A. and Hanna, M. L. (1972). *Arch. Biochem. Biophys.*, **148**, 366
48. Matsuura, A., Iwashima, A. and Nose, Y. (1973). *Biochem. Biophys. Res. Commun.*, **51**, 241
49. Oxender, D. L. (1972). *Ann. Rev. Biochem.*, **41**, 777
50. Oxender, D. L. (1972). *Metabolic Transport*, 133 (L. E. Hokin, editor) (New York: Academic Press)
51. Boos, W. (1974). *Ann. Rev. Biochem.*, (in the press)
52. Boos, W., Gordon, A. S., Hall, R. E. and Price, H. D. (1972). *J. Biol. Chem.*, **247**, 917
53. Kepes, A. and Richarme, G. (1972). *Mitochondria: biogenesis and bioenergetics; Biomembranes: molecular arrangements and transport mechanisms*, 327 (S. G. van den

Bergh, P. Borst, L. L. M. van Deenen, J. C. Riemersma, E. C. Slater and J. M. Tager, editors) (Amsterdam: North Holland)
54. Neu, H. C. and Heppel, L. A. (1965). *J. Biol. Chem.*, **240,** 3685
55. Pardee, A. B. and Watanabe, K. (1968). *J. Bacteriol.*, **96,** 1049
56. Penrose, W. R., Nichoalds, G. E., Piperno, J. R. and Oxender, D. L. (1968). *J. Biol. Chem.*, **243,** 5921
57. Taketa, A. and Kuno, S. (1972). *J. Biochem.*, **72,** 1557
58. Rotman, B., Ganesan, A. K. and Guzman, R. (1968). *J. Mol. Biol.*, **36,** 247
59. Ames, G. F.-L. and Lever, J. (1970). *Proc. Nat. Acad. Sci. USA*, **66,** 1096
60. Medveczky, N. and Rosenberg, H. (1971). *Biochim. Biophys. Acta*, **241,** 494
61. Brown, C. E. and Hogg, R. W. (1972). *J. Bacteriol.*, **111,** 606
62. Berger, E. A. and Heppel, L. A. (1972). *J. Biol. Chem.*, **247,** 7684
63. Rahmanian, M., Claus, D. R. and Oxender, D. L. (1973). *J. Bacteriol.*, **116,** 1258
64. Robbins, J. C. and Oxender, D. L. (1973). *J. Bacteriol.*, **116,** 12
65. Anraku, Y., Naraki, T. and Kanzaki, S. (1973). *J. Biochem.*, **73,** 1149
66. Lengeler, J., Hermann, K. O., Unsöld, H. J. and Boos, W. (1971). *Eur. J. Biochem.* **19,** 457
67. Boos, W. (1969). *Eur. J. Biochem.*, **10,** 66
68. Medveczky, N. and Rosenberg, H. (1970). *Biochim. Biophys. Acta*, **211,** 158
69. Ames, G. F.-L. and Lever, J. E. (1972). *J. Biol. Chem.*, **247,** 4309
70. Boos, W. and Sarvas, M. O. (1970). *Eur. J. Biochem.*, **13,** 526
71. Weiner, J. H., Furlong, C. E. and Heppel, L. A. (1971). *Arch. Biochem. Biophys.*, **124,** 715
72. Rahmanian, M. and Oxender, D. L. (1972). *J. Supramol. Struct.*, **1,** 55
73. Anraku, Y. (1968). *J. Biol. Chem.*, **243,** 3129
74. Pardee, A. B. (1968). *Science*, **162,** 632
75. Lombardi, F. J. and Kaback, H. R. (1972). *J. Biol. Chem.*, **247,** 7844
76. Berger, E. A. (1973). *Proc. Nat. Acad. Sci. USA*, **70,** 1514
77. Parnes, J. R. and Boos, W. (1973). *J. Biol. Chem.*, **248,** 4429
78. Taylor, A. N. and Wasserman, R. H. (1969). *Fed. Proc.*, **28,** 1834
79. Omdahl, J. L. and DeLuca, H. F. (1973). *Physiol. Rev.*, **53,** 327
80. Harmeyer, J. and Deluca, H. F. (1969). *Arch. Biochem. Biophys.*, **133,** 247
81. Corradino, R. A. (1973). *Science*, **179,** 402
82. Gräsbeck, R. (1969). *Progr. Hematol.*, **6,** 233
83. Wilbrandt, W. and Laszt, L. (1933). *Biochem. Z.*, **259,** 398
84. Kundig, W., Ghosh, S. and Roseman, S. (1964). *Proc. Nat. Acad. Sci. USA*, **52,** 1067
85. Tanaka, S. and Lin, E. C. C. (1967). *Proc. Nat. Acad. Sci. USA*, **57,** 913
86. Roseman, S. (1969). *J. Gen. Physiol.*, **54,** 138s
87. Roseman, S. (1972). *Metabolic Transport*, 41 (L. E. Hokin, editor) (New York: Academic Press)
88. Kundig, W. and Roseman, S. (1971). *J. Biol. Chem.*, **246,** 1393
89. Romano, A. H., Eberhard, S. J., Dingle, S. L. and McDowell, T. D. (1970). *J. Bacteriol.*, **104,** 808
90. Phibbs, P. V. Jr. and Eagon, R. C. (1970). *Arch. Biochem. Biophys.*, **138,** 470
91. Saier, M. H. Jr., Feucht, R. U. and Roseman, S. (1971). *J. Biol. Chem.*, **246,** 7819
92. Cirillo, V. P. and Razin, S. (1973). *J. Bacteriol.*, **113,** 212
93. Kundig, W. and Roseman, S. (1971). *J. Biol. Chem.*, **246,** 1407
94. Simoni, R. D., Smith, M. F. and Roseman, S. (1968). *Biochem. Biophys. Res. Commun.* **31,** 804
95. Simoni, R. D., Hays, J. B., Nakazawa, T. and Roseman, S. (1973). *J. Biol. Chem.*, **248,** 957
96. Walters, R. W. Jr. and Anderson, L. (1973). *Biochem. Biophys. Res. Commun.*, **52,** 93
97. Anderson, B., Weigel, N., Kundig, W. and Roseman, S. (1971). *J. Biol. Chem.*, **246,** 7023
98. Simoni, R. D., Nakazawa, T., Hays, J. B. and Roseman, S. (1973). *J. Biol. Chem.*, **248,** 932
99. Hays, J. B., Simoni, R. D. and Roseman, S. (1973). *J. Biol. Chem.*, **248,** 941
100. Gachelin, G. (1970). *Eur. J. Biochem.*, **16,** 342
101. Haguenauer, R. and Kepes, A. (1971). *Biochimie*, **53,** 99

102. Simoni, R. D. and Roseman, S. (1973). *J. Biol. Chem.*, **248,** 966
103. Fox, C. F. and Wilson, G. (1968). *Proc. Nat. Acad. Sci. USA*, **59,** 988
104. Simoni, R. D., Levinthal, M., Kundig, F. D., Kundig, W., Anderson, B., Hartman, P. E. and Roseman, S. (1967). *Proc. Nat. Acad. Sci. USA*, **58,** 1963
105. Epstein, W. and Curtis, S. (1972). *Role of Membranes in Secretory Processes*, 98 (L. Bolis, R. D. Keynes and W. Wilbrandt, editors) (Amsterdam: North Holland)
106. Pastan, I. and Perlman, R. L. (1969). *J. Biol. Chem.*, **244,** 5836
107. Berman, M., Zwaig, N. and Lin, E. C. C. (1970). *Biochem. Biophys. Res. Commun.*, **38,** 272
108. Pastan, I. and Perlman, R. (1970). *Science*, **169,** 339
109. Saier, M. H. Jr. and Roseman, S. (1972). *J. Biol. Chem.*, **247,** 972
110. Richey, D. P. and Lin, E. C. C. (1972). *J. Bacteriol.*, **112,** 784
111. Kashket, E. R. and Wilson, T. H. (1969). *Biochim. Biophys. Acta*, **193,** 294
112. Solomon, E., Miyai, K. and Lin, E. C. C. (1973). *J. Bacteriol.*, **114,** 723
113. Saier, M. H. Jr., Bromberg, F. G. and Roseman, S. (1973). *J. Bacteriol.*, **113,** 512
114. Hochstadt-Ozer, J. and Stadtman, E. R. (1971). *J. Biol. Chem.*, **246,** 5304
115. Kaback, H. R. (1971). *Methods in Enzymology*, **22,** 99 (W. B. Jakoby, editor) (New York: Academic Press)
116. Hochstadt-Ozer, J. and Stadtman, E. R. (1971). *J. Biol. Chem.*, **246,** 5312
117. Meister, A. (1973). *Science*, **180,** 33
118. Eldjarn, L., Jellum, E. and Stokke, O. (1972). *Clin. Chim. Acta*, **40,** 461
119. Fox, M., Thier, S., Rosenberg, L. and Segal, S. (1964). *Biochim. Biophys. Acta*, **79,** 167
120. Hillman, R. E., Albrecht, I. and Rosenberg, L. E. (1968). *J. Biol. Chem.*, **243,** 5566
121. Whittam, R. and Wheeler, K. P. (1970). *Ann. Rev. Physiol.*, **32,** 21
122. Skou, J. C. (1971). *Current Topics in Bioenergetics*, **4,** 357 (D. R. Sanadi, editor) (New York: Academic Press)
123. Dunham, P. B. and Gunn, R. B. (1972). *Arch. Intern. Med.*, **129,** 241
124. Schwartz, A., Lindenmayer, G. E. and Allen, J. C. (1972). *Current Topics in Membranes and Transport*, **3,** 1 (F. Bronner and A. Kleinzeller, editors) (New York: Academic Press)
125. Hokin, L. E. and Dahl, J. L. (1972). *Membrane Transport*, 270 (L. E. Hokin, editor) (New York: Academic Press)
126. Martonosi, M. (1972). *Membrane Transport*, 317 (L. E. Hokin, editor) (New York: Academic Press)
127. Wasserman, R. H. (1972). *Membrane Transport*, 351 (L. E. Hokin, editor) (New York: Academic Press)
128. Skou, J. C. (1957). *Biochim. Biophys. Acta*, **23,** 394
129. Bader, H., Post, R. L. and Bond, G. H. (1968). *Biochim. Biophys. Acta*, **150,** 41
130. Bonting, S. L. (1970). *Membranes and Ion Transport*, **1,** 257 (E. E. Bittar, editor) (London: John Wiley)
131. Glynn, I. M. (1964). *Pharmacol. Rev.*, **16,** 381
132. Charnock, J. S., Rosenthal, A. and Post, R. (1963). *Australian. J. Exp. Biol.*, **41,** 675
133. Albers, R. W., Koval, G. J. and Siegel, G. J. (1968). *Mol. Pharmacol.*, **4,** 324
134. Glynn, I. M. (1968). *Brit. Med. Bull.*, **24,** 165
135. Glynn, I. M., Lew, V. L. and Lüthi, U. (1970). *J. Physiol.*, **207,** 371
136. Glynn, I. M., Hoffman, J. F. and Lew, V. L. (1971). *Phil. Trans. Roy. Soc. Lond.*, **B262,** 91
137. Pitts, B. J. R. and Askari, A. (1973). *Arch. Biochem. Biophys.*, **154,** 476
138. Hafkenscheid, J. C. M. and Bonting, S. L. (1971). *Comp. Biochem. Physiol.*, **39B,** 955
139. Brown, H. D., Chattopadhyay, S. K. and Patel, A. (1967). *Enzymologia*, **32,** 205
140. Sen, A. K. and Post, R. L. (1964). *J. Biol. Chem.*, **239,** 345
141. Baker, P. F. (1965). *J. Physiol.*, **180,** 383
142. Whittam, R. and Ager, M. E. (1965). *Biochem. J.*, **97,** 214
143. Garrahan, P. J. and Glynn, I. M. (1967). *J. Physiol.*, **192,** 217
144. Charnock, J. S. and Post, R. L. (1963). *Nature (London)*, **199,** 910
145. Albers, W. R., Fahn, S. and Koval, G. J. (1963). *Proc. Nat. Acad. Sci. USA*, **50,** 474
146. Post, R. L., Sen, A. K. and Rosenthal, A. S. (1965). *J. Biol. Chem.*, **240,** 1437
147. Post, R. L. and Kume, S. (1973). *J. Biol. Chem.*, **248,** 6993
148. Fahn, S., Koval, G. J. and Albers, R. W. (1966). *J. Biol. Chem.*, **241,** 1882
149. Fahn, S., Hurley, M. R., Koval, G. J. and Albers, R. W. (1966). *J. Biol. Chem.*, **241,** 1890

150. Post, R. L., Kume, S., Tobin, T., Orcutt, B. and Sen, A. K. (1969). *J. Gen. Physiol.*, **54,** 306s
151. Post, R. L., Hegyvary, C. and Kume, S. (1972). *J. Biol. Chem.*, **247,** 6530
152. Siegel, G. J. and Goodwin, B. (1972). *J. Biol. Chem.*, **247,** 3630
153. Nørdly, J. G. and Jensen, J. (1971). *Biochim. Biophys. Acta*, **233,** 104
154. Hegyvary, C. and Post, R. L. (1971). *J. Biol. Chem.*, **246,** 5234
155. Schoner, W., Beusch, R. and Kramer, R. (1968). *Eur. J. Biochem.*, **7,** 102
156. Brinley, F. J. Jr. and Mullins, L. J. (1968). *J. Gen. Physiol.*, **52,** 181
157. Glynn, I. M. and Hoffman, J. F. (1971). *J. Physiol.*, **218,** 239
158. Matsui, H. and Schwartz, A. (1968). *Biochim. Biophys. Acta*, **151,** 655
159. Schwartz, A., Matsui, H. and Laughter, A. H. (1968). *Science*, **159,** 323
160. Schwartz, A., Matsui, H. and Laughter, A. H. (1968). *Science*, **160,** 323
161. Sen, A. K., Tobin, T. and Post, R. L. (1969). *J. Biol. Chem.*, **244,** 6596
162. Tobin, T. and Sen, A. K. (1970). *Biochim. Biophys. Acta*, **198,** 120
163. Lindenmayer, G. E., Laughter, A. H. and Schwartz, A. (1968). *Arch. Biochem. Biophys.* **127,** 187
164. Siegel, G. J., Koval, G. J. and Albers, R. W. (1969). *J. Biol. Chem.*, **244,** 3264
165. Hansen, O. (1971). *Biochim. Biophys. Acta*, **233,** 122
166. Dunham, P. B. and Hoffman, J. F. (1971). *J. Gen. Physiol.*, **58,** 94
167. Garrahan, P. J. and Glynn, I. M. (1967). *J. Physiol.*, **192,** 159
168. Post, R. L. and Sen, A. K. (1965). *J. Histochem. Cytochem.*, **13,** 105
169. Simons, T. J. B. (1974). *J. Physiol.*, **237,** 123
170. Dahms, A. S. and Boyer, P. D. (1973). *J. Biol. Chem.*, **248,** 3155
171. Judah, J. D., Ahmed, K. and McLean, A. E. M. (1962). *Biochim. Biophys. Acta*, **65,** 472
172. Bader, H. and Sen, A. K. (1966). *Biochim. Biophys. Acta*, **118,** 116
173. Rega, A. F., Garrahan, P. J. and Pouchan, M. I. (1970). *J. Membrane Biol.*, **3,** 14
174. Garrahan, P. G. and Rega, A. F. (1972). *J. Physiol.*, **223,** 595
175. Israel, Y. and Titus, E. (1967). *Biochim. Biophys. Acta*, **139,** 450
176. Glynn, I. M. and Lew, V. L. (1970). *J. Physiol.*, **207,** 393
177. Askari, A. and Koyal, D. (1968). *Biochem. Biophys. Res. Commun.*, **32,** 227
178. Kyte, J. (1971). *J. Biol. Chem.*, **246,** 4157
179. Jørgensen, P. L., Skou, J. C. and Solomonson, L. P. (1971). *Biochim. Biophys. Acta*, **233,** 381
180. Jørgensen, P. L. (1972). *Role of Membranes in Secretory Processes*, 247 (L. Bolis, R. E. Keynes and W. Wilbrandt, editors) (Amsterdam: North Holland)
181. Hokin, L. E., Dahl, J. L., Deupree, J. D., Dixon, J. F., Hackney, J. F. and Perdue, J. F. (1973). *J. Biol. Chem.*, **248,** 2593
182. Kyte, J. (1971). *Biochem. Biophys. Res. Commun.*, **43,** 1259
183. Kyte, J. (1972). *J. Biol. Chem.*, **247,** 7642
184. Kyte, J. (1972). *J. Biol. Chem.*, **247,** 7634
185. Kaback, H. R. and Milner, L. S. (1970). *Proc. Nat. Acad. Sci. USA*, **66,** 1008
186. Konings, W. N., Barnes, E. M. Jr. and Kaback, H. R. (1971). *J. Biol. Chem.*, **246,** 5857
187. Kaback, H. R. and Barnes, E. M. Jr. (1971). *J. Biol. Chem.*, **246,** 5523
188. Simoni, R. D. and Shallenberger, M. K. (1972). *Proc. Nat. Acad. Sci. USA*, **69,** 2663
189. Hong, J.-S. and Kaback, H. R. (1972). *Proc. Nat. Acad. Sci. USA*, **69,** 3336
190. Kaback, H. R. (1968). *J. Biol. Chem.*, **243,** 3711
191. Van Thienen, G. and Postma, P. W. (1973). *Biochim. Biophys. Acta*, **323,** 429
192. Prezioso, G., Hong, J.-S., Kerwar, G. K. and Kaback, H. R. (1973). *Arch. Biochem. Biophys.*, **154,** 575
193. Klein, W. L. and Boyer, P. D. (1972). *J. Biol. Chem.*, **247,** 7257
194. Harold, F. M., Baarda, J. R., Baron, C. and Abrams, A. (1969). *J. Biol. Chem.*, **244,** 2261
195. Abrams, A., Smith, J. B. and Baron, C. (1972). *J. Biol. Chem.*, **247,** 1484
196. Lund, E. J. (1928). *J. Exp. Zool.*, **51,** 265
197. Lundegärdh, H. (1954). *Symp. Soc. Exp. Biol.*, **8,** 262
198. Conway, E. J. (1953). *Int. Rev. Cytol.*, **2,** 419
199. Mitchell, P. (1961). *Nature* (*London*), **191,** 144
200. Mitchell, P. (1966). *Biol. Rev.*, **41,** 445
201. Mitchell, P. (1970). *Membranes and Ion Transport*, **1,** 192 (E. E. Bittar, editor) (New York: John Wiley)

202. Greville, G. D. (1969). *Current Topics in Bioenergetics*, **3,** 1 (D. R. Sanadi, editor) (New York: Academic Press)
203. Harold, F. M. (1972). *Bacteriol. Rev.*, **36,** 172
204. Racker, E., Burstein, C., Loyter, A. and Christiansen, R. E. (1970). *Electron Transport and Energy Conservation*, 235 (J. M. Tager, S. Papa, E. Quagliariello and E. C. Slater, editors) (Bari: Adriatica Editrice)
205. West, I. C. and Mitchell, P. (1973). *Biochem. J.*, **132,** 586
206. Kashket, E. R. and Wilson, T. H. (1973). *Proc. Nat. Acad. Sci. USA*, **70,** 2866
207. West, I. C. and Wilson, T. H. (1973). *Biochem. Biophys. Res. Commun.*, **50,** 551
208. Asghar, S. S., Levin, E. and Harold, F. M. (1973). *J. Biol. Chem.*, **248,** 5225
209. Hirata, H., Altendorf, K. and Harold, F. M. (1973). *Proc. Nat. Acad. Sci. USA*, **70,** 1804
210. Jagendorf, A. T. and Uribe, E. (1966). *Proc. Nat. Acad. Sci. USA*, **55,** 170
211. Harold, F. M., Pavlasova, E. and Baarda, J. R. (1970). *Biochim. Biophys. Acta*, **196,** 235
212. Kashket, E. R. and Wong, P. T. S. (1969). *Biochim. Biophys. Acta*, **193,** 212
213. White, S. H. and O'Brien, W. M. (1972). *Biochim. Biophys. Acta*, **255,** 780
214. Harold, F. M. and Papineau, D. (1972). *J. Membrane Biol.*, **8,** 27
215. Slayman, C. L. (1965). *J. Gen. Physiol.*, **48,** 69
216. Schultz, S. G., Wilson, N. L. and Epstein, W. (1962). *J. Gen. Physiol.*, **46,** 159
217. Williams, R. J. P. (1970). *Electron Transport and Energy Conservation*, 7 (M. J. Tager, S. Papa, E. Quagliariello and E. C. Slater, editors) (Bari: Adriatica Edizione)
218. Riklis, E. and Quastel, J. H. (1958). *Can. J. Biochem. Physiol.*, **36,** 347
219. Riggs, T. R., Walker, L. M. and Christensen. (1958). *J. Biol. Chem.*, **233,** 1497
220. Crane, R. K., Miller, D. and Bihler, I. (1961). *Membrane Transport and Metabolism*, 439 (A. Kleinzeller and A. Kotyk, editors) (New York: Academic Press)
221. Schultz, S. G. and Curran, P. F. (1970). *Physiol. Rev.*, **50,** 637
222. *Na-linked Transport of Organic Solutes.* (1972). (E. Heinz, editor) (Berlin: Springer-Verlag)
223. Kimmich, G. A. (1973). *Biochim. Biophys. Acta*, **300,** 31
224. Jacquez, J. A. and Schafer, J. A. (1969). *Biochim. Biophys. Acta*, **193,** 368
225. Potashner, S. J. and Johnstone, R. M. (1971). *Biochim. Biophys. Acta*, **233,** 103
226. Eddy, A. A. (1968). *Biochem. J.*, **108,** 489
227. Kimmich, G. A. (1970). *Biochemistry*, **9,** 3669
228. Schafer, J. A. and Heinz, E. (1971). *Biochim. Biophys. Acta*, **249,** 15
229. Terry, P. M. and Vidaver, G. A. (1973). *Biochim. Biophys. Acta*, **323,** 441
230. Eddy, A. A. (1968). *Biochem. J.*, **108,** 195
231. Gibb, L. E. and Eddy, A. A. (1972). *Biochem. J.*, **129,** 979
232. Rose, R. C. and Schultz, S. G. (1971). *J. Gen. Physiol.*, **57,** 639
233. Lassen, U. V., Nielsen, A. M. T., Pape, L. and Simonsen, L. O. (1971). *J. Membrane Biol.*, **6,** 269
234. Goldner, A. M., Schultz, S. G. and Curran, P. F. (1969). *J. Gen. Physiol.*, **53,** 362
235. Curran, P. F., Schultz, S. G., Chez, R. A. and Fuisz, R. E. (1967). *J. Gen. Physiol.*, **50,** 1261
236. Vidaver, G. A. (1964). *Biochemistry*, **3,** 803
237. Wheeler, K. P. and Christensen, H. N. (1967). *J. Biol. Chem.*, **242,** 3782
238. Koser, B. H. and Christensen, H. N. (1971). *Biochim. Biophys. Acta*, **241,** 9
239. White, J. F. and Armstrong, W. McD. (1970). *Biophys. Soc. Abstr.*, 36a
240. Morville, M., Reid, M. and Eddy, A. A. (1973). *Biochem. J.*, **134,** 11
241. Berger, E. A. and Heppel, L. A. (1974). *J. Biol. Chem.*, (in the press)
242. Cowell, J. L. (1974). *Abstracts Annu. Meeting Amer. Soc. Microbiol.*, 150
243. Curtis, S. J. (1974). *Fed. Proc.*, **33,** 1325
244. Epstein, W. and Kim, B. S. (1971). *J. Bacteriol.*, **108,** 639
245. Bhattacharyya, P., Epstein, W. and Silver, S. (1971). *Proc. Nat. Acad. Sci. USA*, **68,** 488
246. Lombardi, F. J., Reeves, J. P. and Kaback, H. R. (1973). *J. Biol. Chem.*, **248,** 3551

11
Phase Transitions in Model Systems and Membranes

C. F. FOX
University of California, Los Angeles

Abbreviations

DMPC	dimyristoylphosphatidylcholine
DPPC	dipalmitoylphosphatidylcholine
DPPE	dipalmitoylphosphatidylethanolamine
DSC	differential scanning calorimetry
DSPC	distearoylphosphatidylcholine
e.s.r.	electron spin resonance
NDV	Newcastle disease virus
PC	phosphatidylcholine
PE	phosphatidylethanolamine
PG	phosphatidylglycerol
PL	phospholipid
t_h	higher characteristic temperature
t_l	lower characteristic temperature

11.1 INTRODUCTION

The influence of lipid composition and hence lipid structure on the behaviour of catalytic components in membranes can often be assessed from the properties of either the polar head groups or the hydrocarbon chains. The scope of this article will be limited primarily to the effects of hydrocarbon chain composition and temperature on structure–function relationships in membranes. The effects of phospholipid head group composition on the catalytic activity of biomembrane components have been considered in a number of reviews[1-3] and certain aspects are treated in this volume in chapters by Strominger, Eisenman and Epstein.

11.2 THE MELTING BEHAVIOUR OF BINARY PHOSPHOLIPID SYSTEMS

This discussion deals exclusively with binary phospholipid systems, since these are the simplest multicomponent models that can provide a physical rationale for understanding temperature-dependent phase behaviour in membranes. For more thorough recent treatments of the extensive literature on lipid phase transitions, the reader may consult Refs. 4–9.

11.2.1 The phase diagram for an ideal two-component system

The literature on phase behaviour of membrane lipids and the biological consequences of this behaviour is filled with references to 'phase transitions' and 'transition temperatures'. The term phase transition has been used both to describe the beginning or end of the course of melting of membrane lipids and to describe physiological phenomena associated with the melting process. In this review the term phase transition will be used to define the course of melting of a laterally contiguous hydrocarbon phase. Before proceeding further, it is worthwhile to explicitly define the other terms that will be used in this short discourse and to review briefly the events that occur during melting of an ideal, two-component system.

The melting of a single component, such as a pure phospholipid, proceeds by a sharp transition from the solid to liquid state. A mixture of two components, however, is not characterised by a sharp melting point. Instead, the melting of a solid, two-component ideal mixture occurs over a broad temperature range, beginning and ending at distinct lower and upper characteristic temperatures respectively. If the lower characteristic temperatures for all compositions of the two-component mixture are plotted as a function of the concentration of the higher melting component, these points define the *solidus* curve (Figure 11.1). If the upper characteristic temperatures for

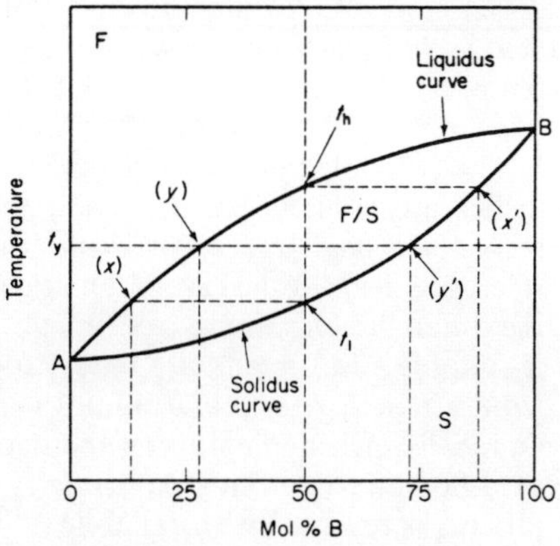

Figure 11.1 Phase diagram describing the melting behaviour of an ideal two-component system composed of A, the lower melting component and B, the higher melting component. Points A and B on the ordinate are the melting temperatures of pure A and pure B respectively. See Section 11.2.1 of the text for details. The regions of the diagram indicating totally liquid or solid solutions are denoted F and S, respectively, and F/S denotes a mixture of liquid and solid solutions

all compositions are plotted in the same way, the points define the *liquidus* curve. The physical state of mixtures of all compositions at temperatures that lie above the *liquidus* curve are liquid or fluid (the region of Figure 11.1 denoted F). Similarly, mixtures of all compositions at temperatures below the *solidus* curve are in a solid or frozen state (denoted S). The physical state at temperatures between the *liquidus* and *solidus* curves will be a mixture of solid and liquid (the area denoted F/S).

In the example shown, a line has been drawn perpendicular to the abcissa at 50 mol % B. This line intersects with points t_l and t_h on the *solidus* and *liquidus* curves respectively, and these slope intercepts are the lower and upper characteristic temperatures of the course of melting of this mixture. The 50:50 mixture of A and B remains solid until heated above t_l, the lower characteristic temperature. At this temperature a liquid phase first forms. The composition of the liquid phase that forms at t_l can be estimated from the phase diagram by drawing a horizontal line through t_l to its intersection with the *liquidus* curve at (x). This corresponds to a 1:7 mixture of A and B in the example shown. When the temperature of the mixture approaches t_h, the composition of the small amount of solid remaining can be estimated by drawing a horizontal line to the *solidus* curve and by dropping a perpendicular line from this point (x') to the abcissa as shown. The compositions of the co-existing liquid and solid phases in mixtures of A and B can be calculated at any temperature between t_l and t_h. For example, the horizontal line extending from t_y intersects with the *liquidus* and *solidus* curves at points (y) and (y'). Thus at temperature t_y, any mixture of A and B will consist of a liquid phase of 73% A and 27% B, and a solid phase of 27% A a 73% B.

11.2.2 Melting behaviour revealed by differential scanning calorimetry

Justification for the use of phase diagrams to describe the melting behaviour of lipid mixtures and membranes evolved initially from the work of Phillips and co-workers who used differential scanning calorimetry (DSC) to study the melting behaviour of a number of binary phospholipid systems in aqueous dispersion[10, 11]. A differential scanning calorimeter measures the transfer of heat to or from a sample when the sample is heated or cooled over a constant temperature programme, e.g. 5–8 °C min^{-1}. In practice, the sample is placed in a small, heat conductive dish, and the sample dish and a reference dish are heated or cooled at the same rate. If the masses of the sample and reference dishes are assumed to be equal, the differential scanning calorimeter will record the heat uptake or evolution of the contents of sample relative to reference dish. In the case of an aqueous dispersion of phospholipid in water, the differential scanning calorimeter will record heat uptake by phospholipid if suitable adjustments are made to correct for heat uptake by or liberation from water. A representation of the melting behaviour of two pure lipids and an ideally melting mixture of the two is shown in Figure 11.2. The mechanics of differential scanning calorimetry are such that only the beginning of a process is revealed with reliability. In an ascending temperature programme (i.e. heating), the temperature at which an endothermic deflection is first observed correlates with the beginning of the melting process. Similarly, the temperature at which an exothermic deflection is first observed during a descending temperature programme correlates with the beginning of freezing (Figure 11.2a and b). For pure phospholipid species such as dipalmitoylphosphatidylcholine (DPPC) or distearoylphosphatidylcholine (DSPC), the initial deflections observed during heating or cooling occur at the same temperature. The DSC curves for the heating or melting of a binary lipid mixture such as DPPC plus DSPC have an appearance similar to the curves shown in Figure 11.2 (a + b). The initial deflections observed upon heating and cooling are different, and the melting (or freezing) occurs over a broad temperature range. As indicated in Figure 11.2 (a + b), the initial endothermic deflection defines the lower characteristic temperature of the phase transition (t_l) for this lipid mixture, and the initial exothermic deflection defines the upper characteristic temperature of the phase transition (t_h). When DSC curves are obtained for a number of compositions of a binary lipid mixture that exhibit ideal melting and freezing behaviour, a plot of all t_l and t_h values as a function of compositions of the mixture (in mol % of the higher melting component) will have the general appearance of Figure 11.1. Phillips, Ladbrooke and Chapman studied the melting and freezing of a number of

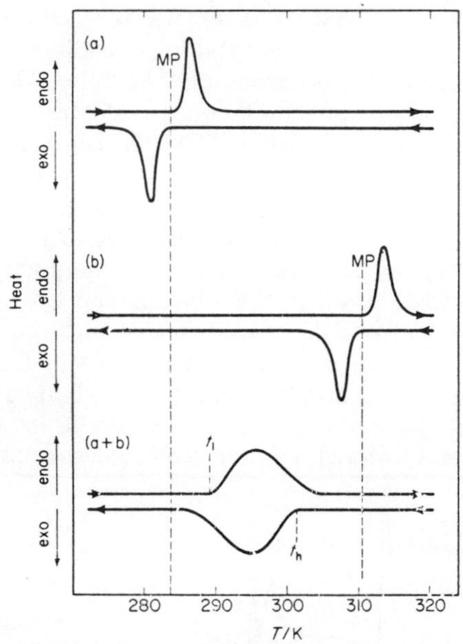

Figure 11.2 Differential scanning calorimetry (DSC) curves for the melting and freezing of two lipids that exhibit ideal mixing properties when a binary mixture of the two is melted or frozen. (a) DSC curves for the melting and freezing of the lower melting lipid. (b) DSC curves for the melting and freezing of the higher melting lipid. Note that in both (a) and (b) the initial endothermic and exothermic deflections from the baseline occur at the same temperatures to define the melting points of these pure lipids. (a + b) DSC curves for the melting and freezing of a mixture of lipids a and b. Note that the initial endothermic and exothermic deflections occur at different and define a broad temperature range over which melting occurs. See Section 11.2.2 for further explanation

binary mixtures of DPPC and DSPC[10]. A plot of the t_l and t_h values *v.* the mol % of DSPC yielded a curve similar to Figure 11.1. Their studies with other binary mixtures of phospholipids, where each phospholipid was substituted with a single fatty acid species, indicated that non-ideal mixing behaviour occurs when the chain length difference of the fatty acids making up the phospholipids is greater than 2 carbon atoms. Non-ideal behaviour was also observed for a mixture of distearoyl- and dioleoyl-lecithins.

11.2.3 Melting behaviour revealed by electron spin resonance (e.s.r.)

The melting behaviour of phospholipid systems can be assessed by a variety of techniques in addition to DSC. The most widely used of these which can give good definition of both the upper and lower characteristic temperatures of phase transitions has been electron spin resonance. This application of e.s.r. is based on the observation by Hubbell and McConnell[12] that the distribution of the nitroxide spin label TEMPO (an acronym for 2,2,6,6-tetramethylpiperidine 1-oxyl) between aqueous and hydrocarbon phases can be determined from its e.s.r. spectrum (Figure 11.3). The high-field line of the first derivative paramagnetic spectrum of nitroxide probes such as TEMPO and 5N10[13] can exhibit hydrocarbon and polar components, and the concentrations of a nitroxide probe in the aqueous and hydrocarbon compartments can be estimated from the amplitudes of these components (Figure 11.3). In their early studies on the partitioning of the spin label probe TEMPO between aqueous and hydrocarbon compartments, Hubbell and McConnell reported that the relative concentration of the probe in water is increased when the temperature is lowered, and this led to the conclusion that TEMPO

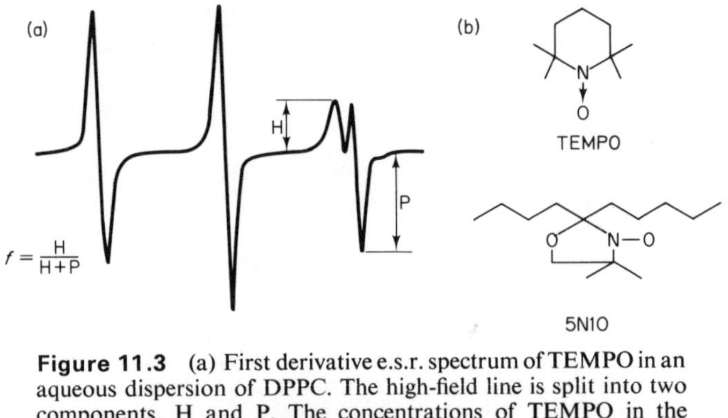

Figure 11.3 (a) First derivative e.s.r. spectrum of TEMPO in an aqueous dispersion of DPPC. The high-field line is split into two components, H and P. The concentrations of TEMPO in the hydrocarbon and polar phases are proportional to the amplitudes of the H and P components and are estimated by measuring the line heights. The fraction of TEMPO in the hydrocarbon phase will thus be proportional to the TEMPO spectral parameter (f) which is the ratio of H:H + P.

(b) Structural formulas of the nitroxide spin resonance probes TEMPO and 5N10.

(Figure 11.3a from Shimshick and McConnell[6], by courtesy of *Biochemistry*)

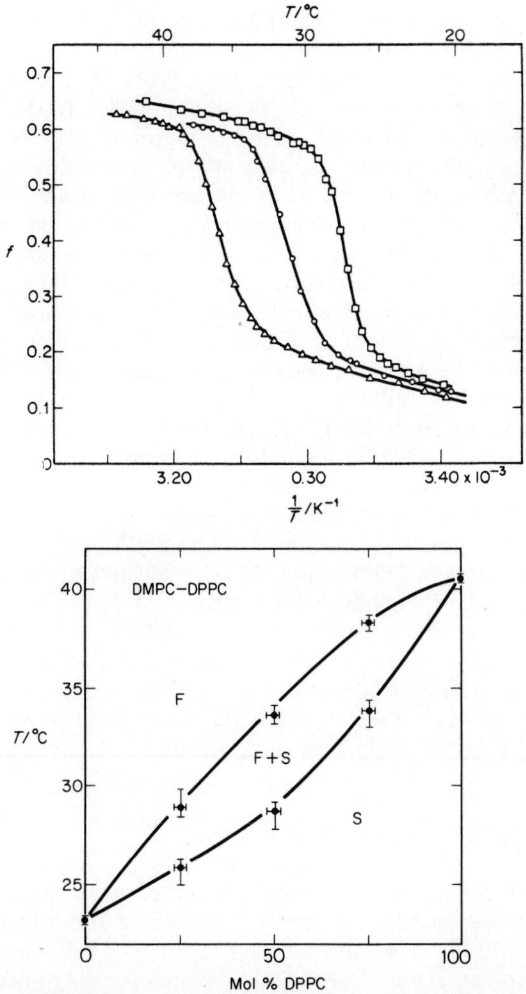

Figure 11.4 Determination of the fluid–solid equilibrium phase diagram for a mixture of DMPC and DPPC by e.s.r.

(a) The TEMPO spectral parameter (f) as a function of $1/T$ for: (△) 76 mol % DPPC, (○) 51 mol % DPPC and (□) 26 mol % DPPC. The upper and lower characteristic temperatures of the phase transition for each can be determined from the slope intercepts obtained when tangents are drawn through the three linear portions of each of the curves.

(b) The fluid–solid equilibrium phase diagram for the DPPC–DMPC binary mixture.

(From Shimshick and McConnell[6] by courtesy of *Biochemistry*) is more soluble in a fluid than in a solid hydrocarbon phase. These studies were extended by Shimshick and McConnell who determined the effects of temperature on the partitioning of TEMPO between water and the hydrocarbon phase of pure, synthetic phospholipids singly or in binary mixtures[6].

Shimshick and McConnell expressed the results of their TEMPO partitioning studies using a spectral parameter that is the fraction of the spin label in the hydrocarbon phase (cf. legend of Figure 11.3). When the localisation of spin label probe is being assessed in a phospholipid–water dispersion of fixed proportions and a decrease in the spectral parameter is observed, this indicates a shift of spin label from the hydrocarbon to the aqueous phase. The change of value of the spectral parameter as a function of temperture was studied with aqueous dispersions of pure phospholipids such as DPPC and DMPC (dimyristoylphosphatidylcholine). At the known melting temperatures of these lipids, the plot of the spectral parameter became almost parallel to the ordinate when the construction of the curves was as shown for the binary phospholipid dispersion in Figure 11.4a. Figure 11.4a describes the TEMPO spectral parameter as a function of temperature for three aqueous dispersions of DMPC and DPPC. The upper and lower characteristic temperatures for the melting process were derived from these data as described in the figure legend and were used to construct a phase diagram (Figure 11.4b). The points on the phase diagram at 0 and 100 mol % DPPC are the melting points of DMPC and DPPC respectively. Studies with other binary phospholipid dispersions in water, such as DMPC–DPPE and DMPC–DSPC, revealed non-ideal behaviour such as that which Phillips *et al.*[10] observed using DSC. The non-ideal behaviour detected by both groups resulted in the generation of phase diagrams where the lower-melting region of the solidus curve lies parallel or nearly parallel to the abcissa, and indicates the existence of immiscible S phase components.

11.2.4 Lateral lipid phase separations

McConnell and his co-workers have concluded that lateral lipid phase separations must occur in membranes[6, 14]. Kornberg and McConnell demonstrated that the lipids of separate monolayers of a bilayer membrane exchange ('flip-flop') at far too slow a rate to account for the rapidity with which equilibrium mixing of lipids occurs during the melting of model lipid bilayer systems[15] (see Chapter 4). Since lipids are constrained from moving vertically between monolayers of a bilayer, it is obvious that if lipids are to move at all they must do so laterally. This movement has been termed lateral diffusion. A number of groups of investigators have demonstrated that lateral lipid diffusion in a bilayer in the fluid state (F state, Figure 11.1) is an extremely rapid process, a molecule of lipid being able to exchange places with a neighbouring molecule at a frequency of greater than 10^7 s^{-1} [14]. This rate of lateral lipid motion within a monolayer can adequately account for the rapid kinetics of melting and freezing of lipids in membranes and aqueous dispersions observed by x-ray diffraction techniques[16], and therefore explains the rapid intermixing between the lipids in the F and S states (Figure 11.1) that must occur when a bilayer composed of a binary or more complicated lipid mixture is heated or cooled within the temperature range defined by the upper and lower characteristic temperatures of the phase transition.

Since Edidin has recently written a comprehensive review on lateral diffusion in membranes and lipid bilayers[17], and since lateral diffusion of lipids and other membrane components has become widely accepted, it will not be further described here. It is nevertheless worthwhile to mention a few unresolved conceptual problems related to the lateral diffusion process. One such problem is the potential lateral motion of membrane components within lipid in the S state. Much of the recent literature in the lectin receptor field contains conclusions based on the assumption that movement of lectin receptors in transformed mammalian cells is severely impeded when membrane lipids are in the S state[18]. In one well-documented study, however, the diffusion of certain cell surface antigens appears to proceed readily below a temperature where lectin receptor mobility may be impeded[19]. A second enigma is the restriction of cell surface receptor mobility characteristic of normal mammalian cells, but not of their oncogenic transformed counterparts when membrane lipids are in the F state[20, 21]. Further details of the experiments that gave rise to these issues will be presented in the following sections. The issues were stated at this point so the reader can deal with additional background information from a position of some understanding of the current research problems in the area.

11.3 LIPID PHASE TRANSITIONS IN CYTOPLASMIC MEMBRANES OF *E. coli*

Up to this point, we have considered the melting of phospholipid bilayer systems composed of either one or two components. The former have a sharp melting point and the latter melt over a broad range bounded by upper and lower characteristic temperatures. Though a biological membrane has components other than lipids, it is possible to treat conceptually its melting behaviour like that of a two-component system. The validity of this type of treatment is apparent from the melting behaviour of membranes from *E. coli*, some of which approximate a binary lipid system.

11.3.1 Fatty acid composition of phospholipids from *E. coli* membranes

In our studies on structure–function relationships in membranes, we have used an unsaturated fatty acid auxotroph of *E. coli* which is incapable of either synthesising unsaturated fatty acids, or of modifying the unsaturated fatty acid that is provided in the growth medium. Furthermore, the phospholipids from membranes of this organism contain little saturated fatty acid other than palmitic acid. Since phosphatidylethanolamine (PE) accounts for nearly 90% of the total membrane lipid in this organism and since phosphatidylglycerol (PG) accounts for most of the rest, the result is a membrane of relatively simple composition (Table 11.1). Compositional analysis of diglycerides derived from cytoplasmic membranes obtained from cells grown with certain supplements reveals what is essentially a binary lipid system on the basis of fatty acid distribution (membranes derived during growth with oleate or elaidate). Thus with the exception of possible contributions from the anionic lipid components, certain of these membrane preparations should have physical properties approximating those of model binary phospholipid systems.

Table 11.1 Distribution of saturated and unsaturated fatty acids in diglycerides derived from cytoplasmic membrane phospholipids of an *E. coli* unsaturated fatty acid auxotroph

Fatty acid supplement for growth	*Disaturated*	*Saturated: Unsaturated*	*Diunsaturated*
Oleate	1.2%	79.1%	19.7%
Linoleate	40.8%	19.8%	39.4%
Elaidate	<0.1%	31.3%	68.7%

An extract of inner membrane phospholipids was digested with *Clostridium welchii* phospholipase C. Comparison of the starting and digested materials revealed complete digestion of phosphatidylethanolamine and phosphatidylglycerol (which account for more than 98% of the phospholipids in this membrane). The diglycerides were resolved by thin layer chromatography on silver nitrate impregnated silicic acid and the fatty acid content in each diglyceride was determined by gas–liquid chromatography. (From Linden *et al.*[13], by courtesy of *J. Supramolec. Struct.*) *E. coli* phospholipids which contain one saturated and one unsaturated fatty acid are substituted with the unsaturated fatty acid at the 2-carbon of glycerol[22]. Thus saturated: unsaturated diglyceride represents a single species

11.3.2 Physical properties of *E. coli* membrane phospholipids *in situ* and in aqueous dispersion

The TEMPO spectral parameter (f) was determined with the cytoplasmic membrane preparations for which compositional analyses were described in Table 11.1. All three membrane preparations yielded plots of f *v.* $1/T$ similar to those obtained with model binary systems (Figure 11.4a). A distinct upper and lower characteristic temperature was revealed for each membrane preparation and the temperature interval between t_l and t_h increased as expected from the fatty acid composition of the membrane phospholipids. Dielaidoyl PL, for example, would be expected to have a melting point closer to that of 1-palmitoyl,2-elaidoyl PL than would dioleoyl PL to 1-palmitoyl, 2-oleoyl PL. It therefore follows logically that the t_l to t_h interval is greater in membranes containing oleic acid than in those containing elaidic acid. The greatest interval between t_l and t_h was detected in membranes derived from cells grown with a linoleic acid supplement. Though the membrane lipids of cells grown with linoleate constituted a ternary system, only two characteristic temperatures were detected in the TEMPO spectral parameters. The results with the linoleate-containing membranes indicate that the rationale developed from binary lipid systems is also applicable to membranes with a more complex lipid composition.

The melting properties of the membranes are similar to those of aqueous

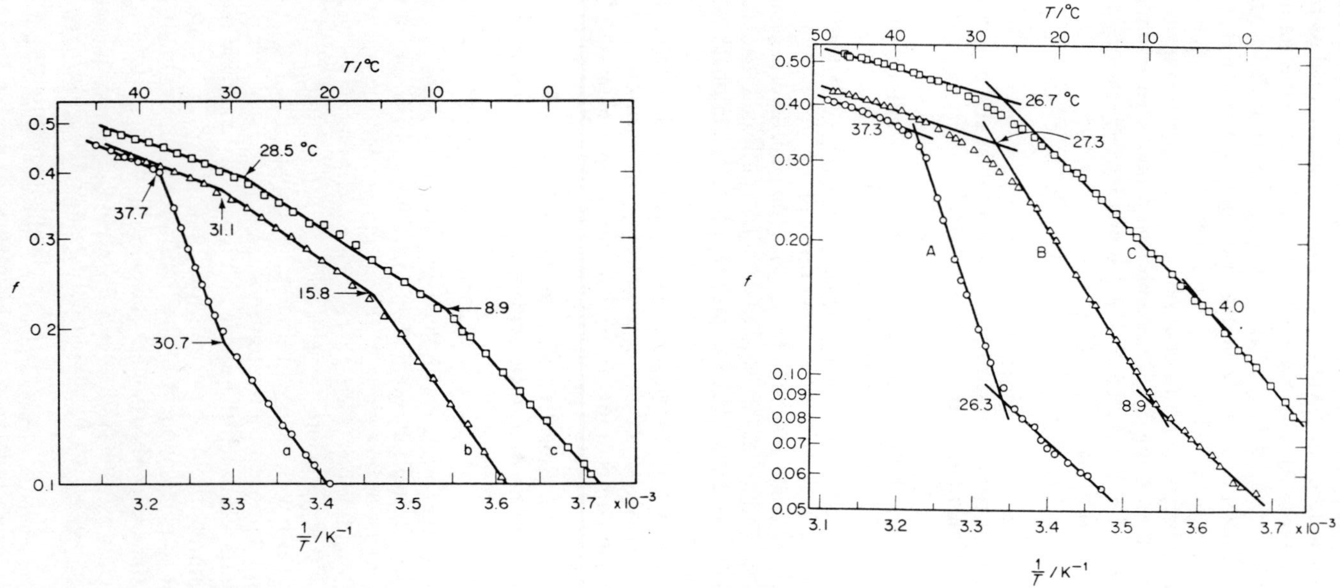

Figure 11.5 (a) The TEMPO spectral parameter (f) as a function of the reciprocal of the absolute temperature ($1/T$) for membranes with phospholipid compositions described in Table 11.1. The cells were grown at 37°C in media supplemented with (a) elaidic acid, (b) oleic acid and (c) linoleic acid. The spectral parameter (f) is an approximate measure of the solubility of the spin-label TEMPO in the *fluid* region of the membrane lipids.

(b) The TEMPO spectral parameter (f) as a function of the reciprocal of the absolute temperature ($1/T$) for aqueous dispersions of phospholipids extracted from these membranes. Lipid extracts of membranes from cells grown with: (A) elaidic acid, (B) oleic acid and (C) linoleic acid.

(From Linden *et al.*[23] by courtesy of *Proc. Nat. Acad. Sci.* (*USA*)) dispersions of phospholipids extracted from them (Figure 11.5). The comparable t_h values are generally within 1–2 °C, but the t_l values for phospholipids in aqueous dispersions are uniformly lower than those exhibited by the phospholipids in membrane. Among the explanations for the consistently lower t_l observed in phospholipid dispersions as opposed to membranes are: (1) The population of vesicles contains some that are sufficiently small that t_l is decreased as predicted by the Kelvin equation of curved surfaces[24]. Since large vesicles would also be present, this could broaden the phase transition by decreasing t_l without affecting t_h. (2) Membrane proteins function to stabilise the membrane lipids. This is the conclusion predicted by the boundary lipid concept developed by Jost, Griffith and co-workers[25, 26].

11.3.3 Correlations between membrane physical state and the effect of temperature on transport rate

Attempts to establish correlations between changes in membrane lipid physical state and function were first approached in the study of galactoside and glucoside transport in *E. coli*[27, 28]. Transport was an obvious starting point since the transport system must extend through the membrane to function and would thus come in contact with the internal, hydrocarbon phase of the membrane. Though earlier studies definitely indicated that such correlations were likely to exist[29-32], the precise physical rationale was recognised only recently[23].

The most common means for studying effects of a change of state in the membrane lipids on a catalytic activity is the plotting of reaction rate *v*. the reciprocal of absolute temperature in the Arrhenius representation. An example of this is shown in Figure 11.6, which describes the effects of temperature on the rate of glucoside transport in an unsaturated fatty acid auxotroph grown with linoleic acid, oleic acid or elaidic acid supplements. The Arrhenius plots for glucoside transport by cells grown with linoleic or elaidic acids reveal two distinct discontinuities in slope or slope intercepts (Figure 11.6a and c)*. These correlate with the upper and lower characteristic temperatures of the phase transition revealed by partitioning of the spin label TEMPO between the hydrocarbon phase of the membrane and the aqueous medium (Figure 11.5a and Table 11.2). There is also a correlation between the lower characteristic temperature for glucoside transport and TEMPO partitioning for cells grown with oleic acid (Figure 11.5b and Table 11.2), but the upper characteristic temperature for transport was not clearly revealed. The fidelity of correlation between the upper and lower characteristic temperatures of the membrane lipid phase transition and glucoside transport leave little doubt that some component of the transport system responds to co-operative changes that occur when patches of S phase lipid begin to form in a fluid membrane (at t_h), or when freezing of the membrane lipids is completed (at t_l). This concept is supported by additional physical and physiological

* Slope intercepts or discontinues in Arrhenius plots for a membrane function will be referred to subsequently as characteristic temperatures for that function, e.g. a characteristic temperature for transport.

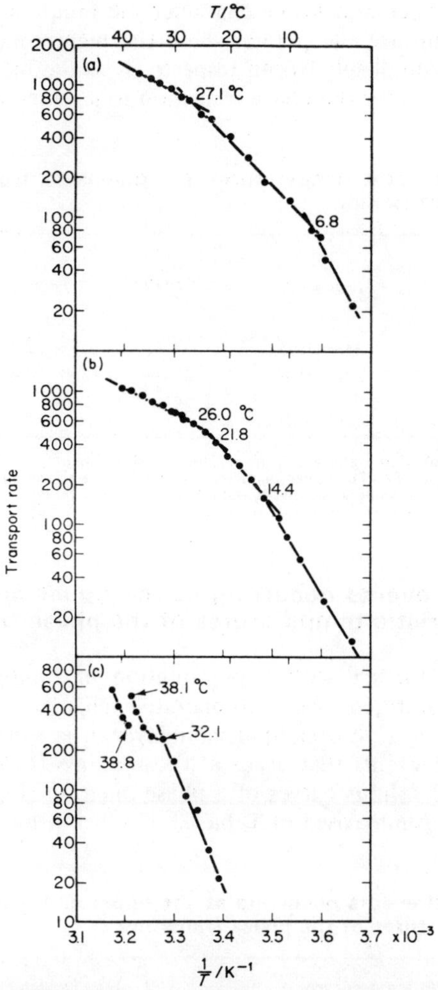

Figure 11.6 Arrhenius plots for β-glucoside transport by cells of an *E. coli* unsaturated fatty acid auxotroph grown in medium supplemented with (a) linoleic acid, (b) oleic acid and (c) elaidic acid. (From Linden *et al.*[23] by courtesy of *Proc. Nat. Acad. Sci.* (*USA*))

data. A second transport system, that responsible for the uptake of β-galactosides, has been studied in the auxotroph described here and the characteristic temperatures for transport by this system also correlate with the characteristic temperatures of the membrane lipid phase transition. The upper and lower characteristic temperatures for the membrane lipid phase transition obtained by TEMPO partitioning for the membrane preparation described in Figure 11.5a have been corroborated by x-ray diffraction studies[33]. Both the low and high angle diffraction patterns obtained at different temperatures indicate that the upper and lower characteristic temperatures revealed by TEMPO partitioning are the points where the membrane lipids begin to freeze and become completely frozen respectively as the membrane is cooled from a temperature above the phase transition to a temperature below it.

Table 11.2 Characteristic temperatures for glucoside transport and spin label partitioning

Fatty acid supplement	t^*_h *Transport*	t_h *TEMPO*	t^*_l *Transport*	t_l *TEMPO*
Elaidate	38.6–38.8	37.7	32.1	30.7
Oleate	26.0,21.8	31.0	14.4	15.8
Linoleate	27.1	28.5	6.8	8.9

The characteristic temperatures for β-glucoside transport and TEMPO partitioning are from the experiments described in Figures 11.5 and 11.6. The upper characteristic temperature of the phase transition is denoted t_h and the lower characteristic temperature, t_l Discontinuities or slope intercepts in Arrhenius plots of β-glucoside transport *v.* $1/T$ which correlate with t_h and t_l are denoted t^*_h and t^*_l

11.3.4 Physical events occurring at the upper and lower characteristic temperatures of the phase transition

Before proceeding further with a presentation of biological phenomena occurring at temperatures where co-operative changes occur throughout contiguous regions of a phospholipid monolayer, it is worthwhile to review briefly the physical events that occur at those points (t_l and t_h) which fall on the liquidus and solidus curves of a phase diagram (Figure 11.1). These physical events are summarised in Table 11.3. When a membrane begins to

Table 11.3 Physical events occurring at the upper and lower characteristic temperatures of the phase transition

Upper characteristic temperature	*Lower characteristic temperature*
Onset of lateral phase separations —begin formation of frozen patches of lipid —lateral phase separation of proteins into F state lipid	End of lateral phase separations —all lipids assume the S state —all membrane protein visualised by electron microscopy after freeze fracture are present in dense patches surrounded by particle free lipid
Onset of enhanced lateral compressibility of membrane lipids (above t_h all lipids are in the liquid expanded state)	End of enhanced lateral compressibility of membrane lipids (below t_l all lipids are in the condensed state)
Begin thickening of membrane (the vertical dimensions of the bilayer increase at regions of S state lipid) (see Chapter 4).	Membrane reaches maximal thickness

The description of events is that which would occur as a consequence of taking a membrane preparation through a decreasing temperature programme proceeding from a temperature above the liquidus curve to a temperature below the solidus curve freeze, patches of S state lipid begin to form. Since bacterial membrane proteins are not attached to an underlying cytoskeletal substratum, they are free to move laterally and are thus likely to remain surrounded by F state lipid at temperatures between t_h and t_l[34]. It is therefore unlikely that alterations which occur at t_h in the rate of transport or other types of enzyme-mediated reactions can be explained by a gradual sequestering of the catalytic membrane proteins in S state lipid. The alterations in transport rate at t_h, including the increase in transport rate observed with decreased temperature in elaidate membranes (Figure 11.5c), are more likely to be caused by co-operative phenomena that result from an onset of the potential for enhanced lipid compressibility in the membrane. This compressibility could lead to cooperative density fluctuations within the membrane lipids and this might facilitate processes involved in catalysis of membrane function, such as the vertical displacement of a membrane protein or changes in protein conformation that would require a lateral displacement of lipids in a monolayer. The alterations observed in transport rate at t_l might also arise from compressibility factors, since the potential for membrane lipid compressibility would reach a minimum when all membrane lipids assume the S state.

Membrane functions that are apparently affected by compressibility considerations, such as transport, can proceed below t_l. There are two ways, however, by which membrane freezing might totally inhibit biological processes. For example, total inhibition might occur in processes which require (1) lateral diffusion of a component of a system or (2) a membrane less thick than that characterised by one in which membrane lipids are in the S state. In the first of these examples, consider a situation in which components must move laterally through the membrane in order to give rise to a biological phenomenon. An example of this might be the agglutination of animal cells which occurs in response to the multivalent plant lectin concanavalin A. This agglutination phenomenon is severely restricted below 15 °C, apparently a lower characteristic temperature of a lipid phase transition in these cells[35, 36]. There are no documented examples of an effect of membrane lipid thickness. Thickening of membrane lipids could interrupt an activity which requires two proteins that lie on opposite sides of the membrane to form a complex which spans the membrane.

11.3.5 Physiological phenomena that respond to events occurring at characteristic temperatures of the phase transition

Summarised in Table 11.4 are a number of physiological phenomena that respond to one or both extremes of the lipid phase transition in bacterial cells. In most cases, i.e. all examples except bacterial growth and membrane assembly, lateral mixing of membrane components and chemotaxis, the response to the designated extreme of the phase transition is a modulation in rate similar to that shown in Figure 11.5 for transport of β-glucosides in *E. coli*. Growth of an unsaturated fatty acid auxoroph of *E. coli* was shown by Overath and his colleagues to have a critical temperature that responds to a change of state in membrane lipids[29]. The growth of another unsaturated fatty acid auxotroph of *E. coli* was studied in the author's laboratory and

Table 11.4 Physiological phenomena responding to events occurring at t_l or t_h in bacterial membranes.

Response to t_h	*Response to t_l*
β-Glucoside transport (23, 37)	β-Glucoside transport (23, 37)
β-Galactoside transport (23, 37)	β-Galactoside transport (23, 37)
	Proline transport (31)
	Succinate-ubiquinone reductase (38)
	Bacterial growth (29, 39, 40)
	Bacterial membrane assembly (39)
	Arrest lateral mixing of membrane components (41)
	Chemotaxis (42)

Certain of the citations given here refer to reports published before the realisation that some membrane functions might have more than one characteristic temperature. The assignment of response to t_l, t_h or both will therefore not be found in all these reports. Some of the pnenomena cited above as responding only to t_l, e.g. proline transport, may also respond to t_h. The failure to detect effects of both extremes of the phase transition on catalytic membrane phenomena was common in studies reported prior to 1973

was observed to have a critical temperature which correlated with t_l [39]. An extensive and thorough study of the effects of lipid composition on bacterial growth rate was conducted by McElhaney who used a Mycoplasma system. The minimum temperature at which cells could grow correlated with the lower characteristic temperature of lipid in membranes from five cultures of cells, each of which had been grown with a different supplement yielding t_l values ranging from 5 to 25 °C [40]. The failure of cells to grow below t_l may arise from defective assembly of essential membrane components. The assembly of the lactose transport system in *E. coli*, for example, is abortive when induction occurs below t_l [39]. Since the soluble enzyme products of genes surrounding the gene for the membrane protein component of the transport system are translated in the same ratio both above and below t_l, some post-translational step in the assembly of this transport system requires fluid membrane lipids. This requirement for fluid membrane lipids is not restricted to the lipids pre-existing in the membrane at the time of transport-system induction, since the transport system can be induced below the t_l of the pre-existing membrane lipids if the membrane lipids formed at the time of transport system induction are themselves fluid[41]. The kinetic properties of the transport system formed when induction occurs under this condition are of interest.

When an unsaturated fatty acid auxotroph is grown sequentially with two different supplements, elaidic and oleic acids respectively, at a growth temperature higher than the t_l values for either of these, the Arrhenius plot for transport induced during a short period of growth with the second supplement (oleic acid) reveals a lower characteristic temperature for transport intermediate between those observed in cells grown with either supplement (Figure 11.7a). This and a similar experiment reported by Overath and his colleagues indicate that membrane components formed before and after the fatty acid shift are capable of mixing by a lateral diffusion process[41, 43, 44]. When induction proceeded at a temperature below the t_l for lipids derived from growth with elaidic acid, but above the t_l for lipids derived from growth with oleic acid, two characteristic temperatures for transport were observed (Figure 11.7b). One of these correlates with the lower characteristic temperature for transport observed for cells grown with oleate alone. This indicates that at least some fraction of the newly formed transport activity is affected by F state oleic acid lipid residing in a matrix of elaidic acid lipid in the S state. When a portion of the cells used for obtaining the data

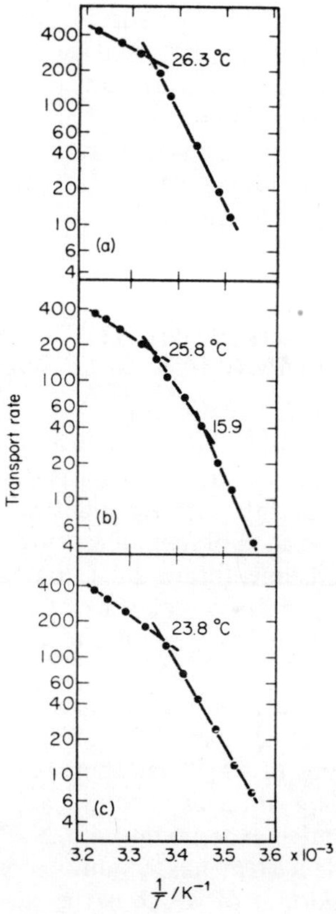

Figure 11.7 The influence of growth temperature on transport induced after shifting from growth with elaidic to growth with oleic acid.

Cells of an unsaturated fatty acid auxotroph of *E. coli* were grown at 37 °C in medium supplemented with elaidic acid, which gives rise to lipid with t_l of 30 °C. The cells were collected by centrifugation and suspended in medium supplemented with oleic acid, which gives rise to lipid with t_l of 14 °C. The cell suspension was then split into two portions. (a) A flask containing one portion of cells was maintained at 37 °C during induction of the lactose transport system. (b) A flask containing the second portion of cells was maintained at 25 °C during induction of the lactose transport system. (c) A portion of the cells grown as described under (b) above was incubated briefly at 37 °C in the presence of chloramphenicol (an inhibitor of further transport system induction) before commencing transport assays. For further details see Section 11.3.3. (From Tsukagoshi and Fox[41] by courtesy of *Biochemistry*)

in Figure 11.7b was incubated above the t_l for the higher melting membrane lipid prior to transport assay, only one characteristic temperature for transport was detected (Figure 11.7c), and this was intermediate between the two shown in Figure 11.7b. This observation further substantiates the conclusion that the experimental conditions described for Figure 11.7b give rise to a membrane containing separate domains of F state lipids derived from oleic acid and S state lipids derived from elaidic acid. Taken collectively, the experiments described in Figures 11.7a–c indicate that lateral diffusion of membrane components formed prior to and after the fatty acid shift did not take place readily below the t_l value for lipids derived from elaidic acid. This suggests that contiguous F state oleate and S state elaidate lipids can form a stable boundary. It is possible, however, that under the conditions of Figure 11.7b the membrane regions derived from oleic acid are incorporated into membrane regions where patches of integral membrane proteins were formed during chilling of the elaidate membranes below their t_l value. The F state oleate lipids and S state elaidate lipids might then be separated by a 'spacer' of membrane protein. The first of these two explanations must be seriously considered, however, because of the results of the complementary experiment in which patches of S state elaidate lipid appeared to form in a matrix of F state oleate lipid[38]. The effects of totally S phase membrane lipids upon the ability of a bacterium to sense and then move in the direction of increasing concentration of an attractant (chemotaxis) is of particular interest because it is the sole phenomenon of an essentially catalytic nature which is totally interrupted at t_l[42]. Since the bacteria are still motile below t_l, the total freezing of the bulk lipid phase apparently interrupts some essential step in the process of attractant recognition.

11.4 CHANGES OF STATE IN LIPIDS OF CULTURED MAMMALIAN CELL MEMBRANES

The lipids in bacterial membranes constitute what appears to be a simple system in which ideal mixing is generally observed and where both monolayers of the bilayer have identical or very similar physical properties. The situation in membranes of cultured mammalian cells is far more complex. The plasma membrane of a mammalian cell appears to exhibit physical asymmetry, each monolayer of the bilayer having a unique phase transition, each of which has a unique t_h and t_l value. The physical properties of the plasma membrane from a mouse fibroblast line and of the membrane of a paramyxovirus, Newcastle disease virus (NDV), have been determined by e.s.r. using the spin label 5N10[45,46]. The characteristic temperatures for partitioning of 5N10 between the membrane hydrocarbon phase and the surrounding aqueous medium are best revealed in plots of h_H/h_p, and a representative experiment with NDV is described in Figure 11.8. (The viral membrane system has a number of advantages for this type of study, the most important of which is the ease with which a virus preparation can be obtained free of contaminating membranes from the host. Since the virus derives its lipids from the host plasma membrane and forms by budding from this membrane, the properties of the viral membrane lipids should be representative of those of the host plasma membrane.) We observe four characteristic temperatures for spin label partitioning both with membranes of NDV and the cytoplasmic membrane fraction from mouse fibroblasts (LM cells). The characteristic temperatures from these two representative plasma membrane fractions are amazingly similar (Table 11.5), suggesting that only minor species-specific variations may occur in cytoplasmic membranes of homeothermic organisms. A fifth characteristic temperature is sometimes observed at approximately 9–10°C. With some membrane

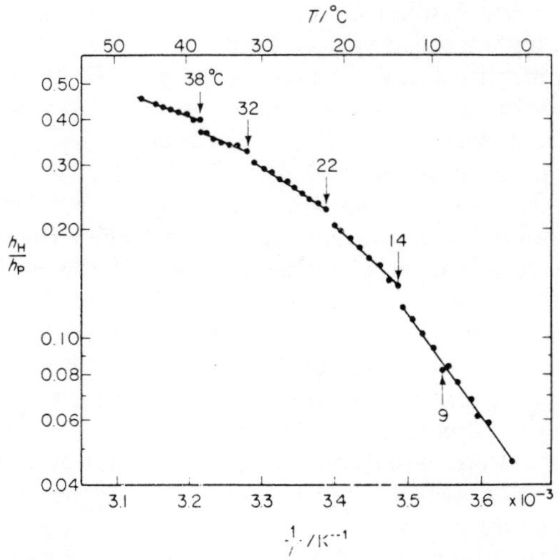

Figure 11.8 Partitioning of 5N10 between the fluid hydrocarbon phase of the membranes of Newcastle disease virus and water as a function of temperature. See Section 11.4 for details

Table 11.5 Physical and physiological characteristic temperatures of animal cell membranes

	Characteristic temperatures (°C)			
5N10 *Partitioning*	$t_{l(o)}$	$t_{l(i)}$	$t_{h(o)}$	$t_{h(i)}$
LM plasma membrane	15	21	30	37
Egg-grown NDV_{HP}	14	22	32	38
Physiological parameters	$t^*_{l(o)}$	$t^*_{l(i)}$	$t^*_{h(o)}$	$t^*_{h(i)}$
a-Aminoisobutyrate transport (LM cells)	16	20	29	37
$Na^+ + K^+ + Mg^{2+}$-ATPase (LM cell plasma membrane)	15	23	28–31	37
$Na^+ + K^+$-ATPase lamb kidney plasma membranes[47]		20		
Con A binding by LM cells[36, 48]	15–19			
Con A binding by 3T3 cells[49]	15			
Con A-mediated haemadsorption by LM cells[36, 48]	13–15			
Con A-mediated agglutination with 3T3 cells[35]	15			
Basal adenyl cyclase†		20		36–38
Hormone-stimulated adenyl cyclase[50]			32	
Rate of mixing of mouse and human cell surface antigens after cell fusion[19]	15‡	21	28–30	

† The characteristic temperature for basal adenyl cyclase at approximately 20°C was observed by J. Keirns (personal communication). The characteristic temperature at approximately 36–38°C can be detected in Figure 1 of the data published by Keirns *et al.*[50]

‡ The rate of antigen mixing, a measure of protein mobility on the cell surface, decreased slowly from 42°C to 30°C. Below 30°C, the rate of antigen mixing declined rapidly to a minimum at 21°C and then increased from 21°C to 15°C. A precipitous drop in the rate of mixing occurred below 15°C. We are grateful to M. Edidin for communicating these data prior to their publication

Data unless otherwise cited are from Wisnieski *et al.*[45]

preparations (LM cells plasma membranes) a change in slope of h_H/h_p is detected at 9 °C, but only a single displaced point is observed at this temperature with NDV membranes. E.S.R. studies with isolated NDV (or LM cell membranes) lipids in aqueous dispersions reveal only two characteristic temperatures. The higher of these is approximately 34 °C, and values for the lower have ranged from 14 to 17 °C.

As stated earlier, the physical and physiological properties of animal cell plasma membranes are consistent with a model where the inner and outer monolayers have independent phase transitions. On the basis of currently available information, we have made assignments as follows: 14(15 °C), lower characteristic temperature of the outer monolayer, $t_{l(o)}$; 21(22 °C), lower characteristic temperature of the inner monolayer, $t_{l(i)}$; 30(32 °C), upper characteristic temperature of the outer monolayer, $t_{h(o)}$; and 37(38 °C), upper characteristic temperature of the inner monolayer, $t_{h(i)}$ (Table 11.5). These assignments were based on the following information. The total amounts of membrane hydrocarbon phase revealed by determination of the change in h_H during freezing was nearly equal over the intervals 14–32 °C and 22–38 °C for NDV membranes. (This was computed by first determining Δh_H/°C over the ranges 32–38 °C and 14–22 °C, and then by multiplying these numbers by the appropriate intervals in °C for the two specified intervals.) Various models in which 9 °C was used as one of four possible characteristic temperatures revealed only one additional set of two intervals (9–22 °C and 14–38 °C) which give rise to roughly equal values for hydrocarbon compartment accessibility to spin label. Since the t_l for extracted membrane lipids is $\geqslant$ 14 °C, the physical data best fit a model of two independent phase transitions occurring in hydrocarbon compartments of roughly equal size with intervals of 14–32 °C and 22–38 °C (Table 11.5). The lower characteristic temperature of one of these intervals (approximately 15 °C) is a critical temperature for cell surface phenomena such as quantitative binding of the lectin concanavalin A (Con A) and Con A-mediated agglutination of cells. Thus 15 °C is likely to be a characteristic temperature of the outer membrane surface. This conclusion is further supported by the properties of adenyl cyclase activity. Basal adenyl cyclase activity (that detected in the absence of hormone) exhibits characteristic temperatures at approximately 20 °C and 37 °C. Since adenyl cyclase is present on the inner surface of the membrane, it is likely that 20 °C and 37 °C define the interval of the phase transition for the inner monolayer of the bilayer. A dramatic alteration in activity is observed near 30 °C for the hormone-sensitive component of adenyl cyclase activity. Since hormones such as glucagon and epinephrine are likely to bind to receptors on the outer membrane surface, it is reasonable to assign a temperature near 30 °C as the upper characteristic temperature of the phase transition for the outer monolayer.

The apparent differences in physical behaviour in the inner and outer monolayers of the membranes of mammalian cells could have their origin either in fatty acid, cholesterol, or polar head group composition. Asymmetry in distribution of cholesterol and phospholipid polar head groups has been reported to occur in erythrocyte membranes[51,52], but we are aware of no reports of asymmetry of fatty acid distribution between the inner and outer surfaces of cell membranes.

11.5 GENERAL CONSIDERATIONS

11.5.1 Boundary lipid

Jost, Griffith and their collaborators have studied the motion of spin labels in cytochrome oxidase preparations containing normal or less than normal amounts of lipid[25, 26]. A spin label was added to the preparation at constant amount of probe per mg of protein and the amount of lipid added to reconstitute the delipidated cytochrome oxidase preparation was varied. In the lipid-poor preparation, the spin label probe was immobilised relative to its mobility in a pure lipid phase. As the ratios of lipid/protein increased, an increased amount of probe partitioned into a more fluid compartment. The same type of observation was made with three different probes, two of which were spin labelled fatty acids, and the third a spin labelled steroid[26]. On the basis of these observations and the amount of lipid remaining associated with the oxidase preparations at which the probe becomes maximally immobilised, the authors concluded that the spin label probes partitioned between the environments for two classes of lipid. These are:

(1) *Boundary lipid.* Boundary lipid exists in a shell immediately surrounding the protein. It is concluded that boundary lipid has decreased mobility compared with bilayer lipid.

(2) *Bilayer lipid.* Bilayer lipid is all lipid other than boundary lipid and is the more mobile of the two species. (This model is strikingly similar to the 'iceberg' hypothesis, in which Klotz attributed unique properties to the shell of water that immediately surrounds a protein in aqueous environment[53].)

The spin label data reported by Jost, Griffith and their colleagues assigns no precise mobility to boundary lipid and can only be taken as an indication that when spin label associates with a lipoprotein species containing no more than sufficient lipid to coat the protein, the spin label binds to sites where it exhibits at least an order of magnitude less motion than if the spin label were present in a phospholipid bilayer containing no protein. Since the more fluid bilayer is known to diffuse laterally at a high rate (sufficient to exchange with a neighbouring lipid molecule 10^7 times s^{-1}), the rate of exchange of bilayer and boundary lipid might still be extremely rapid.

Having ascertained that the lipids which immediately surround an integral membrane protein are less mobile than those more remote from the protein, the next logical consideration is to determine the consequences of this relatively immobile lipid shell. If the lower melting lipid species in a membrane were preferentially associated with protein (i.e. as boundary lipid), this situation might give rise to an increased lower characteristic temperature for the lipid phase transition in the bulk, bilayer lipid, as cited earlier in Section 11.3.2. Whereas this would effect a co-operative phenomenon occurring throughout the entire membrane, some consequences of boundary lipid could be strictly local. Some investigators have determined characteristic temperatures for membrane catalytic activities and have not observed correlations with characteristic temperatures for the membrane lipid phase transition[31, 54]. This has given rise to speculation that certain enzymes might respond to local lipid environments, e.g. where boundary lipid has physical properties different from those of the bulk, bilayer lipid. Though this type of explanation is a viable one, an activity initially thought not to correlate with a characteristic temperature of the lipid phase transition was observed to correlate closely when the enzyme was assayed by an improved technique[31, 38]. Finally, there is no reason why all membrane associated enzymes should be affected by a change of state in membrane lipids. Some membrane proteins have no association with the membrane hydrocarbon phase, and others which do may not be affected by co-operative changes occurring at t_l or t_h.

11.5.2 Lateral diffusion of membrane components in S state lipid

This is an issue of some fundamental importance. The conclusions of many experiments in the cell surface receptor field are based on the assumption that lateral mobility of membrane components such as lectin receptors is essentially arrested below the lower characteristic temperature of the membrane lipid phase transition. One type of experiment which supports the concept that lateral diffusion is arrested in S phase lipid is that described in Figure 11.7. The results of two independent freeze fracture electron microscopic studies are also consistent with the view that lateral motion of membrane components is arrested in S state lipid[55, 56]. Rhodopsin has been isolated from rod outer segments in a form essentially free of lipid and has been reconstituted in bilayers of pure phosphatidylcholine species[55]. When a bilayer containing bleached rhodopsin (rhodopsin that has been exposed to light) is rapidly frozen for freeze cleavage from a temperature where the lipids were initially in the F state, a disperse distribution of particles is observed. If the preparation is first chilled slowly to a temperature below that at which the lipids become totally frozen and is then rapidly chilled as before, a patched distribution of particles is observed. The distribution of particles in dark adapted bilayers, however, is patched if the bilayers are rapidly chilled for freeze fracture from temperatures where the lipids are in the F state. This indicates that dark-adapted rhodopsin tends to aggregate in F state lipid, whereas bleached rhodopsin remains dispersed. Chen and Hubbell studied the dispersal of rhodopsin after bleaching (Figure 11.9). If bleaching occurs when rhodopsin is in F state lipids, it disperses readily, but when rhodopsin residing in S state lipids is bleached, no dispersal is observed. Rhodopsin will disperse, however, in this last case if the sample is heated so that the lipids enter the F state prior to rapid chilling for freeze fracture. Jan and Revel have studied the lateral mobility of rhodopsin in rodent rod outer segments at 4 °C using an antibody specific for rhodopsin[56]. When rod outer segments were treated with an antibody not specific for rhodopsin at either 4 °C or 37 °C and the preparations were fixed prior to freeze fracture, no difference in distribution of intramembranous particles was observed. When rod outer segments were incubated with rhodopsin-specific antibody for 14.5 h at 4 °C prior to fixation and freeze fracture, the appearance of micrographs of replicas was essentially that observed after treatment with non-specific antibody, i.e. there was no evidence for extensive patching. When the rod outer segments were incubated with rhodopsin-specific antibody for

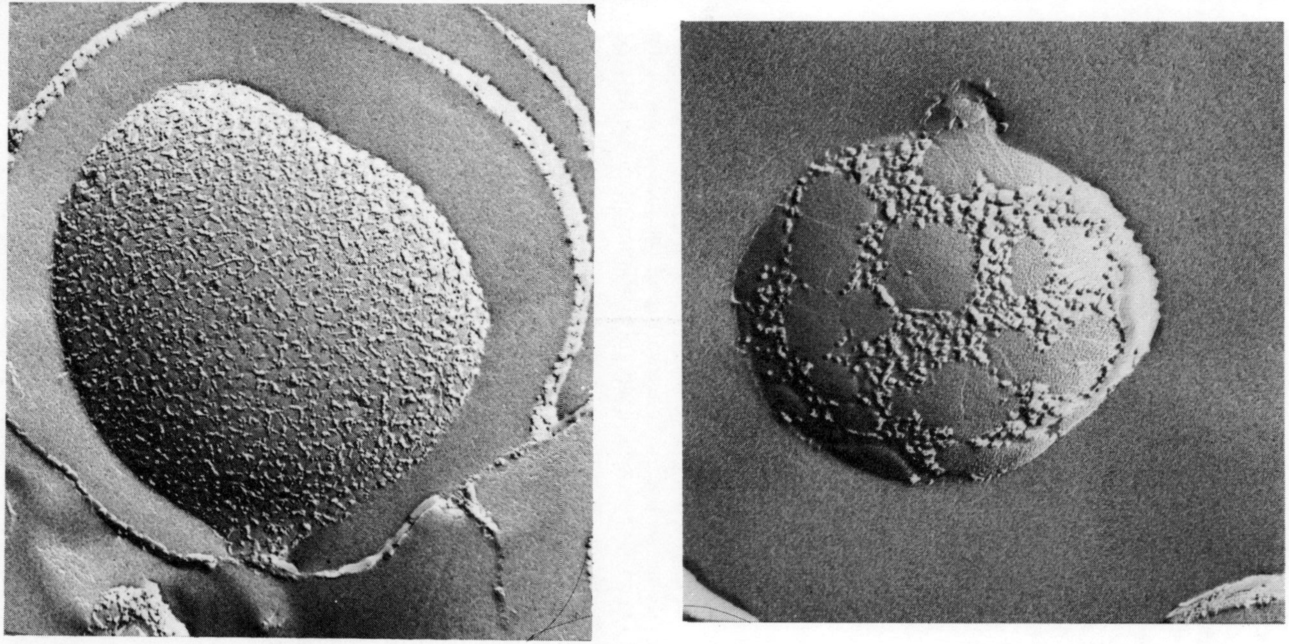

Figure 11.9 Electron micrographs of freeze fracture replicas of model bilayers containing a pure, synthetic phosphatidylcholine species and rhodopsin. The rhodopsin molecules appear as particles in the fracture plane (see Section 11.5.2 for details). The photomicrographs were generously provided by Wayne Hubbell.

LEFT: Fracture plane of a reconstituted phosphatidylcholine-bleached rhodopsin membrane showing a dispersed distribution of particles (×65 000). This appearance is characteristic of that obtained when a preparation containing bleached rhodopsin is rapidly chilled for freeze fracture from a temperature where the membrane lipids are in the F state.

RIGHT: Fracture plane of the same reconstituted membrane preparation showing the patched distribution of particles (×90 000). This appearance is characteristic of that obtained when a preparation containing unbleached rhodopsin is rapidly chilled for freeze fracture from a temperature where the membrane lipids are in the F state, or when a preparation containing bleached rhodopsin is slowly chilled from a temperature where the lipids are in the F state

14 h at 4 °C and then for an additional 30 min at 37 °C prior to fixation and rapid cooling for freeze fracture, extensive patching of intramembranous particles was observed.

Nicolson[20] and Rosenblith *et al.*[21] have studied the temperature dependence of the redistribution of cell surface receptors that occurs in response to the binding of the multivalent plant lectin concanavalin A. When the lectin was added to SV40-transformed mouse fibroblasts at ice bath temperature[20] or 4 °C[21], the distribution of cell surface bound lectin was revealed by a number of staining techniques to be dispersed when the temperature was not increased prior to processing of the preparations for microscopic examination. When the cells were prewarmed to 22 °C[20] or 37 °C[21] prior to staining and assessment of lectin receptor distribution, the lectin receptors appeared to aggregate into patches. Edidin and Weiss have also published a study consistent with the idea that lateral mobility of membrane components is arrested in S state lipids. They reported that capping of the H2 antigen of fibroblasts proceeds at 20 °C in response to specific antiserum, but not at 10 °C[57]. Capping, however, is a complex process and though cap formation was qualitatively arrested at 10 °C, the authors pointed out that the failure to observe capping at 10 °C could just as well have been due to the effects of temperature on cell motility and membrane flow.

From the above discussion, there can be little doubt that lateral mobility of membrane components is arrested in S state lipid. A recent study by Petit and Edidin, however, raises the question as to the extent to which it is arrested[19]. As summarised briefly in Section 11.4 and in Table 11.5, Petit and Edidin have studied the response to temperature of the rate of mixing of cell surface antigens in heterokaryons[19]. Newly formed human–mouse heterokaryons, identifiable by double staining with mouse and human fluorescent antibody preparations, were incubated at various temperatures. The populations of cells were then scored at different times for the proportion of doubly stained cells in which the antigens remained segregated, i.e. where the antigens did not mix. A decrease in the percentage of segregants was interpreted to be proportional to the rate of mixing of these antigens after cell fusion. As described in Table 11.5, the rate of mixing scored in this fashion actually increased when the incubation temperature was decreased from 22 °C to 15 °C. This is consistent with the model for physical asymmetry in the mammalian cell plasma membrane bilayer presented in Section 11.4 and Table 11.5 as follows. At temperatures < 30 °C but > 21 °C, both the inner and outer monolayers should consist of mixtures of F and S state lipid (Table 11.5). If the antigens being scored for mixing penetrated into both monolayers, they would tend to concentrate in areas of the membrane where opposing monolayers consisted of F state lipid. This could give rise to an extraordinarily high temperature coefficient for mixing, since movement of an antigen through a channel of F state lipid in one monolayer could be blocked by a barrier of S state lipid in the opposing monolayer. Below 21 °C, however, this situation could be relieved by displacement of antigen molecules totally from the inner monolayer, into fluid regions of the outer monolayer. This displacement would be most likely to occur abruptly near t_1 for the inner monolayer and, as observed, could give rise to an increased rate of antigen mixing as the temperature is decreased. The precipitous decrease in rate of antigen mixing observed by Petit and Edidin below 15 °C is consistent with the membrane becoming totally frozen at this temperature, but a total arrest of antigen mixing was not observed. The rate did decrease by at least one order of magnitude between 15 °C and 10 °C and by nearly an additional order of magnitude between 10 °C and 6 °C. As a consequence of the unexpected behaviour between 15 °C and 22 °C, however, the apparent rate of mixing at 10 °C was little more than three-fold less than that detected at 22 °C. This result is of concern for a number of reasons, one of which is listed below. As described in Table 11.5, lectin-mediated agglutination of cells is sensitive to a characteristic temperature of a lipid phase transition, probably the t_l of the outer monolayer. One interpretation of the effects of incubation below 15 °C on agglutination assumes a restriction of lectin receptor mobility below this temperature[35, 36, 48]. Though it is conceivable, and indeed highly probably, that lateral mobility of membrane components is not totally arrested in S state lipid, it is possible that the results of Petit and Edidin might have yielded falsely high values for the rate of lateral mobility of membrane components in S state lipids. This could have resulted from any of the following: (1) the procedure for scoring the rate of mixing of antigens in heterokaryons gave falsely high values at low temperatures. If segregants in the doubly antibody labelled cells in the population were more sensitive to destruction, for example, the apparent rate of mixing might be greater than the actual rate. (2) The human cell lipids were not all in the S state below 15 °C. (3) Both the human and mouse cell lipids taken by themselves were in the S state below 15 °C, but heterokaryon formation could give rise to a mixed lipid phase with a t_l of < 15 °C. Statements (2) and (3) are conclusions similar to those stated by the authors[19]. In view of the provocative questions raised by this elegant experiment, the need for a more simple and direct means of measuring rates of lateral mobility in membranes is obvious.

11.5.3 Factors other than S state lipid which arrest lateral diffusion of membrane components

As mentioned briefly in Section 11.2.4, there is an increasing body of evidence to support the view that lateral mobility of certain surface components is arrested by association of these components with submembranous cytoskeletal structures, possibly microfilaments. The information supporting this concept is the subject of an extensive recent review by Nicolson[20]. The logic behind this proposal is compelling and can be stated simply. The membrane lipids of higher cells constitute a fluid matrix in which components should move freely by a lateral diffusion process, but some membrane components are restricted from moving. This concept has been used, for example, to explain differences in the properties of lectin receptor mobility in transformed *v.* normal fibroblasts[18, 20, 21] and the inhibition of lymphocyte surface receptor mobility by concanavalin A[58]. This type of concept can also explain an interesting topographical localisation of certain membrane proteins in cells that normally function in phagocytosis, polymorphonuclear leukocytes (PMN). Ukena and Berlin have observed that neither phagocytosis nor treatment with microtubule disrupting agents alone decreased the rate of lysine or adenine transport in PMN. When PMN were treated with microtubule disrupting agents and the effects of phagocytosis then determined, both transport systems were inhibited to the same extent, approximately 45%. This leads to the conclusion that these membrane transport sites are normally restricted from being incorporated into phagosomes by some elements of the cytoskeleton[59]. This situation, i.e. the binding of transport systems to the cytoskeleton, may be a specialised development in cells which are normally characterised by phagocytosis. Charalampous *et al.* have observed that phagocytosis strongly inhibited the transport of aminoisobutyric acid in a fibroblast line not treated with a microtubule disrupting agent[60].

11.6 CONCLUDING REMARKS

This discourse has focused on membrane structure–function relationships where the response of a physiological phenomenon to temperature change can be correlated directly with a physical alteration in membrane lipids. To this effect, the major emphasis has been placed on the characteristic temperatures which identify the initiation and completion of melting of the membrane lipids. It is at these temperatures where dynamic, co-operative structural perturbations in lipid are initiated or terminated and it is these co-operative perturbations that affect the rates of processes such as transport and enzyme activity. The freezing of membrane lipids affects the distribution of proteins in membranes, resulting in patch formation. The complete freezing of membrane lipid can also arrest lateral motion, but the extent to which it is arrested is not yet clear.

Bacterial membranes, primarily those from *E. coli*, were chosen as the primary model system with which to illustrate these physical and physiological correlations. Another widely studied microbial membrane system is that of Mycoplasma, and the literature on this topic has recently been reviewed by Razin[61]. One point of debate between materials presented in this review and the review by Razin is his statement that cholesterol 'eliminates' the lipid phase transition. Cholesterol certainly reduces the endothermic nature of the phase transition detected by calorimetric analysis[61]. It is obvious from our data, however, that mammalian membranes rich in cholesterol display physical and physiological phenomena that arise from phase transitions in the membrane lipids.

Though bacterial membranes generally exhibit ideal melting behaviour, exceptions have been found[37]. The reader should thus recognise that the emphasis in this review has been on the bacterial membrane systems which apparently exhibit 'ideal' behaviour. This allows analogies to the physical phenomena encountered in the melting of ideal binary phospholipid dispersions in water. For those readers who choose to encounter the intricacies of non-ideal systems, Refs. 4–9 and the literature cited in these should provide for plenty of sleepless nights. Of special interest here are Träuble's penetrating studies into the possible consequences of sequestering contiguous regions of anionic or cationic lipid in membranes, a concept of non-trivial importance in view of our current understanding of membrane asymmetry.

An unprecedented, but perhaps not totally unexpected, observation is the existence of physical asymmetry in the plasma membranes of cells from homeothermic animals. Each of the monolayers of the bilayer appears to have a distinct phase transition, each with unique upper and lower characteristic temperatures. It is of interest that one of the upper characteristic temperatures correlates with body temperature, giving rise to the possibility that circadian rhythms might respond to lipid phase transition-associated phenomena.

In closing, there was cursory reference to a new and exciting area, the role of the cytoskeleton (microfilament–microtubule system) in the maintenance of cell surface organisation. Now that the concept of a fluid membrane and its consequences has been generally accepted, it is time to turn about face and concentrate on those membrane components that tend to remain stationary.

Acknowledgements

Work in the author's laboratory was supported by United States Public Health Service research grants GM-18233 and AI-10733 and by grants BC-79 from the American Cancer Society and DRG-1153 from the Damon Runyon Fund. The author is the recipient of USPHS Research Career Development Award GM-42359. I am grateful to Michael Edidin, Garth Nicolson, Wayne Hubbell, Jean Paul Revel and James Keirns for communicating data prior to their publication, to Harden McConnell for numerous discussions that were both penetrating and entertaining and to Charlotte Miller whose knowledge of literary style exceeds mine.

References

1. Steck, T. L. and Fox, C. F. (1972). *Membrane Molecular Biology*, 27 (C. F. Fox and A. D. Keith, editors) (Stamford, Conn.: Sinauer Ass.)
2. Rothfield, L. and Finkelstein, A. (1968). *Ann. Rev. Biochem.*, **37,** 463
3. Triggle, D. J. (1970). *Recent Progress in Surface Science*, Vol. 3, 273 (J. F. Danielli, A. C. Riddiford and M. D. Rosenberg, editors) (New York: Academic Press)
4. Oldfield, E. and Chapman, D. (1972). *FEBS Lett.*, **23,** 285
5. Chapman, D. (1973). *Biological Membranes*, 91 (D. Chapman and D. F. H. Wallach, editors) (New York: Academic Press)
6. Shimshick, E. J. and McConnell, H. M. (1973). *Biochemistry*, **12,** 2351
7. Overath, P. and Träuble, H. (1973). *Biochemistry*, **12,** 2625
8. Sackmann, E., Träuble, H., Galla, H.-J. and Overath, P. (1973). *Biochemistry*, **12,** 5360
9. Papahadjopoulos, D., Cowden, M. and Kimelberg, H. (1973). *Biochim. Biophys. Acta*, **330,** 8
10. Phillips, M. C., Ladbrooke, B. D. and Chapman, D. (1970). *Biochim. Biophys. Acta*, **196,** 35
11. Phillips, M. C., Hauser, H. and Paltauf, F. (1972). *Chem. Phys. Lipids*, **8,** 127
12. Hubbell, W. L. and McConnell, H. M. (1968). *Proc. Nat. Acad. Sci. USA*, **61,** 12
13. Linden, C. D., Keith, A. D. and Fox, C. F. (1973). *J. Supramolec. Struct.*, **1,** 523
14. McConnell, H. M., Devaux, P. and Scandella, C. (1972). *Memb. Res.*, 27 (C. F. Fox, editor) (New York: Academic Press)
15. Kornberg, R. D. and McConnell, H. M. (1971). *Proc. Nat. Acad. Sci. USA*, **68,** 2564
16. Dupont, Y., Gabriel, A., Chabre, M., Gulik-Krzywicki, T. and Schecther, E. (1972). *Nature (London)*, **238,** 331

17. Edidin, M. (1974). *Ann. Rev. Biophys. Bioeng.*, **3,** 179
18. Nicolson, G. L. (1974). *Int. Rev. Cytol.*, **39,** 89
19. Petit, V. A. and Edidin, M. (1974). *Science*, **184,** 1183
20. Nicolson, G. L. (1973). *Nature New Biol.*, **243,** 218
21. Rosenblith, J. S., Ukena, T. E., Yin, H. H., Berlin, R. D. and Karnovsky, M. J. (1973). *Proc. Nat. Acad. Sci. USA*, **70,** 1625
22. Silbert, D. F. (1970). *Biochemistry*, **9,** 3631
23. Linden, C. D., Wright, K. L., McConnell, H. M. and Fox, C. F. (1973). *Proc. Nat. Acad. Sci. USA*, **70,** 2271
24. Kezdy, F. (1972). *Membrane Molecular Biology*, 123 (C. F. Fox and A. D. Keith, editors) (Stamford, Conn.: Sinauer Ass.)
25. Jost, P. C., Griffith, O. H., Capaldi, R. A. and Vanderkooi, G. (1973). *Proc. Nat. Acad. Sci. USA*, **70,** 480
26. Jost, P. C., Capaldi, R. A., Vanderkooi, G. and Griffith, O. H. (1973). *J. Supramolec. Struct.*, **1,** 269
27. Schairer, H. U. and Overath, P. (1969). *J. Mol. Biol.*, **44,** 209
28. Wilson, G., Rose, S. P. and Fox, C. F. (1970). *Biochem. Biophys. Res. Commun.*, **38,** 617
29. Overath, P., Schairer, H. U. and Stoffel, W. (1970). *Proc. Nat. Acad. Sci. USA*, **67,** 606
30. Wilson, G. and Fox, C. F. (1971). *J. Molec. Biol.*, **55,** 49
31. Esfahani, M., Limbrick, A. R., Knutton, S., Oka, T. and Wakil, S. J. (1971). *Proc. Nat. Acad. Sci. USA*, **68,** 3180
32. Machtiger, N. A. and Fox, C. F. (1973). *Ann. Rev. Biochem.*, **42,** 575
33. Linden, C. D., Blasie, J. K. and Fox, C. F. (to be published)
34. Kleemann, W. and McConnell, H. M. (1974). *Biochim. Biophys. Acta*, **345,** 220
35. Noonan, K. D. and Burger, M. M. (1973). *J. Cell. Biol.*, **59,** 134
36. Rittenhouse, H. G., Williams, R. E., Wisnieski, B. J. and Fox, C. F. (1974). *Biochem. Biophys. Res. Commun.*, **58,** 323
37. Linden, C. D. and Fox, C. F. (1973). *J. Supramolec. Struct.*, **1,** 535
38. Esfahani, M., Crowfoot, P. D. and Wakil, S. J. (1972). *J. Biol. Chem.*, **247,** 7251
39. Tsukagoshi, N. and Fox, C. F. (1973). *Biochemistry*, **12,** 2816
40. McElhaney, R. N. (1974). *J. Mol. Biol.*, **84,** 145
41. Tsukagoshi, N. and Fox, C. F. (1973). *Biochemistry*, **12,** 2822
42. Lofgren, K. W. and Fox, C. F. (1974). *J. Bacteriol.*, **118,** 1181
43. Overath, P., Hill, F. F. and Lamnek-Hirsch, I. (1971). *Nature New. Biol.*, **234,** 264
44. Overath, P., Schairer, H. U., Hill, F. F. and Lamnek-Hirsch, I. (1971). *The Dynamic Structure of Cell Membranes*, 149 (D. F. H. Wallach and H. Fischer, editors) (New York: Springer-Verlag)
45. Wisnieski, B. J., Parkes, J. G., Huang, Y. O. and Fox, C. F. *Proc. Nat. Acad. Sci. USA* (in the press)
46. Wisnieski, B. J. and Fox, C. F. (1974). *J. Supramolec. Struct.* (in the press)
47. Grisham, C. M. and Barnett, R. E. (1973). *Biochemistry*, **12,** 2635
48. Rittenhouse, H. G., and Fox, C. F. (1974). *Biochem. Biophys. Res. Commun.*, **57,** 323
49. Noonan, K. D. and Burger, M. M. (1973). *J. Biol. Chem.*, **248,** 4286
50. Keirns, J. J., Kreiner, P. W. and Bitensky, M. W. (1973). *J. Supramolec. Struct.*, **1,** 368 and *Proc. Nat. Acad. Sci. USA*, **70,** 1785
51. Bretscher, M. S. (1972). *Nature New Biol.*, **236,** 11
52. Verkleij, A. J., Zwaal, R. F. A., Roelofsen, B., Comfurius, P., Kastelijin, D. and van Deenan, L. L. M. (1973). *Biochim. Biophys. Acta*, **323,** 178
53. Klotz, I. M. (1958). *Science*, **128,** 815
54. Mavis, R. D. and Vagelos, P. R. (1972). *J. Biol. Chem.*, **247,** 652
55. Chen, Y. S. and Hubbell, W. L. (1973). *Exp. Eye Res.*, **17,** 517
56. Jan, L. and Revel, J. P. (1974). *J. Cell Biol.*, **62,** 257
57. Edidin, M. and Weiss, A. (1972). *Proc. Nat. Acad. Sci. USA*, **69,** 2456
58. Edelman, G. M., Yahara, I. and Wang, J. L. (1973). *Proc. Nat. Acad. Sci. USA*, **70,** 1442
59. Ukena, T. E. and Berlin, R. D. (1972). *J. Exp. Med.*, **136,** 1
60. Charalampous, F. C., Gonatas, N. K. and Melbourne, A. D. (1973). *J. Cell Biol.*, **59,** 421
61. Razin, S. (1974). *Progress in Surface and Membrane Science* (in the press) (J. F. Danielli, M. D. Rosenberg and D. A. Cadenhead, editors) (New York: Academic Press)

Index